THE RAVEN

To the memory of Ernest Blezard

THE RAVEN

A NATURAL HISTORY
IN BRITAIN AND IRELAND

DEREK RATCLIFFE

With illustrations by
CHRIS ROSE

T & A D POYSER

London

Text © Derek Ratcliffe, 1997
Illustrations © Chris Rose, 1997
ISBN 0-85661-090-9

First published in 1997
by T & A D Poyser Ltd
24-28 Oval Road, London NW1 7DX

United States Edition published by
ACADEMIC PRESS INC.
San Diego, CA 92101

All rights reserved. No part of this book may be reproduced, stored in a retrieval system, or transmitted in any form or by any means, electronic, mechanical, photocopying or otherwise, without the permission of the publisher

A catalogue record for this book
is available from the British Library

Text set in Linotronic Baskerville
Typeset by Phoenix Photosetting, Chatham, Kent
Printed and bound in Great Britain by
The University Press, Cambridge

This book is printed on acid-free paper

ISBN 0-85661-090-9

Contents

Preface		xv
Acknowledgements		xix
Introduction: The Raven: spirit of the wilds		1
1.	THE RAVEN IN HUMAN HISTORY	7
	Prehistory	8
	The Raven in myth and legend	9
	From scavenger friend to predator-foe	16
2.	THE RAVEN'S COUNTRY	27
	Cliff coasts of the agricultural districts	30
	Lowland farmland in the west	32
	Foothills and marginal land	33
	Sheepwalks of the higher hills	34
	Grouse-moors of the eastern uplands	35
	Deer forests of western and northern Scotland	36
	Cliff coasts of the northern and western moorlands	37
3.	DISTRIBUTION AND NUMBERS IN BRITAIN AND IRELAND	40
	South-east England	40
	The Channel Isles	42
	South-west England	42
	East Anglia	46
	Wales	47
	Midlands	53
	Northern England	55
	Southern Scotland	60
	Scottish Highlands	64
	Scottish Islands	68
	Ireland	71
	Grand totals	74
4.	FOOD AND FEEDING HABITS	75
	Studies of food	78
	The Raven as a predator	82
	Other foods	89
	Feeding habits	91

vi   Contents

5. SOCIAL BEHAVIOUR 97
    Territorialism 97
    Flocking and communal roosts 101
    Social behaviour in display 107

6. RAVEN MOVEMENTS 118
    General movements 118
    Daily movements 125

7. ASSOCIATIONS WITH OTHER ANIMALS 127
    Birds 127
    Mammals 138

8. BREEDING: NEST AND NEST SITE 139
    Nest construction 140
    Nest sites 144

9. BREEDING: THE EGG STAGE 160
    Prelaying behaviour 160
    The egg 161
    Clutch size 165
    Laying 168
    Incubation 171

10. BREEDING: THE YOUNG 174
    Nestling growth 175
    Brood size 182
    Breeding performance 188
    Causes of complete breeding failure 189
    Egg-collecting and its effects on breeding performance .. 189
    Regional variations in productivity 191

11. TERRITORIALISM AND POPULATION REGULATION 196
    Surplus non-breeding populations 199
    Rapid remating 200
    The ceiling on numbers 202
    Territorialism and variations in breeding density 204
    Mechanism of territorial spacing 208
    Limitations of unsuitable nesting habitat 210
    Mortality, population turnover and capacity for spread . 211

12. RAVENS IN THE MODERN SCENE 217
    Gamekeeping 218
    Afforestation 225
    Changes in farming practice 232
    Disturbance 236
    Pesticides 238
    Conclusions 240

13. RAVENS ELSEWHERE IN THE WORLD 241
    Subspecies of *Corvus corax* 243
    Other species of Raven 245

14. INTELLIGENCE IN RAVENS 249

Appendices
  1. Distribution of Raven flocks and roosts in Britain and Ireland 256
  2. Calls of the Raven 265
  3. Appearance of the Raven 268
  4. Scientific names of animal and plant species in the text 271
  5. Names of the Raven 274

Bibliography 275

Tables 1–27 284

Index 317

List of Photographs

1.	Raven in a snow-storm	3
2.	Adult Raven	11
3.	Raven flying overhead	24
4.	Coastal nesting haunt, Pembrokeshire	32
5.	Coastal nesting haunt, Ailsa Craig, Ayrshire	38
6.	Raven country, south Wales	49
7.	Quarry nesting haunt, Dinorwic, Caernarvonshire	51
8.	Mountain nesting haunt, Carnedd Dafydd, Caernarvonshire	52
9.	Pennine nesting haunt, Tindale Fells, Cumberland	56
10.	Crag nesting haunt, Borrowdale Fells, Cumberland	58
11.	Cliff nesting haunt, Cairnsmore of Fleet, Kirkcudbrightshire	63
12.	Highland deer forest, Suilven, Sutherland	67
13.	Hebridean cliff nesting haunt, Talisker Head, Skye	69
14.	Coastal nesting haunt, Co. Kerry, western Ireland	73
15.	Sheep in snow, Scottish Borders	77
16.	Sheep carcass, northern Pennines	84
17.	A Lakeland Raven Crag, Westmorland	141
18.	Examination of crag-nest, Westmorland	143
19.	An historic Lakeland Raven site, Cumberland	145
20.	A tree-nest site in shelter clump, Southern Uplands, Kirkcudbrightshire	146
21.	Tree-nest site, Langholm Hills, Dumfriesshire	149
22.	Cottage roof nest, Elan Valley, Radnorshire	152
23.	An overhung crag-nest with eggs, Westmorland	156
24.	Hill-nest with eggs, Langholm Hills, Dumfriesshire	162
25.	Clutch of spotted blue eggs, Elan Valley, Radnorshire	164
26.	Clutch of blue-green eggs, Lake District	164

27. Raven feeding 2- to 3-week-old young, Wales — 176
28. Brood of half-grown young, Southern Uplands — 177
29. Raven feeding month-old young, Wales — 179
30. Raven with near-fledged young, Wales — 181
31. Raven with near-fledged young, Southern Uplands — 183
32. Brood of full-grown young, Southern Uplands — 188
33. Head of recently fledged youngster, Southern Uplands — 191
34. Young raven overhead — 192
35. Demonstrating Raven about to land, Wales — 203
36. Grouse moor, Langholm, Dumfriesshire — 227
37. Blanket afforestation, Southern Uplands, Kirkcudbrightshire — 227
38. Raven site, pre-forest, Cheviots, Northumberland — 230
39. Raven site, post-forest, Cheviots, Northumberland — 231

List of Figures

1.	Regions and districts of Britain and Ireland	31
2.	Breeding distribution of the Raven in Britain and Ireland during 1988–91	44
3.	Change in breeding distribution of the Raven in Britain and Ireland between 1968–72 and 1988–91	45
4.	Variations in breeding density of Ravens in different parts of Britain and Ireland in recent years.	62
5.	Movements of over 100 km by ringed Ravens	122
6.	Direction of Raven dispersal	123
7.	Recovery pattern of Ravens ringed in Cumbria	124
8.	Breeding dispersion of Ravens and Golden Eagles in part of Wester Ross	129
9.	Delay in Raven laying date with increasing altitude in northern England and southern Scotland	169
10.	Advance of Raven laying date with time in northern England and southern Scotland	170
11.	Relationship between brood size and brood age of Ravens	184
12.	The annual cycle of Raven activity in Britain and Ireland	194
13.	Breeding dispersion of Ravens in the Lake District	198
14.	Recovery months of ringed Ravens	212
15.	Survival curve for ringed Ravens	213
16.	Present distribution of grouse moors in Britain	219
17.	Trends in Raven breeding population in different districts of Britain	235
18.	World distribution of Ravens	244

List of Tables

1.	Raven breeding population of Britain and Ireland by county	284
2.	Food of the Raven in various regions of Britain and Ireland	286
3.	Long-distance recoveries of ringed Ravens	288
4.	Distance of Raven movements, from ringing recoveries	288
5.	Distance of Raven movements, according to age, from ringing recoveries	289
6.	Direction of Raven dispersal, from ringing recoveries	290
7.	Raven ringing recoveries, according to month and age	291
8.	Raven tree-nests: choice of tree species by region	292
9.	Raven nest sites on buildings in Britain and Ireland	294
10.	Vertical height of Raven nesting cliffs in several inland regions of Britain	296
11.	Pattern of use of alternative Raven rock-nest sites	297
12.	Use of alternative Raven nest sites according to success or failure	299
13.	Altitude of Raven nesting cliffs in some higher mountain regions of Britain	300
14.	Size of Raven eggs for various regions of Britain and Ireland	300
15.	Raven clutch size for various regions of Britain and Ireland	301
16.	Raven clutch size for first and repeat layings for northern England and southern Scotland	302
17.	Date of Raven first egg for various regions of Britain and Ireland	303
18.	Raven brood size for various regions of Britain and Ireland	304
19.	Raven brood size for first and repeat layings for northern England and southern Scotland	306
20.	Raven breeding performance for various regions of Britain and Ireland	307

21.	Success rate of individual Raven territories since 1945	308
22.	Raven nest-spacing and breeding density for various regions of Britain and Ireland	309
23.	Unusually close nesting and high breeding density of Ravens	312
24.	Life table for Ravens ringed in Britain and Ireland as nestlings	313
25.	Increase in sheep numbers on the Carneddau, north Wales and in Northern Ireland	314
26.	Increase in sheep numbers in various regions of Britain	315
27.	Analyses of organochlorine (o/c) residues in Raven eggs (with Golden Eagle and Buzzard for comparison)	316

Preface

The Raven is a bird that fired my early enthusiasm and it became one of the favourites that especially drew my energies in the field each year. My credentials for writing a book about the bird are no more than that I have watched and admired it for a long time, and in many parts of Britain. My observations have been largely spare time and not particularly systematic, and they have been mainly on the Raven in the nesting season. They have been widely spread, but especially in the Lake District, Pennines, Cheviots, Southern Uplands, north Wales and the Scottish Highlands. Altogether, I have visited over 500 nesting places at least once, and a few of them I have known in over 30 different years. Every year since I was 15 years old, I have been among Ravens at nesting time. When I was a field biologist with the Nature Conservancy, I was in Raven country for a good part of each spring and summer, and with opportunity to take note of their doings even when I was engaged mainly on other work. I have also had the advantage of knowing many other Raven enthusiasts, who have most generously put their information at my disposal. Some have been companions with whom I shared the pleasures of fieldwork. My own information would hardly run to a book, but there is now a large available source of published information on Ravens that I have sought to distil.

When I began looking at Ravens, in 1945, the literature on the bird was quite sparse, apart from mainly rather general references in regional avifaunas. In these, a common theme was the pronounced decline from former abundance and the precarious existence of the species in the remoter parts of our islands. A few writers had taken a particular interest in the species and contributed noteworthy essays in their books.

An early account by the remarkable William Macgillivray (1837), who watched Ravens from hides, is full of information, and many of the Victorian ornithologists wrote interestingly about the bird. R. Bosworth Smith (1905) wrote about the ways of Ravens and their mythology, while the observant George Bolam (1913) spoke of the bird in wild Wales and made the first detailed study of food. John Walpole-Bond (1914) dealt especially with Raven nesting habits and George Abraham (1919) added a rock-climber's experiences from the Lake District. One of the most interesting early accounts was Abel Chapman's (1924) chapter on the Raven in the Northumbrian Borders, while H. A. Gilbert and Arthur Brook (1924), in south Wales, were among the first to photograph and observe Ravens from hides at the nest. Kennedy Orton (1924) set out observations from the mountains of Snowdonia and Ernest Blezard (1928) from the fells of

Lakeland and the adjoining Pennines. Desmond Nethersole-Thompson (1932) gave exciting descriptions of roped cliff descents and tree climbs to nests in an essay on the species in the West Country and Wales. *The Handbook of British Birds* (Witherby et al., 1938) summarised information pre-World War 2.

There was then little more until after the War, but in 1948 Benjamin Ryves gave a most informative account of the Raven on the coast of Cornwall, with much original observation on breeding behaviour. That same year, R. A. H. Coombes produced a paper on the flocking of the Raven in Lakeland. Bannerman (1953) summarised current knowledge of the Raven in Britain and Ireland, but little new work was reported for some time. In 1962, I set out my evidence that, where availability of suitable nesting places was not limiting, breeding density in this species (and the Peregrine) was determined by territorial interaction, which was itself geared to food supply. Interest in the bird grew through the 1960s. In north Wales, Eric Allin published his findings on breeding biology in 1968; he also enthused a group of ringers, with whom he continued his energetic field studies into advanced age. In Ireland, Gabriel Noonan (1971) wrote about Ravens of the south-east and, with a keen team of helpers, began a large-scale programme of ringing of the young during the 1970s and 1980s.

In 1976, Derek Goodwin produced his *Crows of the World*, with a long section on *Corvus corax* containing detailed accounts of voice and behaviour. This was followed in 1978 by Francis Coombs' *The Crows*, with a valuable summary of information about the bird to date. Goodwin published a new and updated edition of his book in 1986. Also in the 1970s, a study of the effects of land-use change, especially blanket conifer afforestation, as a possible explanation of local decline of Ravens in Northumbria and the Southern Uplands, was made by Mick Marquiss, Ian Newton and myself (1978). This was continued by Richard Mearns (1983) and extended to central Wales by Ian Newton, Peter Davis and John Davis (1982). The Davises published a detailed account of breeding biology for their district in 1986 and, in the same year, Peter Dare did likewise for Snowdonia, where he had been monitoring the nesting population for several seasons. John Mitchell (1981) reported a decline in Ravens along the extreme southern fringe of the Highlands and he has continued to monitor this area, as have Patrick Stirling-Aird and his helpers in the adjoining district of southern Perthshire. In 1983, Brian Harle sent me a copy of his unpublished paper on the history of the Raven as a breeding bird in Northumberland.

Tony Cross and Peter Davis repeated the survey of the central Wales study area in 1986, and Tony Cross has continued to monitor the population here and to ring and colour-mark large numbers of their young. Andrew Dixon (in press) has studied the Ravens of the Brecon Beacons farther south in Wales. Other regional studies of breeding biology have been in Orkney by C. J. Booth (1979, 1985) and in Shetland by P. J. Ewins, J. N. Dymond and M. Marquiss (1986) – who also examined food. A further study of breeding performance in Shetland has been made by P. M. Ellis, J. D. Okill, G. W. Petrie and D. Suddaby (1994). The various regional raptor study groups have mostly included Ravens in their activities and the group for south-west Scotland conducted a breeding census in 1994 under the direction of Chris Rollie. Ravens have also been the subject of recent studies for higher degrees. Chris Thomas (1993) made a Ph.D. study of the bird in Argyll, modelling distribution and breeding performance in relation to habitat. In southern Lakeland, Sue King (1995) examined nesting success and breeding density in relation to food supply as part of an M.Sc. submission.

Many other Raven enthusiasts and those routinely involved with surveys of birds of prey, especially the Peregrine, have also contributed information on breeding biology and numbers. Some of their records are incorporated in the now large data base of the British Trust for Ornithology's Nest Records Scheme. Ringing efforts are unified under the BTO's ringing scheme, which covers Britain and Ireland. Some observers have paid attention to the counting of non-breeding flocks. A good deal of information is available in the major BTO publications, the two *Atlases of Breeding Birds* (Sharrock, 1976; Gibbons, Reid & Chapman, 1993) and the *Atlas of Wintering Birds* (Lack, 1986), and in the numerous recent regional avifaunas. *The Birds of the Western Palearctic*, vol. VIII (Cramp & Perrins, 1994), contains a valuable summary of published data on the Raven across this whole region.

There is now a considerable international literature on the various forms of *Corvus corax*, especially in mainland Europe and north America. While this book is essentially about the bird in Britain and Ireland, I have selectively incorporated references to Ravens overseas if these appear to extend or illuminate understanding of the species here. The studies of captive Ravens by Konrad Lorenz (1940, 1971) and Eberhard Gwinner (1964, 1965a, 1965b, 1966) are still the definitive source for much of the detailed information on behaviour. The fascinating recent book *Ravens in Winter* by Bernd Heinrich (1990) is also a new departure, giving a vivid account of the author's field experiments which involved feeding Ravens in the snowy forests of New England in order to gain insight into their social relationships outside the breeding season. Heinrich's findings and conclusions would seem to have general validity and should be noted by Raven-watchers here. His colleague, John Marzluff, is extending this work on social behaviour, and Lawrence Kilham (1989) has also written interestingly about the Raven in New England. Richard Stiehl, Michael Kochert and colleagues have also advanced knowledge of Raven ecology in the western United States.

The Raven is a bird whose ecology is closely tied up with human affairs and is perhaps more intimately entangled with the cultural life of earlier peoples than any other bird in the whole of history. The place of the Raven in superstition, myth and legend is not an easy subject to handle and, as there are good treatments already, I have touched only briefly upon this aspect. My concern has been to present an account of Raven natural history in this part of north-west Europe as a fascinating subject that has drawn many followers. As will be clear from the various gaps and loose ends in the book, there is still a great deal to learn about this bird, which is one of the more spectacular wild creatures of these islands. The relationship between the territorial nesting pairs and the floating population of non-breeders is still obscure and the precise means whereby the breeders space themselves out remains a mystery. The recent work on social behaviour of Ravens in North America poses still further questions. Raven calls and their meanings are still a relatively uncharted field and, despite the innumerable assertions about the high intelligence of the Raven within the world of birds, a great deal more critical work is needed to illuminate this subject.

I have followed my preference for adhering to the old British counties as the basis for dealing with matters of geography. They are, on the whole, more satisfactory units than the new counties for biological purposes, especially in Scotland and Wales, and all but some of the most recent local avifaunas are based upon them. English names are used in the text throughout, but there is an Appendix listing their scientific equivalents. In the interests of easier reading, my use of technical references is on the sparing side and, where names are given as sources but without dates, it should be taken that they are personal communications.

Acknowledgements

My gratitude is due first to those who fostered my early interest in birds and enthused me about Ravens in particular. Ernest Blezard gave freely of his immense fund of self-taught knowledge and fieldcraft, and inspired my early efforts. We later had many memorable trips together into the fell country in search of Ravens. He also expertly examined numerous Raven castings for me, to determine the bird's food by its remains. Ritson Graham and Tom Johnston also encouraged my youthful desire to seek out the hill birds and maintained a lively interest in what I found. I have to thank various field companions and informants with whom I have enjoyed and discussed Ravens: Eric Allin, Dick Balharry, Jim Birkett, Bill Condry, Tony Cross, Peter Walters Davies, John Davis, Peter Davis, Chris Durell, Robin Fisher, Geoff Horne, Stuart Illis, Jim Lockie, John Mitchell, Ian Newton, Ian Prestt, Bill Robinson, Grant Roger, Jean Roger, Chris Rollie, Dick Roxburgh, Ralph Stokoe, Chris Thomas, Des Thompson, Walter Thompson, Donald Watson, Joan Watson and Graham Williams.

Desmond and Maimie Nethersole-Thompson gave me kindly encouragement in the writing of this book, and Desmond spoke vividly of his experiences with Ravens in their nesting haunts. The British Trust for Ornithology has been immensely helpful in providing me with data from the national Nest Records and Ringing Schemes, and I am grateful to the Director, Jeremy Greenwood, for his support. Humphrey Crick kindly gave me computer analyses of Raven-nest record cards and access to the original data. Sue Adams similarly provided print-outs of the ringing recovery records, and Will Peach commented on my interpretation of these data. Rob Fuller has discussed Raven ecology and allowed me to quote from his internal report on upland sheep-grazing and birds. Philip Jackson helped greatly with library facilities. At Monks Wood Experimental Station, then in the Nature Conservancy, Ian Prestt, Tony Bell, Don Jefferies, Arnold Cooke and I worked together under Norman Moore on pesticide and bird of prey problems, and learned about eco-politics as well. Some of the fieldwork was done during my years as a staff member of the Conservancy, and I acknowledge the debt to my former employers. The Nature Conservancy Council funded the studies of Raven populations in relation to land use, in the Southern Uplands and Cheviots, and in central Wales. In his later capacity as Director-General of the Royal Society for the Protection of Birds, Ian Prestt took a keen interest in the Raven story. I thank Barbara Young, Chief Executive of the Royal Society for the Protection of Birds, for use of the Society's Library at the Lodge, where Ian Dawson and his colleagues have been most helpful.

My debt to those who have particularly helped with information on breeding population and nesting biology is very large. Geoff Horne, who has monitored the Ravens of northern Lakeland for many years, has most generously supplied me with his extensive records. Peter Dare has been unstinting in providing detailed data on his study areas in north Wales and Devon. In central Wales, Tony Cross has kindly given me his unpublished information on Ravens and lent his video-tapes of incubation behaviour. I thank Andrew Dixon for allowing me to quote from his manuscript paper on the Brecon Beacon population, and Brian Harle for sending me his unpublished paper on Northumberland Ravens. I am grateful to John Mitchell for information about the birds of the Loch Lomond area and the central Clyde lava plateau over a long period, and for his helpful interest in the book, which has included sending me numerous valuable references and other information. To Patrick Stirling-Aird I am indebted for numerous annual accounts of Ravens in southern Perthshire. In Scotland elsewhere I appreciate much help from Chris Booth, George Carse, Ron Downing, Peter Ellis, Paul Haworth, Mick Marquiss, Wendy Mattingley, Richard Mearns, Bill Neill, Sandy Payne and Chris Rollie. Declan McGrath and Paul Haworth, in the Republic of Ireland, have been especially helpful. Michael Kochert has kindly supplied information on Ravens in the United States.

Colin Harrison informed me about the prehistoric records of Raven remains and kindly cast a critical eye over my treatment of these. For providing me with measurements of Raven eggs, I am grateful to Bob McGowan in the National Museums of Scotland, in Edinburgh; Stephen Moran in the Inverness Museum; and David Clarke and Stephen Hewitt in the Carlisle Museum. Clutch data from other collections were kindly obtained as follows: D. Cockbain, Keswick, by G. Horne; D. Cross, Maybole, by J. Mitchell and I. Newton; and J. Hutchinson, Ayr, by J. Mitchell. For other help with literature and references, I thank A. Cooke and K. W. Brewster. Des Thompson has taken an enthusiastic interest in the project and given invaluable assistance in various ways, which I much appreciate. For permission to use data on sheep numbers in Snowdonia in his report to the Countryside Council for Wales, I thank Alex Turner and the CCW. Alan Fielding has been immensely helpful in undertaking expert statistical examination of some of my data, and preparing figures, and I owe him special thanks. For welcome advice and references on Raven intelligence, I thank Joe Crocker.

I make grateful acknowledgement for help with regional information on Ravens as follows:

South-west England

Peter Dare, Geoff Kaczanow, John Kaczanow, Desmond Nethersole-Thompson, Granville Pictor, Bill Robinson, Dick Treleaven.

Wales

Eric Allin, Bill Condry, Tony Cross, Peter Dare, Peter Walters Davies, John Davis, Peter Davis, Ian Newton, Graham Williams, Iolo Williams, Welsh Raptor Study Group.

Midlands

Shropshire Raven Study Group per Tony Cross, Alan Fielding, Stuart Marsden, Keith Mason.

Northern England

Colin Armitstead, Jim Birkett, Ernest Blezard, Rob Brown, Ingram Cleasby, Pearson Douglas, Ian Findlay, Geoffrey Fryer, Marjory Garnett, Ritson Graham, Brian Harle, Derek Hayward, Geoff Horne, Sue King, Ray Laidler, Brian Little, Percy Parminter, Bill Robson, Paul Stott, Ralph Stokoe, Edward Steward, Edith Steward, George Temperley, Walter Thompson, Terry Wells, North West Protection Group.

Southern Scotland

Chris Cameron, George Carse, Donald Cross, John Hutchinson, Mick Marquiss, Richard Mearns, Ian Newton, Steve Redpath, Chris Rollie, Dick Roxburgh, Ralph Stokoe, George Trafford, Donald Watson, John Young, Dumfries and Galloway Raptor Study Group.

Southern Highlands

Ron Downing, Wendy Mattingley, Don MacCaskill, Mick Marquiss, John Mitchell, Sandy Payne, Patrick Stirling-Aird, Chris Thomas, Roger Broad, Central Scotland Raptor Study Group, Tayside Raptor Study Group.

Northern Highlands

Dick Balharry, Ivan Hills, Stuart Rae, Des Thompson, Jeff Watson.

Scottish Islands

Chris Booth, Peter Ellis, Paul Haworth, Bill Neill, Peter Cunningham.

Republic of Ireland

Paul Haworth, Declan McGrath.

Northern Ireland

Jim Wells, Cliff Dawson, David Knight, Gary Platt, Royal Society for the Protection of Birds.

I thank also the photographers whose work illustrates this book: the late Arthur Brook, Tony Cross, the late Pearson Douglas, Geoff Horne, Bobby Smith and Brian Turner. Chris Rose has most skilfully met my wishes for suitable subjects in his fine drawings, which so well capture the essence of Ravens in various settings. For other help with illustrations, I am grateful to Jean and Grant Roger, John Mitchell, Peter Howlett and David Clarke.

To the numerous landowners on whose ground I have worked, I make collective acknowledgement, and I thank the staff of the Forestry Commission for their interest and help. The wonderful maps of the Ordnance Survey have been a most basic aid at all times, and it is a pleasure to recognise their part in the work.

My earlier fieldwork owed much to the kindness and hospitality of those with whom I stayed among the hills: Margaret and Osborne Jones in Snowdonia; Louisa McGarva, Will and Mary Murdoch in Galloway; and John and Elizabeth Borthwick in Moffatdale.

Andy Richford at Academic Press has been a most tolerant and helpful publisher and I express my gratitude to him. I thank also Tamsin Cousins, who has seen the work through to press.

In my youth, my parents were always supportive and interested in my enthusiasm for birds and, despite their anxieties about my pursuit of those nesting in lonely and dangerous places, were always concerned that I should follow my bent.

My wife, Jeannette, has given constant encouragement and support during the writing of this book, and I am deeply grateful for her help in various ways.

Introduction:
The Raven: spirit of the wilds

The Raven is one of those grander birds whose presence adds so much to the character of the country in which it lives. Once the common and even cherished scavenger of our medieval towns, it became by the late nineteenth century an outcast creature of the lonely places, the sheepwalks, moorlands, mountains and rugged seaboard. Abel Chapman, writing of the bird in his beloved Scottish Borders, said that, 'Today the Raven epitomises the type of a vanished fauna', and he linked it with the wolves, wild boars, bears and wild white cattle of the times when 'Hadrian was drawing his fortified wall from sea to sea'. The Raven is one of our finest and most characterful birds and continues to excite the sensibilities of successive generations of naturalists, especially those who respond to the challenge of the wilds. Its way of nesting before the end of winter, when the weather may be forbiddingly severe, and often in dangerous precipices or lofty trees, adds spice to the quest for close acquaintance in the breeding haunts. Beyond that, it is a bird which – perhaps more than any other – has been invested by humans with symbolic and legendary meaning and mystery.

In appearance it is a striking creature, largest of all the crow tribe, and with a heavy pick-axe bill. The apparent jet blackness of plumage is enlivened in closer view by a glossy iridescence of purple, blue and green. In the air it shows an aquiline spread of wing, with splayed-finger primaries and a large wedge-shaped tail when soaring. When the bird is going places in a hurry its flight is fast and straight, driven by powerful wing-strokes which, in still air or at close quarters, produce a loud beat conveying purposeful strength. In more leisurely mood, the Raven or its partner – for they are often in pairs – will frequently indulge in curious flight antics, rolling over onto its back and tumbling for a moment before reversing the movement. And, advertising its presence to all and sundry, it utters deep, resonant croaks and barks that carry far into the distance. Usually, it is soon gone from sight, leaving the watcher wishing to learn more about it.

I was fortunate enough to grow up in Carlisle, which is second to none as a base for a budding naturalist in Britain and especially favoured by the variety of hill country within working distance. In my youth it also held a dedicated group of field naturalists of an older generation, self-taught and reliant mainly on a combination of bicycle and Shanks's pony for mobility. From them I first learned about the hill country and its birds. They knew all about Ravens and had enthralling tales of adventures among the high crags in quest of their eyries. Fired with enthusiasm, I laid plans for forays of my own in search of the mountain and moorland birds. The Raven soon became a main concern.

The local literature had intriguing snippets of information on the bird and I greatly admired an essay by Canon H. D. Rawnsley, one of the founders of the National Trust and one-time Vicar of Crosthwaite near Keswick. He called it 'After the Ravens in Skiddaw Forest', and beautifully captured the essence of the place, in telling of a long walk with two Keswick men intent on acquiring a young Raven for a pet. The fells of the Skiddaw group are the nearest of the Lakeland massifs to Carlisle and form an inviting skyline to the south of the town. They were the most accessible good Raven country, though the bird was fairly evenly distributed through most of the other Lakeland ranges farther south. I saw my first two Raven nests on the precipices flanking Helvellyn, but they were inaccessible and the first one I actually reached was on a crag of the Skiddaw fells, in a much easier situation. It was a great triumph to look over the edge of the big pile of sticks and see the five blue-green eggs lying in the deep bowl of sheep's wool, as their owners croaked hoarsely in protest and circled overhead.

In a wide arc from north to east are the more massive and gently contoured moorlands of the Borders and Pennines. Ravens were fewer and more widely scattered here, but they nested in picturesque situations. These lonely moors also appealed greatly to my naturalist mentors and had contrasting charms to those of the more spectacular Lake Fells. Then, farther northwards across the Border, were the steep-sided and, in places, craggy hills near Moffat and, to the westward, the desolate rocky moorlands of Galloway. These Southern Upland districts were then good Raven country and the numerous sea-cliffs of Galloway had many more breeding haunts. To all these areas, I duly made my way and then returned time and again.

Part of the fascination of Raven-seeking in the hills was the opportunity for exploration of a modest kind. This was a useful way of spending autumn and winter days, for the Ravens stayed in their breeding grounds all the year round and old nests were quite easy to spot with a little practice. The wonderful maps of the Ordnance Survey were an enormous asset in planning my searches of the uplands for this and other species of birds. In the rugged Lakeland fells it was largely a matter of searching the obvious crags one by one until all the nesting places were found. On the gentler moorlands of the Pennines, Cheviots and Southern Uplands it was often less easy to locate suitable nesting haunts, for fewer of these were marked as crags on the maps. Any waterfalls which might have flanking rocks, or map contourings suggestive of precipitous ground were marked out for future search. Each stream had to be examined to see if there were hidden ravines or hillside outcrops. Other clues and rumours had to be followed up and enquiry of shepherds now and then produced information on hitherto unknown

Introduction: The Raven: spirit of the wilds 3

1. *Raven in a snow-storm,* by W. H. Riddell (The Borders and Beyond, *by Abel Chapman*)

sites. Once in a while there was the thrill of discovery as a new haunt came to light but often search of stream after stream produced nothing in the way of Ravens. Yet it was all experience and there were other things to provide interest: rocky linns with Dippers and Grey Wagtails, promising habitat for Golden Plover and Merlin, wild goats, or simply the pleasure of a 'crack' with friendly hill shepherds.

When the days of winter began to lengthen, plans for the Raven nesting season became firmer and the weather was watched more anxiously. Heavy falls of snow and prolonged frost could mean a change of programme, ruling out the higher-lying and more precipitous haunts in favour of the simpler ones at low levels. March blizzards quite often saw the hill country cut off again – the roads blocked and access to the nesting crags quite out of the question. Even on the gentler moorlands, where the Raven's crags were relatively easy to negotiate, a covering of snow could give a weary plod and then tricky going on the actual rocks. And when there is no snow to contend with, hard frost turns the ground to concrete and glazes any damp spots on the approaches. The rocks become iced and need treating with care. Here is a vivid description, by Ernest Blezard, of Raven forays into the modest hills of the Scottish Borders, just north of Carlisle.

> A complement of eggs was found there on March 9th in a season when the waterfall at the site was still a solid frozen mass, and icicles up to four feet in length hung in rows from the rocks in which the nest was placed. In a later year another haunt in the same district was visited on March 23rd when the nest there proved to hold four eggs on the point of hatching. The site, a rock-walled watercourse deep in the hills and a sunless place with a creepy atmosphere at any time, had this day a deathly coldness which penetrated to the very bones. Despite the exertions of rope work in assailing this stronghold with its uncanny air, the three Raven hunters concerned were half frozen with tears streaming from their eyes during the operations, and were glad to run all the way back down to the low country in order to restore warmth. Cold as it was when these four eggs were hatching, the period in which they had been laid three weeks earlier was marked by much colder weather and frequent snowstorms.

The spring of 1946 saw me in solitary exploration of the lonely hills of Galloway. I had read the Revd C. H. Dick's book *Highways and Byways in Galloway and Carrick* and been enormously taken with his descriptions of these wild uplands. Here is my favourite passage:

> You are here in the heart of the great Cauldron, on an expanse of moor and bog drained by many streams. Although it is almost completely encircled by hills, it gives a wonderful sense of spaciousness. The loneliness is profound for the house is distant about six miles from any road end. If you wander about casting into the burns, you have a feeling of constraint that prevents you from becoming absorbed in your sport. You are here on a precarious sufferance. Something in the wilderness is uneasy and resentful at your presence.

You are glad to hear the croak of the Raven that tells you that you are not quite alone. This is the effect of the place in fine weather. On a sunless day, when the clouds are low, you feel like a lost soul committed to some chill reach of eternity.

I well remember nearing some sunlit granite crags below a high col which gave access to this wonderfully named 'Cauldron of the Dungeon'. A pair of Ravens rose from the biggest face and voiced alarm at my approach. Their nest was plainly visible on a good shelf, and a short, easy climb on clean and sound rock brought me to the ledge and a close inspection of the five eggs that the heather and wool basket contained. Then, a kilometre or so beyond, there was a much more forbidding sweep of granite precipice from which a Peregrine took wing. Another pair of Ravens appeared and hung about the crest of the crags; there were old nests, but I could not spot the occupied one that the birds' behaviour indicated. Finally, yet another range of high cliffs in sight of this one proved to have a third pair of Ravens, behaving as if they had young. Three pairs were evidently nesting, and within a circle no more than 1.5 km across. This was no flash in the pan, for the next year all three pairs were there again, with eggs or young, still within the 1.5-km circle. But there was just the one pair of Peregrines and the next pair was 5 km away.

These were birds that one came to regard as close associates. Where there were Peregrines, there were nearly always Ravens, but rather more than half the Ravens lacked falcon neighbours. The two species were conveniently studied together, for their nests were often within a few hundred metres of each other, and sometimes much closer. They were not harmonious neighbours and constant bickering between them was commonplace. It became clear fairly soon that Ravens and Peregrines had quite similar nesting habits, in resorting to cliff faces for breeding and typically using a selection of different ledges as alternative sites over a period of years. Peregrines often laid in old Raven nests and Ravens sometimes built on ledges previously used by Peregrines. I became intrigued by the way in which both birds appeared to space themselves out regularly when cliffs were plentiful, though the more numerous Ravens bred closer together than the Peregrines. Eventually, I wrote a paper on their breeding density and this aspect of their lives has continued to interest me.

The earlier history of these two species also has common ground, in that both found their status as valued human associates changed to one of outcast enemies when their medieval usefulness faded away. The Raven's story is less complicated and dramatic than that of the Peregrine: there was no pesticide episode and population collapse, and it has not been regarded as a threatened species. Yet it is one of the most fascinating and exciting members of our avifauna, and one whose fortunes are as closely tied to human affairs as any. It has seemed in recent decades to be in the ascendant in many districts, spreading back from the rugged uplands and cliff coasts to softer foothills and lowlands, or even the cities. Yet there are also warning signs that it *could* become a conservation problem. Not a few of the lonely places where once I had the pleasure of acquaintance with the Raven have long been forsaken by the bird, and invariably because human hand has lain more heavily upon these places than previously. Perhaps in our modern times, we should see the Raven as a creature of different and good omen, which – by its survival or disappearance – signals equally the fate of the wilder places of our islands and the richness of their fauna and flora.

CHAPTER 1

The Raven in human history

The Raven is a bird of wide distribution in the northern hemisphere and one adapted to a remarkable range of climate. It is present through much of the Arctic region south of the polar wilderness, and breeds, for instance, all along the bleak coast of northernmost Norway beside the Barents Sea, foraging on the shore or inland on the treeless tundra. There are nesting haunts far north on the coast of Greenland and the islands of the Canadian high arctic. Ravens are also widespread throughout the boreal and temperate forest zones of both Old and New Worlds. Within wooded country, they appear to have some need for open habitats and they flourish in both sub-montane and lowland regions with extensive heaths and grasslands, steppe, and even semi-desert in arid regions. Their present wide occurrence in cold climates suggests that Ravens inhabited Britain and Ireland to the south of glaciation limits during the Devensian Period at the end of the Quaternary Ice Age. They are likely to have spread northwards as tundra and steppe followed the receding ice-sheets some 10,000 years ago and to have become widely distributed as post-glacial forests later advanced to cover most of the land.

PREHISTORY

Remains of the bird in cave earths in south-west Britain have been listed by Harrison (1987), while Bramwell (1959–60) mentions occurrences in Peak District caves. Those at Tornewton Cave in Devon date from the Wolstonian Glaciation (the penultimate of the major ice advances) and show that the Raven was present also during a much earlier cold period of the Pleistocene. Bones from Torbryan and Happaway Caves in the same county are likely to be from a transitional period between the end of the Devensian and the onset of the Holocene, i.e. from 13,000 to 9,000 years ago. Climatic conditions would then have ranged from arctic to cold temperate. Ossom's Cave in the Manifold Valley, Staffordshire, contained Raven bones along with those of Ptarmigan, Eagle Owl and Reindeer; these were also of late-glacial age, though Bramwell interprets the conditions as subarctic/boreal rather than arctic: open ground with grasslands, moorland, marsh, water and scattered bushes or trees.

Remains in Port Eynon Cave in Gower, south Wales, are Early Holocene (9,000–6,000 years ago), while those at Dowel Cave, near Buxton, are Late Neolithic and accompanied by bones of other species, suggesting a bird fauna similar to that existing today. Near Bakewell, Raven bones were found in Bronze Age food refuse, while remains at Woodbury Settlement, Devon, are Iron Age and others, at Hucclecote Villa near Gloucester, Silchester in Hampshire and Colchester in Essex, are Roman. Excavations of still later remains in the Dark Ages have shown the presence of Ravens in settlements at Lagore in Co. Meath and at Jarlshof in Shetland that were occupied by the Vikings and Norsemen (Fisher, 1966). While Colin Harrison suggests that, for various reasons, Ravens are one of the birds least likely to have remains preserved from the distant past, the patchy record shows that the species has probably been continuously present in these islands since the end of the last Ice Age.

There is no telling just when Ravens began their long connection with humankind. Perhaps even earlier in the evolution of primitive anthropoids, they learned to follow our early ancestors, as they followed wolves, polar bears, coyotes and other carnivores, in order to scavenge from their leavings of food. While the Raven shows some tendency towards an omnivorous life-style – since vegetable matter figures frequently in its diet – this is a bird of somewhat vulturine habits. It has evolved primarily as a carrion-feeder, though it is well able to take smaller fry among the living, including birds, mammals, reptiles and amphibians. Humans also provided the bird with a food supply more directly, for, seemingly, when the opportunity arises, the Raven finds a meal of defunct human flesh as acceptable as that of any other large animal.

Whether the Raven existed precisely in its present form before the evolution of proto-humans is a moot point but the bird evidently has a very long history of association with *Homo sapiens* and has developed a way of life that remains inextricably interwoven with human activities through much of its range. As human occupation of Europe became more complete, Ravens became more permanently established as camp-followers and settlement-scavengers, in close familiarity with people. The bird came to have a special significance for those with whom it lived, and to be more than a casual companion.

THE RAVEN IN MYTH AND LEGEND

Perhaps more than any other bird, the Raven has from ancient times held a notable place in the minds of people in the northern hemisphere and the way in which they regarded their world. It has figured widely in the primitive attempts to make sense and meaning of the mysterious and bewildering environment within which *Homo sapiens* evolved, and has acquired a powerful symbolism as an earthly agent of the supernatural forces which were presumed to run the show. Those interested in this aspect of the bird cannot do better than to read the learned accounts by the Harrow schoolmaster R. Bosworth Smith, in his book *Bird Life and Bird Lore* (1905), and by the parson-naturalist Edward Armstrong, in his New Naturalist *The Folklore of Birds* (1958). Others, including Swainson (1886), Schufeldt (1890), Forbes (1905), Lloyd (*c.* 1925), Kennedy and Walker (1988) and Heinrich (1990) have summarised some of Raven mythology. I have drawn heavily upon Smith's account and continue with a relevant quotation.

> A bird whose literary history begins with Cain, with Noah, and with Elijah, and who gave his name to the Midianite chieftain Oreb; whose every action and cry was observed and noted down, alike by the descendants of Romulus and the ancestors of Rolf the Ganger; who occurs in every second play of Shakespeare; who forms the subject of the most eerie poem of Edgar Allen Poe, and enlivens the pages of the *Roderick Random* of Smollett, of the *Rookwood* of Ainsworth, of the *Barnaby Rudge* of Dickens, is a bird whose historical and literary pre-eminence is unapproached.
>
> This was the favourite bird of fable, in which 'he serves to point many a moral and adorn many a tale'.

It seems to have been the ease with which people could see human attributes or meaning in the Raven's appearance and behaviour, and project

Barnaby summoned Grip to dinner.
(Drawing by F. Barnard in Barnaby Rudge *by Charles Dickens)*

upon the bird their own inner feelings, that gave it such a pre-eminent place in human affairs. As Kennedy (1988) said, it was the bird most like ourselves. A major significance of the Raven in these human associations was as the bearer of omen – the message from on high to foretell the outcome of important events. The omen could be good or bad, and the bird in general was either revered or reviled, according to circumstances or to the particular culture. From this the Raven also became a symbol or allegory and, sometimes, a messenger with instructions as well as tidings. It was regarded as prophet, soothsayer and magician, and invested with miraculous powers, including transmutation into human form, or vice versa. The sombre blackness of the plumage, the sepulchral voice, the ability to mimic human speech and the knowing demeanour of the bird added up to a connection with the dark forces and death in the minds of some. On the other hand, the boldness of the Raven in captivity, its evident wiliness and its appearance of formidable strength and dominance among other birds appealed to another side of primitive thought, so that the bird also became a talisman of success in war and conquest.

Perhaps both these superstitions come together in the continuing association of Ravens with the Tower of London, that place of grim memory

2. *Adult female Raven near the nest in Wales (photo: A.V. Cross)*

from medieval England. The ceremonial and legendary role of the bird there, dating from the time of Charles II, is most earnestly maintained, as the newspapers remind us from time to time, with reports that a new recruit has been supplied from some wild location in order to keep up the required number of at least six, lest the White Tower crumble away and the Crown itself be at risk of demise. These are the only Ravens that many people in this country ever see and they must help perpetuate the ominous connection. The legends about the bird, which originate from many parts of the world, are seemingly countless, and some of them persist to this day in places where superstition dies hard. Common themes run through many myths from different regions, such as the connections with Deluge and Creation stories, as well as death. There was also a widespread belief in the talismanic powers of Raven relics, such as bones or beaks, and the mythical 'Raven stone', supposedly brought by the bird from the sea to its nest.

Bird figures identifiable as Ravens or Crows occur among the Palaeolithic cave-paintings at Lascaux, in the Dordogne, France (Armstrong, 1958). They are also among the group of 12 bird species identified by Willoughby Verner in the Neolithic paintings (6,000–8,000 years old) in the cave of Tajo Segura in Cadiz Province, Spain (Gurney, 1921). Some of the earliest allusions to the Raven are in the Old Testament, such as the feeding of Elijah in the wilderness and the flight from the Ark to reconnoitre the Flood. From the latter comes the meaning of 'Corby Messenger' – an unreliable emissary, who fails to return or does so too late.

He send furth Corby Messingeir
Into the air for to espy
Gif he saw ony mountains dry
Sum sayis the Rauin did furth remane
And came not to the ark again. (*Warkis*, Lindsay, 1592)

The bird was revered as friend and companion of the early Christian saints and martyrs, rendering service and protection. Bosworth Smith remarks that the Raven also figures in the Koran. To the Ancient Greeks it was Apollo's bird and both they and the Romans elevated its prestige as an adjudicator of human affairs.

> The Raven was placed by the ancient Romans at the head of all the birds of omen, the *oscines* (os cano), as they were called: birds, that is, which, by their weird and startling cries, possessed the curious and enviable privilege of prescribing every detail of the public and social life – commanding this or forbidding that – of even so severely practical a people. (Smith, 1905)

[The song-birds were later known as the Oscines.]

The frequency of its bones in the remains at Silchester and other Roman settlements points to the Raven having lived there in a captive or semi-domesticated state (Ritchie, 1920). Some were evidently kept as pets or mascots. The term *Corvus*, or *Corax*, was also given to certain Roman implements of war.

In Anglo-Saxon times, the legendary King Arthur was equated with a Celtic warrior-god Bran (a Cornish name for the bird), who was accompanied by a symbolic Raven. Arthur was believed not to die, but to undergo magical transformation into a Raven and, for centuries afterwards, it was taboo to kill a Raven in the West Country (Penhallurick, 1978). Ravens are mentioned in the *Saxon Chronicle*, of around AD 700.

The Raven had famous associations with the Norsemen, to whom it was the sacred bird of their religion. In the oft-repeated legend, their war-God Odin had two Ravens named Hugin and Munin (Thought and Memory); each day they flew all over the world to gather information but, in the evening, they sat on his shoulder and whispered in his ears what they had seen. And so Odin was Ravneguden – the Raven God. The war-banner of the Norsemen was fashioned in the form of a large Raven. On windy days the banner appeared to be flying and the omen for battle was good, but on calm days, when it hung drooping and lifeless, the portents for victory were bad. Following in the vein of the Ark story, the Norse sea-rovers carried Ravens on their voyages of discovery because of the bird's supposed ability to sense invisible land. The rover Flokki was said to carry three Ravens; when released the first bird returned to the Faeroes which he had just left; farther out the second simply came back to his ship; but after he steered still farther to the north-west, the third disappeared in the same direction. Continuing, Flokki came to a large barren island which he named Iceland.

William the Conquerer is depicted on the Bayeux Tapestry behind the

Raven standard at the Battle of Hastings. It was mentioned in connection with Ireland in the writings of Giraldus Cambrensis (1183–86). During the Middle Ages the bird apparently lost its place as a fateful henchman in matters of warfare and exploration, but remained very widely a supernatural purveyor of omen and tidings, especially about death, pestilence and other ill fortune. Smith (1905) says it is mentioned – with Crows – at least 50 times by Shakespeare, far more than any other bird, in allusions mostly of a sombre kind. For example:

> A barren detested vale, you see, it is;
> The trees, though summer, yet forlorn and lean,
> O'ercome with moss and baleful misseltoe;
> Here never shines the sun; here nothing breeds,
> Unless the nightly owl or fatal raven. (*Titus Andronicus*, Act 2 Scene 3)
> Or
> It comes o'er my memory
> As doth the raven o'er the infectious house,
> Boding to all. (*Othello*, Act 4 Scene 1)

In more recent times, Edgar Allen Poe's weird poem 'The Raven' evidently uses the bird as a metaphor for the malign workings of Fate.

It is interesting how the observation of the Raven's appearance on the battlefields of olden time became turned back to front in meaning. Because the birds had learned to associate the assembling of armies with prospective carnage and abundance of food, they heeded the signs and turned up in anticipation. We can now see it as a good – though unsavoury – example of the conditioned learning of a habit. This response was, however, then regarded as an omen of imminent slaughter. Bolam (1913), speaking of the Borders of both Scotland and Wales, commented that:

> The expression "Corby Messenger" had a more sinister meaning, associated with coming strife and bloodshed, with its attendant glut of carrion; and in the days of Howell Dda or Owain Glyndwr [warrior chieftains of the Welsh Borders], the Ravens of the Dee valley must have chanted their hoarse requiem over many a stricken field, of which they had, perhaps, been looked upon as the forerunners.

Much more recently – at least up to Edwardian times – the Raven's habit of showing interest in Highland deer-stalking parties, in anticipation of a feast on the 'gralloch', was regarded by some stalkers as the portent for a successful day's sport! Such is the human propensity for preferring magic to rationality.

Smith (1887) said that 'The North American Indians honour it as unearthly, and invest it with extraordinary knowledge and power, and place its skin on the heads of their officiating priests as a distinguishing mark of their office'. The Raven is a frequent totem bird in Native American culture. It figures much in the beliefs of the Inuit, as in the legend of the Raven's creation of light by flinging mica chips in the air. A tale of cunning and treachery is that in which a Raven told a party of Inuit to camp under a cliff, which he then brought down on them so that he could eat their bodies in the spring. Ravens are frequent subjects in Inuit carvings. They and especially their young are nevertheless considered good eating by some Eskimos (Freuchen & Salomonsen, 1960).

Human flesh as a food source is a strange and repulsive idea in these days, though Ravens no doubt remain alert to the possibilities. In 1870 two were seen on the body of a drowned sailor under Boulby Cliff on the Yorkshire coast (Nelson, 1907). Borrer (1891) said that the skeleton of a man was found in a gorse thicket on the South Downs after Ravens had been noticed regularly about the spot. In rural Wales, it was reported that a murder was discovered through a pair of Ravens croaking over the corpse (Forrest, 1907) and the body of a person drowned in Bala Lake was found through Ravens assembling there (Bolam, 1913). Carrion Crows are also attracted to human corpses; Blezard (1954) recorded dissecting a male in Cumbria and finding a human eye lens, fingernails and tendon in the stomach. These remains were confirmed by a medical pathologist and the drowned body from which they came was found floating tangled in wreckage in the River Eden near where the crow was obtained.

The anonymous old ballad 'The Twa Corbies' shows insight into the gruesome attentions that Ravens could give to human bodies. There are different versions of this and an English one has been dated to the sixteenth century. I give below the best known and Scottish version, quoted by Sir Walter Scott (1806).

> As I was walking all alane
> I heard twa corbies making a mane;
> The tane unto the t'other say,
> 'Where sall we gang and dine the day?
>
> In behint yon auld fail dyke
> I wot there lies a new slain Knight;
> And naebody kens that he lies there,
> But his hawk, his hound, and his lady fair.
>
> His hound is to the hunting gane,
> His hawk to fetch the wild fowl hame,
> His lady's ta'en another mate
> So we may mak our dinner sweet.
>
> Ye'll sit on his white hause-bane,
> And I'll pike out his bonny blue een;
> Wi' ae lock o' his gowden hair
> We'll theek our nest when it grows bare.
>
> Mony a one for him makes mane,
> But nane sall ken where he is gane;
> O'er his white banes, when they are bare,
> The wind sall blaw for evermair.

[*mane* = moan, *fail dyke* = turf wall, *hause-bane* = collar-bone, *theek* = thatch; the rest should be decipherable]

Although Bolam found superstition about Ravens to be prevalent in rural Wales at the beginning of this century, it has tended to die out in Britain. It has had its advantages for the bird. In the 1920s, one gamekeeper at Langholm, Dumfriesshire, had no compunction whatever in killing Peregrines and Hen Harriers, but would never touch the Ravens, which bred every year on his ground. It is surely significant that greater tolerance is shown to Ravens in the Nordic countries where 'the cult of Odin once held supreme sway'. The extreme tameness of Ravens reported in Icelandic fishing villages and Aleutian settlements bespeaks not only an absence of persecution, but also a respect for the bird. In Finnmark, the northernmost county of Norway, the bird is welcomed as a scavenger and well regarded by all; not so the Hooded Crow (Gunnar Henriksen).

I close this brief sketch with a summarising quotation from Armstrong (1958), who was concerned to understand the evolution of myths as a global anthropological phenomenon.

> Raven mythology shows considerable homogeneity throughout the whole area [northern regions of the northern hemisphere] in spite of differences in detail. The Raven peeps forth from the mists of time and the thickets of mythology, as a bird of slaughter, a storm bird, a sun and fire bird, a messenger, an oracular figure and a craftsman or culture hero. First, perhaps, it achieved prominence as a death bird, then as a storm and rain bird it became associated with creation and deluge. With the anthropomorphisation of supernatural powers and the invention of more elaborate tool- and weapon-making techniques men thought of it as an artificer and messenger of the gods.

FROM SCAVENGER FRIEND TO PREDATOR-FOE

In the Middle Ages the Raven was a common inhabitant of our towns and villages, living there off the refuse and discarded food so freely available in these once unhygienic places. In this it seemingly shared an ecological niche with the Red Kite in some towns. Ritchie (1920) has shown that the Raven was once actually encouraged and protected as a disposer of garbage and cleanser of cities. He quotes two continental travellers in Britain who remarked on this town-cleansing function. The Venetian Capello, after spending the 1496–97 winter in England wrote, 'Nor do they dislike what we so much abominate, i.e. crows, rooks and jackdaws; and the raven may croak at his pleasure, for no one cares for the omen; there is even a penalty attached to destroying them, as they say that they will keep the streets of the town free from all filth'. Then, too, the German traveller Von Wedel wrote, ' on the 6th [of September 1584] we rode to Barwick [Berwick-on-Tweed]. There are many ravens in this town which it is forbidden to shoot, upon pain of a crown's payment, for they are considered to drive away bad air'. Glegg (1935) noted that, in London, there was a penalty for destroying Ravens in 1500, and probably for some time afterwards; they and the Red Kite thrived on the streets of the city during the reign of Henry VIII. Kirke

Drawing from The Twa Corbies *in* Scots Ballads *by Robert Burns (1939)*

Drawing from The Twa Corbies *in* Scots Ballads *by Robert Burns (1939)*

Drawing from The Twa Corbies *in* Scots Ballads *by Robert Burns (1939)*

Nash (1935) said they were abundant and protected in Edinburgh around 1600. Yet statutes of Henry VIII and Elizabeth I outlawed Crows and bounties were paid for their destruction – a farthing a head in one parish.

As the towns and villages became cleaned up, so the Raven lost its congenial urban habitat through lack of food, as well as lack of need for its services. Inland it remained a widespread bird in the agricultural scene, where livestock management was general, yet primitive by the standards of today, with a mortality of animals old and young which must have supplied the scavenger birds and mammals with a reliable source of food. Possibly it was always less welcomed in rural areas than in the towns and it doubtlessly acquired the reputation of a killer of lambs, chickens, rabbits and game, while its value as a scavenger of dead livestock became ever less appreciated. Records below show that the persecution of Ravens had begun by the mid-1600s in some country areas and, by the eighteenth century, the Raven was evidently an object of dislike quite widely, with a price on its head in many areas. Glegg (1935) observed that the earlier protection under penalties in London had been replaced in 1768 by bounties for killing them. The written testimony of churchwardens' records shows the widespread practice of paying bounty rewards to encourage destruction of Ravens. In Ireland, Ussher and Warren (1900) said that, 'Rutty, in 1772, mentioned that Ravens frequented in numbers the neighbourhood of great towns and were held in some veneration for devouring carcasses and filth', which implies that the bird was well regarded there for longer than in Britain.

The Revd H. A. Macpherson, author of *The Vertebrate Fauna of Lakeland* (1892), delved into these old records for the district and gave an illuminating detailed account. He begins by quoting a passage from Wordsworth, who made early comment on the bounty scheme in reference to a Raven seen at Ullswater:

> Friday, November 9th [1805]. A Raven was seen aloft; not hovering like the Kite for that is not the habit of the bird; but passing onwards with a straight-forward perseverance, and timing the motion of its wings to its own croaking. The waters were agitated; and the iron tone of the Raven's voice, which strikes upon the ear at all times as the more dolorous from its regularity, was in fine keeping with the wild scene before our eyes. This carnivorous fowl is a great enemy to the lambs of these solitudes. I recollect frequently seeing, when a boy, bunches of unfledged Ravens suspended from the churchyard of Hawkshead, for which a reward of so much a head was given to the adventurous destroyer.

The last comment evidently alluded to the raiding of the dangerously placed crag-nests.

The earliest church records were for the parish of Orton, Westmorland, in 1636–37, when the price per head was $1d$. By 1649 the price had risen to $2d$, but that for young birds remained at a penny. The first Hawkshead records were for 1731, 'Four Ravens killing, $4d$ p. piece, $1s\,4d$', and this sum appears to have been adopted throughout the following half-century of accounts, during which Raven rewards were recorded in 26 separate years.

The total number paid for in those years was 174 Ravens, giving an annual average of over six and a maximum of 16 in 1772. Prices varied locally, for while $4d$ was the rule at this time (and half-price for youngsters), Cartmel Priory paid only $2d$ by an order decreed in 1751. The longest run of records was for Greystoke parish, over the 90-year period 1752–1842, when churchwardens accounted for 966 Raven payments. Analysed into separate decades, the figures were: 1752–61, 154 Ravens; 1762–71, 204; 1772–81, 135; 1782–91, 116; 1792–1801, 134; 1802–11, 80; 1812–21, 41; 1822–31, 63; 1832–41, 39. The largest number killed in one year was 48 in 1766/67. A constant rate of $4d$ a time during the whole period points either to zero inflation or a dwindling sense of urgency about the Raven problem. Macpherson noted that it was only in the last three decades (i.e. before 1892) that he could find single years in which *no* Ravens were paid for by the Greystoke churchwardens.

The practice seems to have been general throughout Lakeland during the eighteenth century and Macpherson found records for the parishes of Patterdale, Martindale, Crosthwaite, Under Skiddaw, Borrowdale, Wythburn, St Johns and Kendal. It also extended into adjoining parts of Yorkshire, as in the parish of Dent, where the church account books recorded annual payments, usually of $2d$ per head, for Ravens between 1713 and 1750. Numbers were usually five to eight but in 1737 there were no fewer than 35, for which $5s$ $10d$ was paid. Farther south, in Cheshire, Coward (1910) remarked that, 'So plentiful indeed were Ravens at the beginning of the 18th century that only one penny per head was paid for their destruction by the churchwardens of Stockport'. Bounty payments were also made in Wales. P. Hope Jones (in Lovegrove, Williams & Williams, 1994) details annual totals of 15–120 Ravens killed and rewards paid, in Llanfar parish, Merioneth, during 1720–57.

No doubt a more extensive examination of church records would show the bounty system to be widespread in Raven country for, in those days, this was a general means of encouraging the destruction of 'vermin'. The practice of hanging up the bodies of the victims was equally general and lingers today as an atavistic gesture of some gamekeepers, crow-shooters and molecatchers. The rewards were an evident inducement to the locals to seek out and pillage Raven eyries after the young had hatched, and doubtless many nesting places were regularly raided over the years.

Scotland has supplied a wealth of information on predator destruction in the form of parish and estate records from the late eighteenth century onwards. Some of this was subsidised by bounty payments to all and sundry but, as concern in many areas extended or passed from the rearing of sheep and cattle to management for game, so the job of gamekeeper evolved to deal with the problem of vertebrate competitors with man. Increasingly, the large estates employed special custodians to put down all flesh-eating creatures, even those which were only suspect, and there was again often the inducement of headage rewards to promote results.

In Scotland, 'vermin lists' quoted by Ritchie (1920) chronicle that, in the Deeside parishes of Braemar, Crathie, Glenmuick, Tulloch and Glengarden, 1,347 Ravens and Hooded Crows were killed during 1776–87.

A table given by Harvie-Brown and Buckley (1887) lists 1,962 Ravens and 2,647 Crows/Magpies killed in the county of Sutherland and the two Caithness estates of Langwell and Sandside during 1819–26, all paid for by the Sutherland Association for the Protection of Property. On the estates of the Duchess of Sutherland, 936 Ravens were killed during 1831–34, for a reward of 2s each. The famous (or infamous) list from Glen Garry, Inverness-shire listed 475 Ravens destroyed in the 4 years from 1837 to 1840 (Pearsall, 1950). Harvie-Brown and Buckley (1887) knew of at least 662 Ravens destroyed in Sutherland, including 134 by a single Assynt keeper, during 1870–80, for a bounty of 2s 6d each.

Assynt continued to be cleansed of its 'vermin' for, on the Inchnadamph shootings, 149 Ravens were destroyed in 1880–89 and 91 in 1890–99 (Harvie-Brown & Macpherson, 1904). At Inverewe, in Wester Ross, 27 were killed in 1897–1902 and, at Ullinish on Skye, 25 between November 1885 and July 1886, while similar numbers were put down in some years on the Portree shootings. Harvie-Brown also noted that, in Perthshire, the Atholl estates office paid for 117 Ravens killed between 1894 and 1904, and the Breadalbane estates office for 576 between 1891 and 1901 (the latter ranging from 36 to 99 in separate years). The higher rates of bounty in the Highlands are intriguing. A letter from the Marquis of Bute in 1808 proposed 'a list of rewards' for the killing of predators on Bute, in which the Raven was rated highest of individual species, at 6s 6d, though 'Game Hawkes' nests' (Peregrine) were worth 10s 6d – considerable sums in those days. Ravens were clearly regarded as serious nuisances in these northern areas.

In north Wales, Forrest (1907) reported that 464 Ravens were killed in the 28 years ending 1902 on Lord Penrhyn's estate, which included part of the northern mountains of Snowdonia. In the West Country, Hudson (1915) was told by the head keeper of the Forest of Exmoor that, when he took the post, in around 1890, he shot and trapped 52 Ravens in one year. After a quarter of a century of continual warfare against the bird, the number rapidly diminished and for the past several years none had been killed.

Even though it is now thought distinctly possible that the estate vermin lists became inflated through manipulation of the bounty system by the beneficiaries, they still point to a relentless slaughter of all birds and mammals with a price on their head. Those quoted are simply some of the examples which have found their way into the literature and there is little doubt that this kind of destruction was virtually countrywide.

Many Ravens were shot or trapped, either at the nest or at baits left out in a conspicuous open place. In common with the raptors, Ravens perch by preference in some prominent place and so the pole or cairn of stones surmounted by a trap was often their downfall, despite the wariness of some individuals for jaw-traps. The sitting female can quite often be surprised on the nest at fairly close quarters and then offers an easy target to the gunner. Even when the nest cannot be reached, it is often within range of some safe vantage point, whence a shot or volley of stones will put paid to the eggs or young. Harvie-Brown and Buckley (1887) recorded that a keeper wiped out a pair and brood on a Caithness sea-stack, by determined rifle practice and much expenditure of ammunition, at a range of 230 m.

But probably the Raven's greatest weakness is its vulnerability to poison. Baits of chicken eggs or animal flesh laced with poison have been used for a very long time by gamekeepers as a means of destroying such enemies. Harvie-Brown and Buckley (1887) referred to the use of strychnine to kill vermin in 1870 and noted that, in the winter of 1885/86, one keeper poisoned a dead salmon on the bank of the R. Naver, thereby accounting for 20–30 Ravens. Saxby (1874) said the bird was relentlessly persecuted on Shetland, including by poison. Rawnsley (1899) mentioned a Raven nailed up at Skiddaw House in Lakeland which had succumbed to a 'hollow or poisoned and quite irresistible egg'. In his fascinating book about the wildlife of a then remote district in Merioneth, Bolam (1913) said:

> Old Ravens, however, soon become very wary of a trap, and are not easily taken in that way, while against the guns of the keepers they could probably also hold their own, but the insidious poisoned bait is beyond the highest intelligence yet developed among birds, and before that the race is gradually disappearing.

This last comment fairly sums up the position of our Raven population by that time, i.e. the onset of World War 1. It is instructive, though somewhat depressing, to read through the accounts of the species in the ornithological works of the nineteenth and early twentieth centuries. The Raven's fall from grace began around 1650 and thereafter it was in trouble. The common theme is of its retreat under the onslaught of almost universal and incessant persecution, especially by shepherds and gamekeepers. Few species could withstand this level of persecution without showing some ill effects and there are limits to the Raven's resilience. Beyond doubt, the unending destruction, of which I have quoted examples from the dismal catalogue, gradually told on the bird, cutting back its distribution and reducing its numbers. In combination with declining suitability of habitat in many areas, first in the towns and cities, and then in the more developed and arable farmlands, this warfare had, by the early 1900s, greatly curtailed its geographical range and its overall population. The bird simply became extinct over much of the lowlands, where its former wide occurrence remains denoted by place-names. It vanished from the sheep-grazed chalk downs, where it had persisted locally until the mid-1800s, and in the lowlands often hung on longest in parks of the large estates. Even in the uplands the extensive grouse-moors became inhospitable terrain from which the Raven was generally banished as a nesting bird.

Ravens found refuge mainly in districts where there was little or no preservation of game. Even here, the bird's ability to hold its own against the hostility of sheep-farmers probably depended on enough pairs nesting in remote or inaccessible situations where they and their young stood a good chance of escaping destruction. Ravens remained relatively numerous on the precipitous coastline of western and northern Britain, where safe nesting places abutted land on which game was generally of little interest. Inland they lost nearly all their lowland tree-nesting haunts and held out in the rockier uplands of Wales, northern England and Scotland, where they nested mainly on crags. The Raven typically had the company of Peregrine

24 *The Raven in human history*

3. Raven flying overhead (photo: Pearson Douglas)

and Buzzard, two other outlawed predators largely banished to the wilder parts of Britain, and in central Wales was neighbour to the Red Kite, lingering here as a far more depleted remnant in a last refuge within our islands. After this great retreat, the Raven managed in general to hold the line and then to readvance locally. It became tolerated on many of the sheepwalks and many shepherds took a more benign view of it than the Carrion Crow. Slowly, its fortunes began to improve again.

Killing of birds was not the only misfortune for the Raven. From about 1850 onwards, egg-collecting became an increasingly fashionable pursuit among the gentlemen ornithologists of the time. Its popularity remained high through the first half of this century and, even now, it dies hard, despite general condemnation and the high risks of prosecution. Any species which was both uncommon and laid particularly colourful and variable eggs became a special target for the 'brotherhood of oologists', as they have liked to call themselves. Ravens lay quite attractive eggs, in varying hues from blue to green, variously spotted and blotched with a range of brown, black and grey. They are the biggest of the Crows' eggs, though still small in relation to the size of the bird. Part of their attraction was in the difficulties of taking them – from tall trees and sheer cliffs – and in their association with wild and beautiful country. The Raven has also, during the last 150 years, been sufficiently uncommon for its eggs to have acquired a certain scarcity value. For some eggers, too, there was in addition the hope that sooner or later they would chance upon the rare erythristic, or pink, variety.

I shall return to egg-collecting and its effects in Chapter 10, but another kind of human attention deserves brief notice here. During Victorian times especially, the Raven found favour as a domestic pet. Its readiness in taking to captivity, with its engaging behaviour, ability to imitate the human voice and other sounds, and seemingly knowing and playful nature, gave the bird a certain character and status in this respect. Some of the nest-plundering in those days was to supply this demand. Bosworth Smith, who himself kept tame Ravens for many years, remarked that the eyries of the Dorset coast were regularly raided in the late nineteenth century and the young sold to bird-dealers in Leadenhall Market at 10 to 15s each. Penhallurick (1978) noted that dealers in Cornwall received nearly 20 young Ravens from the coast there in 1872, and 19 in 1874, and that four lives were lost in this pursuit. Ryves (1948) knew an old man in Cornwall who had helped supply this trade in his youth. It is no longer fashionable to keep such birds as pets and so this practice has faded away as a factor of any possible importance to the population. Probably it was never more than a minor and local issue. A number of localities scattered over England, Wales and Scotland are claimed to have provided Ravens to the Tower of London, and renewal of the captive stock there now and then receives press comment.

The early bird protection legislation did little for the Raven. The Wild Birds Protection Act 1880 aimed to give all wild birds (but not their eggs) protection between 1 March and 1 August. This seasonal restraint excluded any land-holder or person authorised by them, except for a list of specially protected species that did not include the Raven. With responsibility for bird protection subsequently delegated to county authorities, with power to make County Orders and by-laws, the Raven and its eggs became somewhat variably protected within its range, but such law was little subject to enforcement. The Protection of Birds Act 1954 was intended to standardise legal protection countrywide, under the principle that all wild species and their eggs were protected except for a small number regarded as pests or quarry. The Raven received blanket protection at first, but in 1956 a separate Order allowed the killing or taking of Ravens in Argyll by authorised persons, and this was followed in 1965 by a similar Order for Skye – both the result of lobbying by sheep-farmers over alleged depredations by non-breeding flocks. The Wildlife and Countryside Act 1981 finally removed these local exclusions and restored blanket protection to the Raven throughout Great Britain. Its provisions, which made possession of wild birds' eggs an offence, unless they were proved to be acquired legally, were a further clamp-down on egg-collecting.

The indiscriminate use of poison to kill predators remains a problem. Foxes and Carrion/Hooded Crows continue to be regarded by both gamekeepers and shepherds as prime enemies. Since the control of both by other means is labour intensive, there is a great temptation to resort to the simplest method: putting out poisoned baits, which also account for Ravens, Buzzards, Golden Eagles and Red Kites (RSPB, 1980; RSPB & NCC, 1991). Laying of poisoned baits in the open has long been illegal, but it is often difficult to detect (except when egg-baits are used) and proving individual responsibility is still more problematic. Campaigns against it have had some effect locally, but it still continues to be a common practice in some areas.

During the twentieth century, the Raven has gradually been restored to a place of general esteem within the public perception of wildlife. It still has a few detractors and outright enemies, but they mostly hold their tongues nowadays; any persecution is covert and illegal. The bird is cherished as one of our most spectacular and interesting species. It has, in parallel, recovered its range and numbers in some parts of Britain and Ireland, to give an overall status which is probably the healthiest for well over 100 years. Some writers describe it as almost common in certain districts. It evidently has the potential to continue this recovery, yet some districts have shown an appreciable recent decline. In considering its natural history in these islands, I want to examine the bird's prospects more closely. The more significant aspects of human influence on the Raven have been indirect, through land use and management practices, and these will form an important part of the story that I have to tell.

CHAPTER 2

The Raven's country

After its retreat to the wilder parts of our islands, away from close contact with humanity, the Raven has come to be regarded as one of our hardiest birds, not least in its habit of nesting before the end of winter, almost regardless of situation or weather. This is a creature conditioned to harsh environments, because there it finds both security and a sufficient supply of essential food. The breeding season itself has become timed so that the young hatch when availability of food is at its greatest, though the penalty is that nesting must begin often when conditions of arctic severity prevail, at least in some mountain breeding areas. The young are launched upon the world before some of our summer visitors have even reappeared, often to face notably inclement weather.

The mountains and moorlands are both windswept and misty on many days of the year, but they have the further disadvantage of being the coldest and snowiest parts of these islands. Most species of hill bird leave the high ground before the onset of winter and the Raven is one of the few which are able to subsist in the uplands the whole year round. In areas where the nesting haunts are in the higher uplands, it takes the heaviest snow and hardest frost to drive it down to the foothills and marginal land of the hill farms. The bird's relatively large size gives it some advantage during times of enforced fast, since bigger birds have a lower heat and energy loss than small ones. Yet winter blizzards and ensuing deep snow must largely preclude foraging activity for a time and, even though they may cause heavy mortality among sheep flocks still on the hill, this boost in food supply does not become available until the thaw. It is, indeed, something of a mystery how the bird keeps going in the hill country through the worst of winter stresses.

The weather is often especially adverse when the Raven begins nest-building, but only deep snow packed onto the eyrie ledges seems to delay these operations. And, once the eggs are laid, hardly anything deters incubation or the feeding and care of the young. The Raven-seeker may be faced with dauntingly severe conditions that make it difficult or impossible to penetrate the nesting grounds at all. Drifts may choke the hill roads and cut off the last farms and shepherds' cottages for many days, while the actual precipices remain so plastered with snow and ice as to afford a challenge to even the most experienced alpine climbers. Fresh blizzards may suddenly overtake the hill-wanderer and birds alike, posing difficulties for both. Abel Chapman (1924) has well portrayed the capriciousness of the weather so typical of Raven nesting time on the lonely moorlands of the Borders between Scotland and England.

> On 18th March (1915), a cold bright spring morning, still and sunny, we – to wit, George Bolam and the author – had set forth on a week's survey of our Realm. After a long morning's work devoted to other objects, by 2 o'clock we had reached a Raven's eyrie which already at that early date contained four new-hatched ravelets – almost naked, their flesh-pink bodies scantily clad in slate-blue down, and with yellow gapes. At that hour – as by magic – the day changed. A furious blast blew up from the north-east, bearing indigo-black clouds, all snow-laden, that enshrouded the hills and shut off the light of Heaven. Five minutes later snow swirled round in blinding drift – there was no mistaking the portent. At once, we decided, and hastened to retire ere retreat should be definitely barred by snow. It was a lucky decision since this snow-blizzard speedily developed into one of the wildest days

of many a wild winter. We succeeded in regaining our base – though only just in time; and when morning dawned found both highways and byways – even the railway – snow-blocked, and communication with the outer world temporarily suspended. Far worse, that sudden cataclysm had cost two human lives. On the moors above Wark a stalwart shepherd had been overwhelmed in the drifting snow – both he and his faithful collie being found dead next day, and a poor little school-girl had also perished on her way home from school. But how had our callow ravelets fared? Well, they simply never turned a hair, and all four, on the last day of April, were beginning to practise aviation!

I have known of many occasions when incubating Ravens have been overtaken by snowstorms which plastered the crags and the whole of their feeding grounds to some depth. Yet, when the thaw came and examination of the nesting grounds again became feasible, their breeding was in most cases unimpaired. One nest, built on a stunted bush on a cliff face, was brought crashing down by the weight of accumulated snow and ice, but it appeared to have held an incomplete clutch. I have a photograph of another eyrie in which snow lies thickly all around, including on the rim of the nest, but the cup and eggs were kept clear by the sitting bird. In 1987, Chris Rollie saw a Galloway nest with the eggs completely buried under a layer of snow, but this had melted 2 days later to reveal a clutch of six, which produced a healthy brood of six fledged young. The tremendous snowfalls of February/March 1947 nevertheless delayed many hill Ravens in their nesting, but this was exceptional.

In the passage quoted above, Chapman was making the point that the Raven was well able to cope with snow alone, but he went on to show that extremes of frost could prove fatal to the young of even this hardy mountain bird. After the unusually low temperatures of March/April 1919, he found a nestful of half-grown young that had been frozen on the Cheviot. In the hard winter of 1941, Ernest Blezard found a hatching brood all dead in their shells – the result, apparently, of the sitting bird having been kept off the nest for a critical time during the preceding days of icy weather. Deserted eggs are often split by frost, so that the female needs to cover incomplete layings during hard weather. Yet the Raven is able to live and breed far north of the Arctic Circle, in frigid regions that far eclipse any part of Britain in severity of cold. It endures what Chapman called 'the thraldom of the Arctic winter'.

By comparison, many of the coastal haunts of Ravens in these islands are benign and hospitable. The warming effect of the sea in winter produces relatively mild conditions, especially in the south and west. Nesting is early on some of the sea-cliff ranges, where many Raven eyries have full clutches well before February is out. While low temperatures are not a problem here, the weather can be hostile in different ways. The ferocity of the gales which rage so frequently against the Atlantic seaboard must be a hindrance, if not a hazard, to at least some coastal birds. A storm-lashed cliff coast is an awesome spectacle, as great breakers pound in and surge far up the precipice, to fling clouds of spray over its crest and then fall back in cataracts of white water. Care is needed even in approaching the cliff edge, where sudden gusts and eddies can make seeking acquaintance with Ravens at their home a more than usually dangerous business. These storms must test the safety of the big stick-nests and cause the most sheltered of sites to be chosen on the rocks. Yet, many Raven-watchers have remarked how the birds themselves seem almost to revel in testing their powers of flight against the tempest.

Contrasting with the days of storm are others when the whole range of cliffs lies fog-bound, with chill and clammy mist creeping along the half-seen faces. The waves lap invisibly below at the cliff base and the fog-horn blares and booms through the murk. If Ravens are to be seen at all, it is usually as a flurry of wings as a surprised bird beats a rapid retreat from a cliff-top perch. Croaks come loudly through the mist, as though the bird can see what is going on, but these are not conditions for learning about its nesting either. Ravens seem able to make their way about the cliffs when these are shrouded in sea-fog, but they can hardly be unimpeded, especially in their search for food. Perhaps they forage particularly along the shore at such times, for sea-fogs often stretch far inland. Some sea-cliffs drop into deep water, but others have a beach or undercliff where the birds can glean and satisfy their needs for a time.

Ravens are content enough to inhabit places where the conditions are easier, provided that food supply is adequate and human attitudes not too unfriendly. We tend, in this country and in Ireland, to think of the species as a wilderness bird of mountains and rugged coasts, and forget that it once flourished in comfortable surroundings of urban tameness. It is not a bird like the Dotterel, which summers only in really hard environments. On the contrary, it is an adaptable opportunist that can make the most of a wide range of habitat choices. Many of the more southerly coastal breeding haunts abut ordinary farmlands of pasture and arable that are the habitat of familiar lowland birds. And it has tended increasingly in the West Country and Wales to return to the soft lowlands where it was once so much at home.

I shall briefly describe some of the distinctive habitats which the Raven inhabits in Britain and Ireland.

CLIFF COASTS OF THE AGRICULTURAL DISTRICTS

These coasts extend from the south-easterly limits of distribution, in the Isle of Wight, westwards to Land's End and then along the Atlantic coast of Cornwall, Devon and Somerset. They occur at intervals along the coasts of Wales, the Isle of Man, Galloway and Ayrshire to the southern fringe of the Highlands in the Firth of Clyde. The few breeding haunts on the east coast of Scotland mostly fall into this category. In Ireland, probably the majority of coastal nesting places belong here, except in parts of the far west (see Photo 14). Ravens here resort almost exclusively to rock nesting sites on the sea-cliffs.

Many sea-cliffs have adjoining areas of untilled rough grassland or

Figure 1. Regions and districts of Britain and Ireland as referred to in the text.

heathery heath behind their crests, and less precipitous slopes running down to the sea have similar vegetation, often thickly grown with scrub of brambles, blackthorn, elder and even patches of woodland. Rough ground of this kind has diminished greatly in recent decades, as ploughing or pasture improvement have pushed to the very cliff edge in many areas, and its loss may have been significant to some Ravens. Most of these more southerly sea-cliffs nowadays abut a hinterland of productive lowland agricultural areas, mainly of stock-rearing but also of arable (see Photo 4). The birds forage a good deal inland on adjoining farm pastures or uncultivated ground, as well as on the shore, and food items here commonly include the rabbit and mole, those ubiquitous denizens of the farmland scene.

4. *The Pembrokeshire coast; extensive sea-cliffs backed by small farms with cattle and sheep pastures, and some unimproved heath (photo: D. A. Ratcliffe)*

LOWLAND FARMLAND IN THE WEST

In places, the Raven has greatly extended its range inland on the lowland farms of the West Country and Wales, where the rearing of beef cattle and sheep prevails, and rather little of the land is arable. In Devon, Somerset, south and central Wales, Denbighshire, Herefordshire and Shropshire, it is now widespread in quite ordinary farmland country with little uncultivated ground other than that occupied by trees. Here, it nests in woodlands and shelter belts, or even hedgerow trees, but also occasionally in quarries or on man-made structures. In such terrain, the Raven consorts with many familiar farmland and woodland birds, and seemingly holds a niche not obviously different from that of the Carrion Crow.

In earlier times, there were many nesting places on chalk downland and in lowland parks, in various districts, but these were lost during the great nineteenth-century decline. Parkland, where livestock rearing continues, has again become important Raven habitat in Shropshire.

FOOTHILLS AND MARGINAL LAND

This includes ground just above and below the limits of cultivation on the upland farms. Much of it has improved grassland, characterised by the intense greenness that NPK (nitrogen-phosphate-potassium) fertilisers produce, but some of it is unimproved pasture, with bracken on dry ground and rushes in wetter places, and includes that intermediate zone known as the *ffridd* in Wales. Sheep and cattle are pastured widely, but some of the land is used for production of hay crops, and there may be a little arable. There are often scattered small woodlands and shelter belts, and clumps, or sometimes lines, of trees, or – in more southerly districts – hedgerows with occasional bigger trees. Farmsteadings are usually not far away and, in Wales especially, there are often deserted houses that date from earlier days of transhumance. These frequently have an adjoining clump or fringe of trees.

These habitats are much favoured by Ravens in the West Country, e.g. around the edges of Dartmoor and Exmoor and in many parts of Wales and the Welsh Borders. Although extensive in northern England and parts of Scotland, they are less used by the species there. Many Ravens frequenting the eastern and central lowlands of Ireland live in this kind of terrain. The birds breeding in such country are predominantly tree-nesters. Their associates here are often Buzzards and, in central Wales, Red Kites, as well as the ubiquitous Carrion Crows.

SHEEPWALKS OF THE HIGHER HILLS

The foothills and marginal lands grade into the extensive open sheepwalks that occupy so much of the hill ground in Wales, northern England, southern Scotland, the southern Highlands and the main massifs of the Irish mountains. These are predominantly green and grassy, but the swards are mostly unimproved fescue-bent on steep dry slopes, mat grass and heath rush on damper ground, and flying bent or cotton grass where drainage is still more impeded. The dry ground below 450 m is often extensively invaded by bracken. There is a variable extent of low shrub vegetation, especially of heather, bilberry, crowberry and, particularly in the West Country, Wales and Ireland, western gorse. The dwarf shrub heaths were evidently once much more extensive, or even dominant, and have been steadily retreating under the combined effects of grazing by domestic stock and repeated moor-burning.

The main variations between the sheep hills are in the extent of bedrock exposed as crag and scree. On the softer sedimentary formations of Ordovician and Silurian rocks that compose so much of the Welsh hills and the Southern Uplands, there is a prevalence of steep but smooth-sided mountains with deep, water-worn valleys, but few cliffs of any size. Only here and there do precipitous escarpments or stream ravines occur. The Howgill Fells, between the Lake Fells and Pennines, are a typical example of such terrain. The Carboniferous rocks of the Pennines and Cheviots also have large areas of the same kinds of vegetation, but the more massive forms of these moorlands, with broad plateaux and gentle slopes, have given rise to extensive areas of blanket bog dominated by cotton grasses and heather, sometimes with abundant bog mosses. The Pennines also have large areas of limestone grassland locally, carrying sheep at higher density and with a richer flora than on the prevalent base-poor, acidic soils.

In places, ice-carving among these sedimentary hills has left more plentiful scarps where Ravens can find nesting places, as in the hills around Rhayader and Llangynog in Wales, and near Moffat in Scotland. However, it is in the extensive areas of harder sandstones and igneous or metamorphic rocks where high crags are most numerous – in Snowdonia, Lakeland, Galloway, the Trossachs, Wicklow, Antrim, Down, Waterford, Cork and Kerry. Ravens here reach saturation as cliff-nesters and are often found in company with Peregrines (Photographs 8, 10 & 11).

While some of the sheep uplands are important for other large raptors, such as the Buzzard and (in central Wales) the Red Kite, their bird fauna tends to be rather limited. The Meadow Pipit and Wheatear are usually everywhere and the Kestrel is a constant, smaller predator. The waders are distinctly patchy though, and Grouse, both Red and Black, even more so. The Chough is a distinctive inhabitant of the sheepwalks very locally in Snowdonia, the Isle of Man and western Ireland. On the whole, though, the Raven's bird associates on the grassy uplands are rather few.

Most of these sheep hills have only scattered fragments of native woodland, though in a few areas the hanging oakwoods are so placed as to give the impression of extensive cover, as in the Vale of Ffestiniog in Snowdonia, Borrowdale in Lakeland, Loch Lomondside in the southern Highlands and the Killarney district in Kerry (Photograph 6). These uplands have, since 1940, been especially the focus of large-scale afforestation programmes, almost entirely with the non-native conifers: Sitka and Norway spruce, larch, lodgepole pine and Douglas fir. In many parts of Wales, the Southern Uplands, parts of the southern Highlands and various ranges of the Irish hills, vast blankets of trees have obliterated former sheepwalks up to the planting limit at around 500 m. Over most of this ground the sheep have gone and a woodland bird community has replaced that of the open fell.

Cliffs are the preferred nesting habitats among the higher hills, but among the less rocky sheepwalks of uplands in Wales and the West Country, Ravens have become widespread and numerous as tree-nesters.

GROUSE-MOORS OF THE EASTERN UPLANDS

In eastern Britain and Ireland the combination of low rainfall and gentle contouring has favoured the management of moorlands for Red Grouse. The bird's principal food-plant, ling heather, thrives best under a dry climate, while level or gently sloping ground is best for driving the grouse over guns. This kind of terrain is also conducive to peat formation and many moors have a good deal of blanket bog as well as dry heather ground.

Few managed grouse-moors now remain in Wales, but they were formerly widespread from Montgomery through Merioneth to Denbigh and even eastern Caernarvon. The Berwyn Mountains and Denbigh Moors were notable areas. The Pennines still have important areas of grouse-moor, from the Peak District northwards and including the western outlier of the Bowland Fells. The North York Moors also have extensive grouse-moors, but large expanses of this habitat in the Cheviots have been lost to coniferous afforestation. Scattered areas survive amid the sheepwalks and conifer forests of the Southern Uplands between Wigtown and Ayrshire in the west and Berwickshire in the east. The main Scottish grouse-moors lie in a broad belt around the eastern fringes of the Highlands, from eastern Perth through Angus, Kincardine, Aberdeen, Banff, Moray and Nairn. North of the Great Glen there are further areas in the east of Inverness, Ross, Sutherland and Caithness. Afforestation has encroached extensively in all these counties in recent years. In Ireland there is little grouse-preserving nowadays, but the Wicklow Mountains were once a noteworthy area (see Fig. 16).

While most grouse-moors lie on acidic rocks and soils, they are often more productive habitats than the typical sheepwalks for other upland birds. This is partly a matter of topography and some of the best Curlew areas are on sheep ground with gentle terrain. Yet species such as Golden Plover, Dunlin, Merlin, Short-eared Owl, Hen Harrier, Ring Ouzel and Stonechat are usually more numerous on heather ground than on the grasslands. Not that the Raven finds a congenial habitat here, for it is unwelcome to grouse-preservers and usually discouraged from making its home in what is otherwise potentially very suitable ground. Numbers are thus generally low and on many moors the bird is absent. Good crags for nesting are often in short supply, but trees are usually very widely available (see Photo 36).

DEER FORESTS OF WESTERN AND NORTHERN SCOTLAND

Much of the Highlands and Islands from Arran and Kintyre northwards are managed especially for red deer and the hillier parts of the region are mostly divided up into deer forests. This animal is able to subsist with little attention in some of the most barren hill country in Britain. Some of the 'forests' were virtually treeless after 1850, though they also included much of the remaining areas of native Scots pine woodland, as well as variable amounts of native oak and birch. In recent decades there has been a good deal of coniferous afforestation, especially in Argyll, Sutherland and Caithness, and the new plantations have little food-value for deer, though they provide good shelter. Most of the deer forests also have some sheep and the two animals are between them causing the same kind of conversion of dwarf shrub heath to grassland that has been so widespread south of the Highlands. Under the generally wet and cool climate, there is a prevalence of peaty soils with low carrying capacity and the species which tend to become dominant under heavy grazing and repeated moor-burning are flying bent and deer sedge.

Although Red Grouse are widespread through the deer-forest country, they are mostly in low numbers nowadays, along with some of the other typical upland birds, such as Golden Plover and Curlew. The most distinctive deer-forest species are the Golden Eagle, Greenshank and Black-throated Diver (see Photo 12). It is believed that, under the adverse conditions of climate and soil, extractive land management has caused a serious loss of fertility across the region, though the evidence for associated bird declines comes mainly from the general statements of earlier writers, such as Osgood MacKenzie (1924) and Harvie-Brown and Macpherson (1904). Though widespread, the Raven is mostly at low density, for it not only has a more limited food supply than in more productive uplands, but also faces both competition and predation from its Golden Eagle neighbours. It is almost exclusively a crag-nester among the deer forests.

Although red deer occur among the mountains of Donegal and Kerry, Ireland can hardly be said to have deer forests on the Scottish pattern, and its uplands are virtually all sheepwalks. Many of the mountain massifs of western Ireland nevertheless have an essential ecological similarity to those of the western Highlands, with a prevalence of almost identical wet grasslands, heaths and bogs. The only marked vegetational difference is that the black bogrush tends in places to replace deer sedge as one of the most characteristic species. Carrying capacity is again low and some Irish ecologists believe that there has been a run-down in soil fertility comparable to that in the western Highlands. Moorland bird populations are again sparse and Ravens are at low density here, even though they have not had to contend with Golden Eagles for nearly a century.

All the British and Irish mountains with appreciable areas above the climatic tree-line (around 610 m, but lower in the far north and west) can be said to have a montane or alpine zone. In regions outside the Highlands, this is often distinguished largely by the disappearance of heather, the lower stature of plants in communities otherwise not very different from those at lower levels, and the presence of 'alpines', such as mountain sedge and least willow. Only within the Highlands is there a substantial area of distinctive montane plant communities, including types owing their character to prolonged snow-cover. Sheep and deer find palatable summer grazing on the high tops and so Ravens range widely across the montane zone in search of carrion and smaller fry. Bird numbers tend to decrease with altitude and only in the Highlands is there an important new species in terms of biomass – the Ptarmigan, which varies considerably in numbers between different massifs and is given to cyclical fluctuations.

CLIFF COASTS OF THE NORTHERN AND WESTERN MOORLANDS

From the southern fringes of the western Highlands and Islands northwards, and along much of the western Irish seaboard between Cork and Donegal, large and continuous areas of uncultivated moorland and mountain extend to the very edge of the sea-cliffs. Farmed land here is either completely absent or extremely restricted in the Raven's coastal haunts. The carrying capacity of these moorlands away from the sea may be no greater than those of interior sheepwalks and deer forests. The sea nevertheless appears to exert an extra and beneficial influence for Ravens dwelling on these rocky coasts. Sea-spray tends to produce a zone of soil enrichment along the crest of the cliffs and there may be other adjacent maritime habitats in which to forage, such as sand-dunes, machair,

5. *The Gannet cliffs of Ailsa Craig, Ayrshire; ravens in these large sea-bird breeding stations usually have a copious supply of dead birds and eggs on which to feed (photo: D. A. Ratcliffe)*

salt-marsh and rocky shores. The cliffs themselves may have sea-bird colonies, as well as giving way to steep slopes with rough vegetation which supports good numbers of animals that figure as Raven food. For whatever reason, these sea-cliff haunts tend to have Ravens breeding at higher density than in the barren interiors of the same moorlands and so are regarded as a distinctive habitat worthy of separate recognition (see Photo 13).

On places on the Isle of Man, the east coast of Scotland (notably in Caithness), in parts of Orkney (but not Hoy) and locally on the Irish coast, the sea-cliff Raven haunts abut a mixture of moorland and rather low-grade farmland. They are thus intermediate between the two northern and southern coastal types. Most of the Shetland breeding places back onto uncultivated moorland, though virtually the whole of this island group is under a strong maritime influence and probably has a higher carrying capacity than the deer forests of the Highland mainland. This maritime effect is even stronger on the small, storm-swept oceanic islands of the west, such as St Kilda, North Rona, the Shiants and Handa, and, in western Ireland, Aranmore, Achill, Clare Island, the Aran Islands and the Blaskets.

The Raven could thus be said to belong to a range of bird communities, insofar as this is a meaningful concept in the avian world. It is one of the most accommodating of species and few others can match its ability to adapt to almost the whole range of conditions that Britain and Ireland can offer. The Carrion and Hooded Crows do not penetrate as high into the mountains but, unlike the Raven, are able to occupy the richer arable farmlands of the low country. In the other bird groups it is surpassed perhaps only by the Peregrine in adaptability to different environments, both in these islands and elsewhere in the world.

CHAPTER 3

Distribution and numbers in Britain and Ireland

The extremely wide occurrence of place-names containing the word 'Raven' suggests that the bird once had a similarly widespread distribution across Britain and Ireland. Bartholomew's *Survey Gazetteer of the British Isles*, ninth edn lists 61 entries of localities beginning with 'Raven', in 39 different counties, and there are many more in which the word appears other than as a prefix – but these are more difficult to identify and count. Many such place-names have no other record of the bird and lie in areas from which it has long disappeared. The regional avifaunas contain an interesting history of the Raven, as it has chanced to be recorded in local knowledge, and this is summarised in the following pages. Where accurate historical information is available for different periods, this is included to illustrate more detailed population trends.

SOUTH-EAST ENGLAND

Although abundant in London in 1500, the Raven was almost gone by 1800. One pair nested in Hyde Park in 1826 and another in Bush Hill Park before 1840. A pair bred regularly at Enfield during 1840–50, but this was the last record for Middlesex (Glegg, 1935). Sage (1959) noted that Ravens were common in the Hitchen area of Hertfordshire up to 1700 and bred near Welwyn as late as 1846, after which there was only a handful of sightings up to 1955. The bird's history in Surrey is vague; Bucknill (1900) said, 'Undoubtedly a resident in the early part of the century [nineteenth] but there is very little definite information about it at all'. Possibly it bred up to 1850. In Kent, Ticehurst (1909) knew of only two certain inland breeding stations: at Lullingstone Park up to 1844 and at Cobham Park, where nesting ceased before 1868. There was a possible nest near Beckenham in 1885. On the coast, Ravens nested on the chalk cliffs either side of Dover, there being three pairs on the South Foreland section in 1841. At least one pair bred on the Kent coast up to 1890, but the species was extinct before 1900.

Walpole-Bond (1938) has catalogued the history of the Raven in Sussex. He listed the following 18 tree-nesting haunts, with latest dates of nesting: Bramber Castle (1843), Burton Park (1849), Petworth Park (1849), Danny Park–Wolstonbury Hill (1855), Cooden (1865–70), Glynde (1868), Sheffield Park (1870–75), Heathfield Park (1876), Devil's Dyke (possibly 1876), Harting (1866), Parham Park (1877), Firle Park (1878), Ashburnam Park (1877), Lancing–Cissbury–Chanctonbury–Findon (1880) and Pippinford Park, Ashdown Forest (1881). On the coast there were nine sea-cliff-nesting areas: Rottingdean (1878), Cheyne Gap (1894–95), Newhaven (1893), Seaford Head (1895), Seven Sisters (1892), Beltout (1886), west Beachy Head (1894), east Beachy Head (1894) and Ecclesbourne–Fairlight (1895). Ravens thus hung on along the sea-cliffs slightly longer than they did inland, but they had certainly ceased breeding by 1900, though birds lingered for a few years. Walpole-Bond continued to see Ravens sporadically in various parts of Sussex up to 1937, mainly from August to April. He believed some birds, at least, were from the Isle of Wight. Shrubb (1979) recorded that a pair bred at Seaford Head and then Beachy Head between 1938 and 1945, but at least one of the birds had escaped from captivity. The species remained a scarce autumn and winter visitor, with nine records after 1946.

The Raven has maintained a tenuous foothold on the Isle of Wight. Cheverton (1989) describes it as a scarce resident, with irregular nesting records over the last 40 years. Culver Cliff had two or three pairs during 1913–15, but none since 1958. A pair bred at Main Bench, east of the Needles, from 1966 to 1969, but from then until 1988 even single birds were unusual. The *New Atlas* recorded breeding in this western area in 1988–91.

In Hampshire, Kelsall and Munn (1905) said that the last two nests in the New Forest were in 1858, at Burley and Puckpits, and that several former 'Ravensnest Woods' were deserted. Other former nesting places, scattered across the county, with last known dates, were the Hackwood estate (1855), Rooksbury Park (1865), Tangley Clumps (1866), near Whitsbury (1872–73), Longwood (1884), Avington Park (1885), Andover Road (1877) and Ashdown Copse on the Wiltshire border (1887). There are no recent breeding records for Hampshire and the bird is only occasionally seen. In Berkshire, Ravens still nested in Windsor Great Park in 1868, while, in the adjoining Oxfordshire, the last pair bred for some years on top of a column in Blenheim Park up to 1847; nesting had ceased earlier at Crowsley Park, Henley (Radford, 1966). There have been only rare sight records in Oxfordshire since 1900.

THE CHANNEL ISLES

Dobson (1952) reported the following figures for Raven breeding population on the sea-cliffs:

Jersey	Three to four pairs annually, but only one in 1951
Guernsey	Only occasional nesting (1909 and 1939)
Alderney	Two pairs in 1923–38, one in 1946
Sark	Three pairs in 1938
Herm	A nest in 1939

The first *Atlas of Breeding Birds* (Sharrock, 1976) showed Ravens breeding on all five islands during 1967–72, in five different 10 × 10-km squares, and the *New Atlas* (Gibbons, Reid & Chapman, 1993) also recorded the same breeding distribution for 1988–91, but with the addition of a sixth square, on Jersey. The species thus appears to maintain its rather meagre status in the Channel Isles.

SOUTH-WEST ENGLAND

The Dorset coast is the most easterly part of southern England still retaining a breeding population of Ravens. Prendergast and Boys (1983) state that the number of pairs here has fluctuated between two and eight. At that time, 1972 was the last good year, with up to seven pairs between Ballard Down and Lyme Regis. Three Ravens were found poisoned in Purbeck in 1979–80, evidence of continuing hazards to the species. Seven pairs nevertheless nested again on the Dorset coast in 1994 (G. Pictor). Most of the numerous former inland nesting places, such as that at Badbury Rings, are long forsaken, but a pair has nested inland in west Dorset since 1992.

In Devon, Moore (1969) noted that the Raven had decreased considerably during the second half of the nineteenth century, becoming scarce and mainly coastal, but that an increase was reported by 1906. The Reports of the Devon Bird Watching and Preservation Society said in 1929 that Ravens 'are renewing the tree-nesting habit' and, in 1933, that they were 'increasing around Ilfracombe in the north' and 'now nesting inland in quarries and trees for want of cliff space'. The uplands of Dartmoor and Exmoor remained an inland stronghold thoughout and, by the late 1960s, the species was widely distributed as an inland breeder in the western half of the county.

The tetrad map and accompanying text by P. J. Dare, in Sitters (1988), for the period 1977–85, suggest a substantial increase since 1945. Ravens are now numerous in many inland areas, but especially concentrated around the fringes of Dartmoor and Exmoor, and thin out in the eastern half of Devon. Dare notes that they are most frequent in livestock- (sheep and beef cattle) rearing districts and rather uncommon where arable and dairy farms predominate. Most inland pairs are tree-nesters in secluded woodlands, copses, shelter belts, and even hedgerow trees, on both marginal hill farms and in the lowlands, and there are several nesting places in woodlands beside estuaries. Much of the country they inhabit has a lowland rather than an upland character, with enclosed and highly improved grass fields but also numerous woodlands and trees in a rolling landscape of low hills. Breeding on natural inland rocks, especially tors, has declined since 1945, since these are mostly small and now much disturbed by recreation-seekers. However, Peter Dare and Geoff Kaczanow tell me that five to ten pairs still use such rocks, mostly outcrops on steep valley sides in the Dartmoor area. Use of sites in quarries has increased and these now hold 10–15 pairs. A few man-made structures are also occupied. Dare estimates the inland population at 200–300 pairs.

The Devon sea-cliffs have long been a breeding refuge, contiguous with those of Cornwall. Dare (1988) estimates 20–28 pairs on the north mainland coast and another 23–33 pairs on the south coast. Comparison with earlier records shows that the north-coast population has remained stable over the last 50 years and the south-coast numbers are also believed to be little changed. On Lundy Island, from two to four pairs annually bred from 1914 to 1965 (Moore, 1969), but no fewer than nine pairs were reported to have nested in 1966 and 1967; in 1970, the extraordinary total of 11 pairs was claimed (Lundy Field Society Annual Reports). It is unclear how many of these 11 pairs were proved to breed. The overall coastal breeders can be given as an estimated median figure of 55 pairs.

The tetrad map in Sitters (1988) showed confirmed breeding in 236 of these Devon 4 × 4-km squares, plus probable breeding in another 154. Dare puts the total Devon breeding population at 250–350 pairs, placing it with Argyll as the largest for any county of either Britain or Ireland. After Wales, it is the area with the biggest increase in numbers during this century. There are also more tree-nesters than in any other county.

Penhallurick (1978) reviewed the history of the Raven in Cornwall, quoting early writers who found them to be frequent sea-cliff-nesters, although decreasing in numbers by the mid-nineteenth century. Churchwardens' accounts show them to have been much persecuted in previous decades. Clark, in 1906, believed the Raven was increasing locally and Harvey, in 1915, noted that it was holding its own on the coast. Nethersole-Thompson (1932) found Ravens breeding at high density on the Atlantic coast of Cornwall and Devon. Ryves (1948) estimated the breeding population at possibly 50 pairs, mainly in the northern half of Cornwall and stable in numbers. He knew of several inland-nesting pairs but noted that tree-nesting was uncommon.

Penhallurick (1978) said that no complete census had ever been attempted, but felt that the breeding population was greater than 50 pairs, though unlikely to exceed 100 pairs. He commented that they were scarce on the south coast east of the Lizard. Dick Treleaven has found that, on the

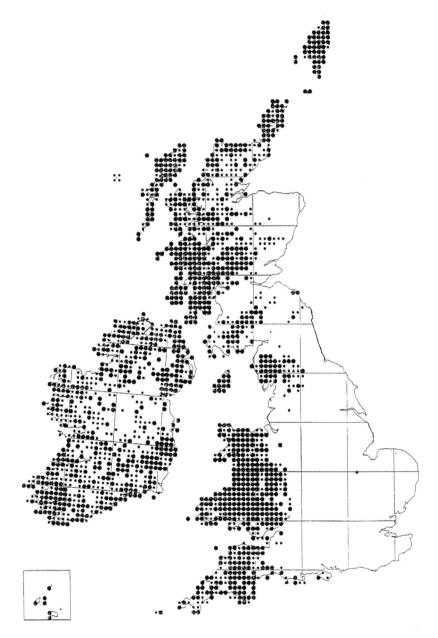

Figure 2. Breeding distribution of the Raven in Britain and Ireland during 1988–91. (Reproduced from The New Atlas of Breeding Birds in Britain and Ireland: 1988–1991 *(Gibbons, Reid and Chapman, 1993) with permission from the British Trust for Ornithology.)*
NOTES: Symbols show 10 × 10-km grid squares in which Ravens bred during the survey period 1988–91. A large dot refers to a 10-km square with evidence of breeding; a small dot refers to presence during the breeding season but no evidence of breeding.

Distribution and numbers in Britain and Ireland 45

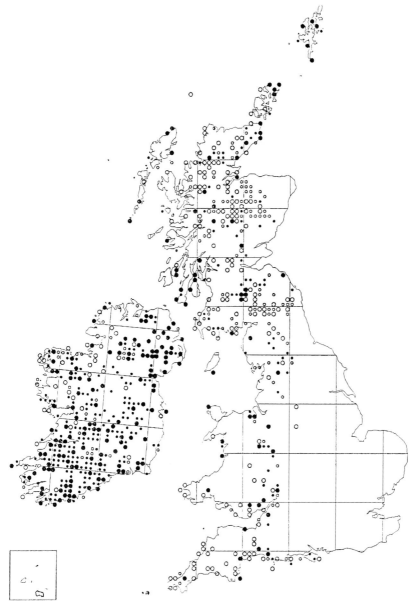

Figure 3. Change in breeding distribution of the Raven in Britain and Ireland between 1968–72 and 1988–91. (Reproduced from The New Atlas of Breeding Birds in Britain and Ireland: 1988–1991 *(Gibbons, Reid and Chapman, 1993) with permission from the British Trust for Ornithology.)*

NOTES: Symbols show 10 × 10-km grid squares in which Ravens bred during the survey periods 1968–72 and 1988–91. A black dot refers to a 10-km square with gains; an open dot refers to a 10-km square with losses; a large dot refers to a 10-km square with evidence of breeding; a small dot refers to presence during the breeding season but no evidence of breeding.

Atlantic coast of the north, breeding pairs are distributed more or less regularly along the lengthy cliff sections, in numbers probably exceeding those of Peregrines by 25–50% – which suggests a population of 40–45 pairs for this sector. Inland sites include the granite tors of Bodmin Moor, quarries and open-cast china-clay workings, old mine engine-houses and stacks, and other structures. Tree-nests were evidently under-recorded. Two pairs bred on the Scillies in 1995 (A. & R. Colston). Bearing in mind the large numbers in adjoining Devon, a population of 100 pairs seems reasonable.

Palmer and Ballance (1968) recorded that, in east Somerset, most former inland breeding places, such as the Glastonbury Tor tower and Willet Hill Folly, were deserted by 1850, though Cheddar Gorge remained intermittently occupied. On the coast, Sand Point, Brean Down and Steep Holm stayed tenanted, though not continuously. Pairs bred on the Somerset side of the Avon Gorge in 1940 and in a tree at Shapwick in 1958. Farther west, a few pairs bred on the sea-cliffs of the Quantocks or in trees inland. Former abundance on Exmoor was greatly reduced by 1900, but World War 1 saw a marked recovery and Ravens became generally spread across Exmoor and along the cliff coast west of Minehead. More recently, the Somerset Ornithological Society (1988) has confirmed this picture of widespread breeding in the west of the county and noted occasional nesting in the Blackdown Hills. Inland nests are mainly in trees, but a few are in quarries. No precise population figures have been published, but the high density of Ravens on Exmoor in Devon is likely to continue into western Somerset, so that a total population of 50 pairs is estimated.

Gloucestershire is now on the eastern limits of Raven distribution. Swaine (1982) reported that the bird was still fairly numerous in the Wye Valley and Forest of Dean until the mid-nineteenth century, when decline became noticeable; probably all had gone by 1880. There was also scattered breeding in the Cotswolds up to 1875. Nesting was again recorded in the Wye Valley and Forest of Dean from 1952, and four pairs have bred regularly up to the present, in rocks and trees. By 1982, there had been only one recent nesting record east of the Severn, but breeding Ravens have now returned to the Avon Gorge. Farther east, in Wiltshire, Smith (1887) named 22 nineteenth-century breeding places, nine of which were situated on Salisbury Plain; only two remained occupied when he wrote, at Wilton Park and Compton Park, where they were protected. All these Wiltshire nest sites were in trees, especially conifers and elms, and many were in estate parkland. Savernake Forest had at least three pairs. Ravens disappeared from the county soon after and Buxton (1981) said that there were only 12 records of birds seen during the last 30 years. In 1994 a pair nested successfully in southern Wiltshire and birds were seen in two other localities: two pairs bred in Wiltshire in 1995–1996 (G. Pictor).

EAST ANGLIA

In Cambridgeshire, Ravens were recorded at Bottisham in the early 1800s, and as breeding near Fen Ditton in 1828, but vanished from most parts of the district soon after (Bircham, 1989). Newton (1864–1902) records that an egg 'probably from Madingley or Quy, Cambridge, was taken not later than 1843' and a clutch came from Copt Hall, near Epping, Essex, in 1846. For Essex, Cox (1984) says that the last nests were at Mundon, Goldhanger and North Fambridge in 1890 (a rather late date for this region), and that there were sporadic sightings of birds up to 1909. Five records of Ravens were reported in Essex during 1958–82.

Newton (1864–1902) catalogued the history of 'what I believe to have been the last pair which inhabited East Anglia' [i.e. Suffolk and Norfolk] over 15 years from 1848, in the Breckland of west Suffolk. These birds, which represented more than two individuals over the period, were tolerated on the sheep-grazed rabbit warrens around Icklingham on condition that their eggs were taken every year. They were said to feed especially on moles. The qualified protection evidently allowed this pair to survive after all others had been eliminated from the region. They moved their nesting places between a Scots pine plantation belt on Icklingham Heath and a similar belt at the Duke's Ride on the Elveden estate, some 2 or 3 km away. The last nesting was in 1863, after which the pair disappeared, believed poisoned. Newton added that, ' The only tradition of other Ravens in west Suffolk I ever heard was of a pair at Ickworth, and another at either Fakenham or Sapiston'.

For Norfolk, Stevenson (1866) quoted Sir Thomas Browne in 1662 as saying Ravens 'were in great plenty near Norwich' and a Mr Hunt who asserted in 1829 that they were 'found in woods in every part of the county'. He also mentioned that Gurney and Fisher, in 1846, said the bird was 'still breeding in Norfolk, but in small and decreasing numbers', and that Gurney said a pair nested on the ruined steeple at South Walsham, probably in the mid-1800s. Newton noted that, ' About 1848 Mr Newcome had a live bird from Middleton, near Lynn, in Norfolk, and I think there had been a nest at Weeting not long before'. The last known nest was in a plantation at Beechamwell on the Swaffham brecks in 1859.

WALES

Lovegrove, Williams and Williams (1994) have summarised changes in the status of the Raven for the whole of Wales and I have quoted their figures for recent breeding population size. There was steady decline from 1745 and the bird had disappeared from many lowland and foothill areas by the second half of the nineteenth century. Numbers reached their lowest ebb around 1910. Recovery took place from 1914 onwards, slowly at first, but then uninterruptedly over several decades, some lost areas, such as the Marches, being reclaimed and numbers increasing in the favoured upland heartlands. This has involved a large population build-up in many areas, with a widespread return to tree-nesting, which is now prevalent in the lowlands and the less rocky uplands.

In the far south, Heathcote, Griffin and Salmon (1967) said that the Raven had decreased markedly by 1900, but a few pairs still bred on the crags of the coalfield and along the rocky coast of Glamorgan. The bird began to re-establish during World War 1 and to occupy tree sites in the Vale of Glamorgan, as well as increasing on inland and coastal cliffs. World War 2 saw a further increase and by 1967 there were about 30 pairs, in several groups. Thomas (1992) said that, 'Ravens nest throughout West Glamorgan' and 'They are a common sight on the cliffs of south Gower and in the northern uplands, but are also found in most of the eastern wooded valleys'. He estimated the population at 50–60 pairs. For the whole of the old county of Glamorgan, Hurford and Lansdown (1995) reported confirmed breeding in 52 tetrads and probable breeding in another 83. Their map shows breeders to be concentrated mainly in the hilly industrial valleys of the north, and on both sea-cliffs and in inland areas of the Gower peninsula in the west; there were very few in the south, with only two or three pairs on the Vale sea-cliffs. This suggests another 30 pairs besides those in west Glamorgan, giving a total of 80–90 pairs overall. Andrew Dixon believes that this is an under-estimate, and that the county total is more likely to be 110 pairs, consisting of 15 pairs in Gower; 40 pairs in the Lliw valley district north of Swansea, Vale of Neath, and around Hirwaun and Merthyr Tydfil; 40 pairs in the eastern valleys, Tawe valley area, Rhondda valleys, and around Neath and Maesteg; and 15 pairs in the Pontypridd and Vale of Glamorgan area. There have been two recent nests on buildings.

Tyler *et al.* (1987) noted a similar pattern of decline and recovery in Gwent (formerly Monmouth). By 1939, Ravens were thought to nest only in the extreme north and north-west, and then in a very few places. They then expanded and by 1975 were reported from most parts of the county, nesting predominantly in trees. The population was recently put at 100–150 pairs, though the tetrad map showed a similar picture to Glamorgan, with breeding confirmed in 46 tetrads and probable in 78. Andrew Dixon suggests that the true number is closer to 75 pairs, with 15 pairs in the Old Red Sandstone region of the Black Mountains in the far north; 30 pairs in the industrial valleys overlying the Coal Measures in the north-west; 25 pairs in the agricultural lowlands of the centre and east; and three pairs on the estuarine levels in the south. Nesting on electricity pylons is reported from two areas.

Gilbert and Brook (1931) regarded Pembrokeshire as an important Raven county; they found 39 nests in one year on the mainland coast, remarking that there were many others inland, though only one in a tree. Lockley (1949) said that Ravens were increasing in Pembrokeshire and estimated their population at 80 pairs, of which 60 nested on coastal and island cliffs and the rest inland, often in trees. In 1946 four pairs nested on Skomer and there have since been 'several'. Two pairs usually nest on Ramsey, one on Skokholm (two in 1966) and one regularly on Grassholm. Donovan and Rees (1994) report 12 pairs nesting on the islands, 65 pairs around the outer coast and another 60 pairs inland, giving a total breeding strength for Pembrokeshire of about 140 pairs. Tree-nesting has increased considerably.

The adjoining Carmarthenshire has a few coastal pairs, but the bulk of the population nests inland, the most recent estimate being about 100 pairs

6. *Raven country in a wooded cwm of the south Wales hills; extensive grassy sheepwalks lie above the woods (photo: Nature Conservancy)*

– a very large growth since 1940. Breconshire is another county to have shown a remarkable increase: an estimated 20 pairs in 1914 (all in rocks) had doubled by the 1920s, and the population has recently been put at 200 pairs (Peers & Shrubb, 1990). Andrew Dixon studied the area closely in 1993–95 and found that breeding density varies widely between the low ground of the main valleys, the intermediate massif of Mynydd Eppynt and the high ground of the Brecon Beacons and Drygarn Fawr, with 8, 41 and 110 pairs respectively, i.e. a total of approximately 160 pairs. In both Carmarthen and Brecon, this expansion has involved a great increase in tree-nesting, which now includes the majority of nests. In Cardiganshire, Ingram, Salmon and Condry (1966) said that the Raven 'has increased very considerably during the past 50 years and is now well distributed as a resident breeding species all along the coastal cliffs, on crags in the hills to the north and east, and there and elsewhere since 1927 has reverted freely to its ancient tree-nesting habit'. The most recent figure is 165–190 breeding pairs. Mid-Powys (Radnor with the northern part of Brecon) had an estimated 30 pairs in 1955, but this had increased to nearer 50 pairs by 1985 (Peers, 1985).

In north Wales, Forrest (1907) said the Raven, though much persecuted by sheep-farmers and gamekeepers, was still widespread in the

wilder disticts, both coastal and inland, but he was vague on numbers and reliable figures for earlier years are generally lacking. Tree-nesting had virtually ceased by 1900. Breeding has clearly increased in Montgomeryshire during recent decades, and numbers are reckoned at 120–140 pairs, the majority of which are tree-nesters. Merioneth is a rugged county which has probably always held good numbers. Bolam (1913) mentioned the bird as a widespread but much harried breeder around Llanuwchllyn in the early 1900s. During visits to the Arenigs, Rhinogs, Arans and the Cader Idris area in 1951–53 and 1961, I saw 20 nesting pairs and found that in rugged areas the bird was as numerous as in Snowdonia. The recent total of *c.* 100 breeding pairs must represent a considerable increase since Bolam's time.

During 1951–53, at least 14 pairs nested on the sea-cliffs of Anglesey, but this figure was based on an incomplete survey. The most recent estimate of the breeding population for the island is about 27 pairs, some being inland, with regular tree-nesting in at least one locality (Lovegrove, Williams & Williams, 1994). Although two dense clusters on the north coast have been reduced, there has evidently been an appreciable increase overall. On the Caernarvonshire coast, probably at least 15 pairs breed on the cliffs of Creuddyn and Lleyn.

7. *A quarry nesting place in north Wales. Dinorwic, Caernarvonshire (photo: D. A. Ratcliffe)*

The mountains of Snowdonia, with their abundance of lofty crags, probably remained a stronghold throughout the period of earlier decline. During 1950–57, I located 37 pairs in the mountain ranges north of the Traeth Bach (Caernarvonshire and north Merioneth). Only one of these pairs nested in trees. Other evidence suggested there were probably another nine pairs in localities that I had not visited within this area and, in 1961–67, I found a

8. *Raven haunt in a rugged recess of the mountains of Snowdonia; sheep numbers have risen markedly here over the last 20 years or so and Ravens also have increased dramatically. Carnedd Dafydd, Caernarvonshire (photo: D. A. Ratcliffe)*

further three pairs. This gave a possible total of 49 territories; although some of these were visited only once, there appeared to be only two where no breeding was occurring at the time. I believe these represented a minimum regular breeding population of 42 pairs. During 1978–84, Dare (1986a, 1986b) surveyed the whole of Snowdonia north of Portmadoc and found 97 nesting territories, with a 91% occupancy, representing an average of 88 pairs during this period. Pairs were spread fairly evenly over the whole district, but the south-east had an especially dense cluster around Blaenau Ffestiniog. Dare found only three regularly tree-nesting pairs, but another four pairs alternated between rocks and trees.

Peter Dare has kindly given me a copy of his data, which show that he covered a rather larger area than my original survey but, when comparison is based on the same ground that we both covered, his total is at least 63 territories, of which four were occupied irregularly, i.e. a minimum population of 59 pairs. This suggests that the increase in breeding population between 1950–67 and 1978–84, for the area we both covered, was at least 40% (i.e. from 42 to 59 pairs, taking conservative figures). The increase for the whole of Snowdonia could have been appreciably larger than this, depending on whether some of the numerous quarry-nesting places found by Dare in the more peripheral parts of Snowdonia were first occupied in more recent years. Julian Driver has found continuing increase in the massif of the Carneddau, in northern Snowdonia, since 1987, with another four new territories regularly occupied in addition to the 19–25 pairs found by Dare. One is a tree-nesting pair. This suggests that the Snowdonian population is now around 100 pairs.

Dare (1986a, 1986b) also studied a contiguous area to the east of Snowdonia, mainly in Denbighshire and consisting of gentler moorland and enclosed hill farms. Suitable crags for Ravens were few here and, of the 20 pairs, 13 nested only in trees. By reference to earlier data provided by E. K. Allin, he regarded this group as having been stable since the 1950s. The most recent estimates give a breeding population of 50–55 pairs for the whole of Denbighshire and another 10–12 for the adjoining county of Flint. The majority of pairs nest in trees, with quarries as the next preference.

Lovegrove, Williams and Williams (1994) estimate the total Raven breeding strength in Wales in the early 1990s at 1,250–1,500 pairs, which is by far the largest regional segment of the whole British and Irish population. This is where recovery of the formerly depressed population has been strongest since around 1914.

MIDLANDS

The record of the Raven in the Midlands is mainly pre-1900. The large Welsh population spills over into the Border counties of Hereford and Shropshire, but then fades out. Bull (1888) said, 'It is not uncommon in the more wild and hilly districts of Herefordshire', mainly along the marches with the Black Mountains and the Radnor hills. He mentioned earlier tree sites at Whitfield, Croft Ambrey, Brampton Bryan, Wigmore Rolls, Sellack and Dinmore Hill but implied that tree-nesting had ceased. The *New Atlas* (Gibbons, Reid & Chapman, 1993) shows breeding Ravens as now widespread in western Hereford, with a scatter of records in the east. There are no recent census figures, but it is described as a common breeding resident in suitable habitat, and was recorded 'throughout the year' in 29 localities in 1991 (Hereford Bird Reports). A population of 30 pairs thus seems probable, with up to 50 a distinct possibility.

In Shropshire, Deans *et al.* (1992) state that the last nineteenth-century nest was on the Long Mynd in 1884. A pair returned to the identical spot in 1918 but breeding was sporadic up to 1939, after which it increased. By 1960, about 20 pairs were nesting in the Clun Forest upland and the west-central hills, and by the 1980s there were 30–35 pairs, mainly in south-west Shropshire, all nesting in trees. The occupied parts include the Wrekin, Clee Hills and Millichope area, but pairs also nest east of Clungunford and in Loton Park. The Shropshire Raven Study Group located 57 nesting pairs in 1996, and Leo Smith estimates that numbers in unsurveyed areas could bring the total to at least 80 pairs. From his knowledge of the county, Tony Cross feels that the present figure is close to 75 pairs.

Lord & Munns (1970) and Harrison *et al.* (1982) describe the Raven as a very scarce breeding species, still not firmly established, in the counties of Worcester, Warwick and Stafford. Birds frequently appear, especially in the Malverns area, but do not usually settle to nesting. Old breeding records are at Needwood Forest up to 1863, on the crags of the Roaches, around Swythamley and in the lowlands, in Staffordshire; at Stanford-on-Teme in the late 1840s, near Upton-on-Severn in 1856 and Bewdley in 1859, in Worcestershire; and in Warwick Park in the 1850s.

Leicestershire had breeding Ravens in Charnwood Forest up to the mid-1820s, but they had gone by the 1840s and the bird has been only a rare vagrant in recent times (Hickling, 1978). Lilford (1895) said it was formerly common and well known in Northamptonshire, but authenticated records had become few after the last certain nesting, near Wigsthorpe around 1850. A prospecting pair seen around an old tree-nest site in 1880 were not proved to breed. Earlier evidence of Ravens in Lincolnshire is sparse: a pair bred on Louth church in 1693 and they were said to be common before 1750. At the end of the 1700s a pair still nested near Scunthorpe, but extirpation in the early nineteenth century is implied (Lorand & Atkin, 1989). The history of the bird in Nottinghamshire is also vague. Whitaker (1907) spoke of it being seen at times in Sherwood Forest up to *c.* 1855, with occasional sightings up to 1906, but gave no breeding records.

In Derbyshire Whitlock (1893) said that the Raven was described as common in 1789 and was well known as a resident on the crags of both the Low Peak and High Peak in the early 1800s. Raven Tor, adjoining Overton Park at Ashover, and Castles Rock were regular haunts. In 1844 it was 'not uncommon' in Dovedale, nesting in both trees and rocks. Breeding continued on the Derwent Moors of the High Peak up to 1863 (though the last known site was just in Yorkshire), but the Raven was long extinct at the time of writing. Frost (1978) mentions other nineteenth-century nesting places at Lathkill Dale, Ashbourne and Alport Castles, and probably in Cheedale and near Edale; and said it was still occasionally seen on the grouse-moors in 1905. He recorded that a pair returned and bred at Alport Castles in 1967–68, but they may have been released birds. Ravens have been seen increasingly in the High Peak in recent years and at least one pair bred within the Derbyshire portion in 1996 (S. Marsden).

For Cheshire, Coward (1910) said that the Raven was an extinct but once common resident, and mentioned a reference by Brockholes in 1874 to its former abundance in winter on the Dee marshes. There was an unsuccessful nest on Hilbre Island in 1857, and the scar called Raven Stones on Saddleworth Moor doubtless identified a former Peak District nesting place. Bell (1962) said there were only occasional records of passing birds during the twentieth century. Since then, the increase and spread of the Raven in north Wales along the Dee estuary has spilled over into Cheshire. The first nest was in 1991 and fledged three young and, by 1994, four pairs were holding territory (though only one was proved to breed) and there were increased sightings of birds from a new area of the county (Wells, 1995). A pair reared three young on Chester Town Hall in 1996 (Chester Tourist Information Office).

NORTHERN ENGLAND

Yorkshire

Nelson (1907) quotes Thomas Allis who in 1844 wrote, 'F. O. Morris observes that twenty-five years ago this bird was common in most parts, but has since been gradually getting scarce'. Nelson noted the wide occurrence in Yorkshire of place-names containing the word 'Raven' and mentioned widely scattered, former lowland tree-nesting sites from which it had long vanished: at Wycliffe-on-Tees, Walton Park, Bishop's Wood near Selby, Gowerdale, Scampston deer park and Hornsey Mere, and also on buildings at Beverley Minster and Castle Howard Mausoleum. Ravens bred on the seacliffs of Flamborough until 1889 and there were earlier nesting places at Filey Brigg, Scarborough, Ravenscar, Whitby, Boulby Cliff and Saltburn. The North York Moors had several crag-nesting places in the Helmsley, Danby, Pickering and Guisborough districts until 1860–75. The bird also bred in the Yorkshire section of the High Peak until 1863 (Seebohm, 1883). By the time Nelson wrote, however, the Raven was reduced to a few pairs nesting annually in the wildest parts of the Pennines and outlying Howgill Fells (now in Cumbria). He named other haunts in the upper parts of Nidderdale, Wharfedale, Wensleydale and Swaledale which were occupied up to 1880 but have since been deserted. Upper Teesdale still had a single pair which sometimes nested on the Durham side of the river.

This parlous state of affairs appears to have continued for the next few decades after Nelson wrote. In 1955, I saw Raven nests in four different localities in the Yorkshire Dales, plus a fledged brood in a fifth area. Suitable cliffs in at least another dozen localities bore no trace of Ravens but nests were subsequently seen on three of these during the late 1950s and 1960s. The Ornithological Report of the Yorkshire Naturalists' Union (YNU) for 1970 said that three pairs bred in Watsonian vice-county 65, but these were probably the three territories in the Howgill and Dent Fells where Ravens have nested regularly from 1900 right up to the present; in the subsequent new county system this area was transferred to Cumbria, but I count them here as belonging to the old Yorkshire. The YNU Report for

9. A northern Pennine Raven nesting place on a streamside crag amidst grouse moors; breeding has not occurred here for some years. Tindale Fells, Cumberland (photo: D. A. Ratcliffe)

1972 recorded three successful pairs in vice-county 64, which includes the Craven district. Since then, the Raven has almost disappeared from the Yorkshire Pennines. The YNU Reports for 1984–92 give no breeding records at all in 4 years, and a maximum of one certain and one probable breeding (1991–92). It appears that sporadic nesting continues in the Craven area and a single pair attempts, with varying success, to occupy the ancient territory in Upper Teesdale. It is many years since even attempted nesting was reported from the Yorkshire coast or the North York Moors. Two pairs were, however, reported to be nesting in the south Yorkshire part of the Peak district in 1996 (S. Marsden).

Lancashire

Mitchell (1892) noted the wide occurrence of Raven place-names in hill districts, and former breeding places in Wyresdale and Cliviger. He remarked on the hostile treatment of a pair which had recently attempted to return to the Bowland Fells and feared that extinction of the few pairs still breeding in the north of the county was imminent. About half a century later, Oakes (1953) reported that the high fells of Furness Lancashire held the only breeding Ravens in Lancashire (this district is covered along with

Cumberland and Westmorland, which together form the Lakeland area. More than 40 years later, the bird has failed to recolonise the grouse-moors of the Bowland Fells and western Pennines, though it is frequently seen there. An apparently successful nest was reported in Bowland Forest in 1987 (BTO Nest Records), but another nest that year had the eggs smashed; when the birds began building the following year, they soon disappeared, suspected killed (P. Stott). These are the only recent breeding records. In the Lancashire lowlands, an isolated pair nested successfully in an old quarry near Blackburn in 1994 and 1995 (P. Stott).

Cumberland, Westmorland and Lancashire north of the Sands

These counties together form the faunal area of Lakeland as defined by Macpherson (1892) and also correspond to the new county of Cumbria, apart from the present inclusion of the Howgill and Dent Fells (formerly in Yorkshire). The region includes not only the Lake District proper but also the Bewcastle–Gilsland Moors at the south-west extremity of the Cheviots and the highest section of the Pennines. The Raven population of the last two districts amounts to very little. Although there were two occasional recent nesting places just across the Northumberland border, breeding on the Cumbrian side of the Bewcastle–Gilsland Moors is unknown during this century. In the Lakeland Pennines, Blezard (1946) knew nine nesting places, of which seldom more than two were occupied in any year during 1920–40. During 1950–70, there were four regularly occupied territories, plus two irregular ones, but the breeding population has since declined to one or two pairs and, in some years, not a single attempted nesting has been reported. Tree-nesting was only occasional during this century, apart from a run of tree-nesting in one territory near Kirkby Stephen in the late 1960s.

The Lake District has one of the best-documented Raven populations in Britain, although nineteenth- century records are scanty. Macpherson and Duckworth (1886) believed that Ravens were widespread in lowland districts as tree-nesters in the early 1800s. When they wrote, nesting was known in only about a dozen localities but they felt that, despite persecution, Ravens remained as numerous as half a century before. Only three nesting places were named, including St Bees' Head on the coast. Macpherson (1892) detailed the catalogue of former destruction in churchwardens' records (see pp. 20, 21) but said little about current status, other than to suggest that there were more Ravens in Westmorland than Cumberland. The bird probably remained more abundant than these earlier writers realised.

During 1900–19, the records of E. B. Dunlop, E. S. Steward, P. W. Parminter, D. Scott, W. C. Lawrie and G. D. Abraham named 43 nesting places of separate pairs, of which 33 have remained regularly occupied ever since. This list represented coverage of only about half the district and, during 1920–39, the records of J. Coward, R. J. Birkett, D. Cockbain and P. Douglas added another 24 pairs, of which 18 proved to be regular nesters. Complete coverage was probably not achieved until 1940–69, during which time a further 14 pairs became known, probably 12 of them regulars (R. J.

Birkett, G. Horne, D. A. Ratcliffe). Adding these together, there were, during 1900–69, 81 known territories, of which two became permanently deserted, four were occasional, 12 were irregular and 63 were regularly occupied. The population evidently ranged from 64–74 pairs, with a mean of 69.

From about 1970 onwards pairs of nesting Ravens began to appear in places where breeding was previously unknown and several formerly irregular territories became regularly used. Second pairs also appeared in several previously regular territories. This amounted to a slight but distinct increase in population, though some pairs eventually dropped out again or became irregular in their nesting. Numbers during 1980–95 can be put at 82–92 pairs, with a mean of 87. This amounts to an increase of about 26% in average population over the earlier period 1900–69 (G. Horne, D. Hayward, P. Stott, D. A. Ratcliffe). The earlier stability in breeding population was noteworthy and numbers appear to have stabilised again at the higher level. Tree-nesting is still uncommon and mostly sporadic throughout the district, but may be overlooked in the south. There are post-1900

10. *Ravens have an abundance of nesting crags on the rugged Borrowdale Volcanic Fells of central Lakeland. Borrowdale and Derwentwater, Cumberland (photo: D. A. Ratcliffe)*

records from only 12 localities; one regular pair on Great Mell Fell has dropped out, and the only presently regular pair of tree-nesters is in the Shap area (D. Hayward).

Northumberland and Durham

Temperley (1951) recorded that, in Durham, the Raven bred on the coast at Dalton and Marsden near Sunderland up to *c.* 1834 but had lost inland rock sites in Wolsingham Park and along the Derwent by 1840, and was virtually extinct by 1905. Breeding thereafter was confined to the years when the Upper Teesdale pair nested on the Durham side, and to occasional attempts in Weardale. Bolam (1912) said of Northumberland that, 'About a hundred years ago the Raven was numerous'. He mentions a traditional tree-nesting place at Ilderton and stated, 'up to 1837, these birds had not been driven from nesting upon trees, nor altogether banished from the low country; though the latter event must have taken practical shape within the next ten or 20 years, and had begun some time previously'. Bolam then knew of only three annual nesting pairs in the Cheviots, and three occasionals, all on cliffs. Chapman (1907) also listed only five haunts known to be occupied during a long familiarity with the county. Bolam's MS records for 1915 gave nine Northumbrian nesting localities, though not all were reoccupied annually. Later, Bolam (1932) said, ' During the last 20 or 30 years the number of occupied eyries may sometimes have run to half a score'.

Bill Robinson and I saw nests in 11 localities during the early 1950s. Although not all were visited in the same year, they mostly appeared to be regularly occupied. Other information suggested 3–4 more pairs. The Ornithological Report for Northumberland and Durham for 1967 said, 'Of some 25 pairs present in the north and north-west of the county, 11–13 pairs bred successfully, a pleasing increase'. This surprising figure seems now to be regarded as an over-estimate, and Galloway and Meek (1983) said at least 11 pairs 'may have bred' in 1967, the highest certain annual total since 1900. Numbers then declined rapidly in the early 1970s. A full survey in 1974 showed only three pairs and one certain nest and by 1977 the Raven was almost extinct in Northumberland. Brian Harle thought that three pairs bred again in 1982 and the county bird reports since then indicate that the species just hung on as a breeder, with from one to three pairs in most years (also Day, Hodgson & Rossiter, 1995). Brian Little tells me that one Cheviot pair has remained constant throughout, and that there have been increased sightings of birds during the last few years, with a successful tree-nest in 1995; and perhaps incipient recovery, with three pairs rearing young in 1996.

In 1983, Brian Harle compiled a list of 41 recorded Raven nesting places in Northumberland, though 11 had been used only occasionally and another seven were deserted by 1960. Nearly all were in the Cheviots or the outlying Simonside and Wanney Hills. Scarcity of suitable crags severely limited the scope for nesting within the south Tyne and Allendales and, although nesting occurred in the 1930s on the abrupt Whin Sill crags along the Roman Wall, recent disturbance levels have been inhibitory. Tree-nesting was occasional, there being at least six records since 1927.

Isle of Man

Cullen and Jennings (1986) have provided an admirable historical account of the Raven on the Isle of Man. It was evidently outlawed quite early on and headage payments were recorded for all parishes except Ballaugh during 1689 and 1690, pointing to a widespread distribution at that time. Substantial decline by the late 1800s was apparent. In the earliest survey, Ralfe knew of about 15 nesting places in 1903, all in rocks and all but one on the extensive Manx sea-cliffs. Cowin (1941) reported that 33 pairs nested in that year, there being a substantial increase in inland nestings. In 1977, there were again 33 pairs, but the 1986 breeding population was assessed at about 40 pairs, with 15–18 of these at coastal locations during the past decade. Even on the assumption that the whole area of the island is utilised by Ravens for feeding (and the northern lowlands look rather unsuitable), this amounts to the fourth highest nesting density in Britain (Table 22). Tree-nesting was unknown until 1932 but has increased slowly since then, with six out of 17 inland nests so placed in 1977 (Cullen, 1978), and nine or ten of the inland nests in trees during each of the most recent years. One pair alternated between a coastal sand-cliff and a tree just inland. Tree-nests are mainly in woodland areas. Other inland nests are in quarries, gorges or low cliffs produced as a result of mining, and nests on buildings are very rare. Manx Ravens are little persecuted nowadays.

SOUTHERN SCOTLAND

Evans (1911) and Bolam (1912) indicate that the Raven was widespread and relatively numerous 'in olden days' within the Tweed area, encompassing much of the counties of Berwick, Peebles, Selkirk and Roxburgh. From its medieval abundance in Edinburgh it was reduced to a single pair nesting on the crags of Arthur's Seat by the early 1800s (Macgillivray, 1837). By the early twentieth century, only a few pairs survived in the wilder hill country, where cliffs provided nesting sites. Muirhead (1889) recorded that several pairs bred on the high sea-cliffs of the Berwickshire coast between Fast Castle and Burnmouth until 1868, but regular nesting had evidently ceased well before 1900. Several inland haunts were also deserted. Only occasional nesting occurred in the Lothians by this time, though there were former nesting places on the Lammermuirs and the islands off north Berwick; Baxter and Rintoul (1953) state that the Bass Rock was the only Forth island on which Ravens had nested.

Information from this district during the period 1910–40 is scanty. Nesting in the Lammermuirs was reported around 1910 and a handful of pairs evidently continued to lead a tenuous existence. The Moorfoot Hills in Midlothian had a regular pair up to the 1960s, but this disappeared after 1976 (Andrews, 1986), and the Pentland Hills had an occasionally used nesting place up to 1966 (Munro, 1988). During 1949–60, I saw three regularly occupied territories and a fourth irregular one in Peeblesshire, and later heard of a fifth haunt. I failed to locate any nesting Ravens in Selkirkshire but found the vestiges of an ancient nest on one small rock. In Roxburghshire, there was a once-off tree-nest near Newcastleton in 1947 and an occasionally occupied rock site on the Hermitage Water. In all these Southern Uplands areas, a general scarcity of good crags appeared to be a major factor in holding Ravens down to these low numbers, though there was ample scope for tree-nesting. The Scottish Cheviots are likewise almost devoid of suitable nesting rocks and had no earlier records. By the 1990s, numbers were similar, with three or four pairs (perhaps none regular) in Peeblesshire, a single pair which returned to the Pentlands and bred successfully in 1993, and a single pair once more in the Moorfoot Hills. A flying brood, seemingly reared on the Scottish Cheviots, was also seen in 1994, and a pair bred here in 1996 (B. Little). Two more pairs nested in the Hermitage area in 1996 (S. Redpath). A single pair hung on along the Berwickshire coast for many years and the *New Atlas* recorded breeding in two 10 × 10-km squares here in 1988–91. A pair was still present in 1994, though it was not proved to breed.

The western half of southern Scotland has held the bulk of the Raven population during the present century. Gladstone (1910) said that 12–16 pairs of Ravens nested in Dumfriesshire annually. During the period 1949–70, the records of W. Murray, D. Cross and myself gave 23 territories with rock-nesting Ravens, of which 16 were evidently regularly used. At least three of these territories had alternative sites in trees and there were also three regular tree-nesting pairs, plus at least another four occasional or once-off tree-nests in Dumfriesshire. During 1994 a total of 19 territories were found occupied, with seven tree-nests (C. Rollie 1994 survey). A new tree-nest was found near Langholm in 1996 (S. Redpath). Parts of the hill ground of this county have few cliffs, so that the Raven population was somewhat unevenly distributed. The rockier hills of Kirkcudbrightshire to the west had more scope for crag-nesters; there were few earlier records but during 1946–70 I saw 30 territories with rock-nesting Ravens, and other observers reported another four. Of these, six had dropped out by 1960, but 18 of the rest were regular up to 1970. In three territories there were alternative tree sites and at least another four pairs nested only in trees. A full survey, organised by C. Rollie in 1994, showed 27 pairs present, eight nesting in trees.

Paton and Pike (1929) said that Ayrshire annually had 11 breeding pairs, including one or two in Kyle. During 1950–70, at least 15 inland Raven territories were known in Ayrshire, of which two were in trees; occasional tree-nests were reported elsewhere. Most of these were in the district of Carrick, but at least two were in the far north of the county. In 1994, Raven pairs were known in 13 inland territories (C. Rollie). Wigtownshire has few inland cliffs that could hold Ravens but Dickson (1992) mentions two occasionally used rock sites on the moors, and reports sporadic tree-nesting during recent years. Three occupied territories were known in 1994, one in a quarry. Renfrewshire had a single inland pair in 1994 (C. Rollie).

On the cliff-bound sections of the Galloway and Ayrshire coast, Ravens

Distribution and numbers in Britain and Ireland

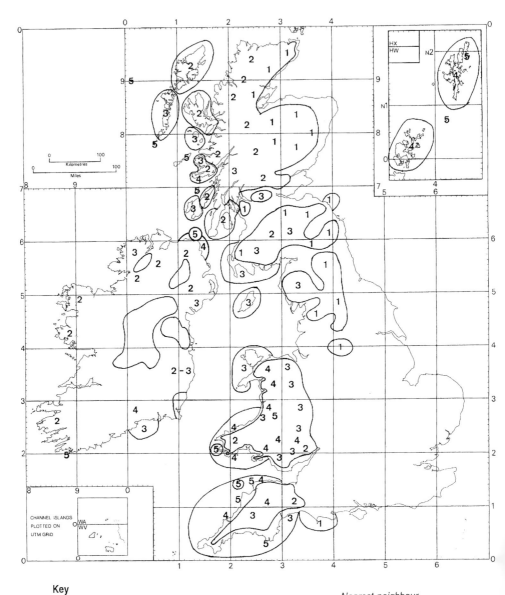

Key

		Pairs/100 km²	Nearest neighbour distance (km)
5	Very high	>15.0	<2.0
4	High	9.0–14.9	2.0–3.0
3	Moderate	5.0–8.9	3.0–4.5
2	Low	1.0–4.9	4.6–8.0
1	Very low	<1.0	>8.0

Notes: Figures are given only for those areas where reliable data on breeding density are available.
Raven breeding areas shown by encircling lines in Britain; but in Ireland, where Ravens are widespread, areas appearing as "islands" have few if any breeders.

11. Granite cliffs in the Galloway uplands, where Ravens have continued to nest despite much robbing of eggs. Cairnsmore of Fleet, Kirkcudbrightshire (photo: D. A. Ratcliffe)

have bred in 50 different territories since 1945. Of these, 11 were in Kirkcudbrightshire, 29 in Wigtownshire (20 in the Rhinns, 9 in the Machers) and 10 in Ayrshire. In 1961–62, pairs were seen in at least 27 territories and, of these, 21 certainly bred, but at least 10 formerly regular territories were not visited, so that the population could have numbered up to 37 pairs. In 1974–75, all but four territories were checked and at least 19 found occupied, with 13 definite nests, i.e. a maximum population of 23 pairs. In 1981, all territories except one (formerly regular) were examined and only 12 pairs located, of which 10 may have bred. Allowing for a possible thirteenth pair in 1981, the figures showed minimum decreases of 32% since 1974–75, 52% since 1961–62, and >55% since the pre-1961 period. Breeding success in this population had declined still further, by at least 66% in 1981 compared with 1961–62. There has since been a recovery, with 33 pairs holding territory on the Galloway and Ayrshire coast in 1994, of which at least 18 bred successfully. Records up to 1962 are mainly from R. Stokoe, W. Hughes, D. Watson and myself; those for 1974–75 are from M. Marquiss, for 1981 from R. Mearns (1983), and for 1994 from C. Rollie. Dickson (1992) summarises figures up to that date.

Figure 4. Variations in breeding density of Ravens in different parts of Britain and Ireland in recent years.

The Dumfries and Galloway Raptor Study Group conducted a complete survey of both coastal and inland Ravens in southern Scotland in 1994. I am indebted to the group, and especially to C. Rollie, for the results. Altogether, 79 pairs were proved to breed in southern Scotland in 1994, 19 of them in trees, and birds were present in another 30 localities. Chris Rollie informs me that, in 1995, Ravens bred in at least another five new localities (three in trees) and pairs reoccupied two more, long-deserted haunts. By 1996, another three pairs bred and two more held territory in the Langholm–Newcastleton area (S. Redpath).

SCOTTISH HIGHLANDS

Southern fringe

I have included the Clyde Islands within this region. McWilliam (1936) said that there were about eight pairs of Ravens nesting on Arran and about three more on Bute. Gibson (1953) gave a larger total, with eight to ten pairs on Arran plus another two on Holy Island. For Bute he said two pairs normally nested, with a third pair either there or on Inchmarnock and another one or two on Little Cumbrae. Other records added that two pairs nested on Inchmarnock in 1952–53, and that a pair occasionally nested at the north end of Great Cumbrae. The Arran Bird Reports give 17–18 pairs for the island in 1982–84, decreasing to 13 pairs in 1987–89, but with 15 pairs again in 1991.

The Raven population of Stirlingshire and Dunbartonshire has been monitored by John Mitchell since the mid-1960s. In the south of this district, the low Kilpatrick, Campsie, Fintry and Gargunnock Hills held at least six pairs in the 1960s, but these were reduced to three pairs by 1981 and only two in 1987. Yet, by 1995, numbers had climbed back to a higher level than before, with eight pairs breeding successfully and two other pairs occupying old territories, though without nesting. Monitoring in the northern part of the district is complicated by the sharing of four territories between adjoining counties: two between Dunbarton and Stirling across Loch Lomond, one between Stirling and Perth near Inversnaid, and another between Stirling and Perth in the Ochil Hills. There are in total 14 territories, of which only five were occupied in 1987; in 1995 eight territories were occupied in northern Dunbartonshire but only four by breeding pairs, while in northern Stirlingshire four territories were occupied, all but one by breeding pairs. An additional tree-nesting pair was discovered in 1995 in Dunbartonshire, and bred again in 1996.

Central and south-west Perthshire, south of the Glen Dochart–Glen Lochay watershed and Loch Tay, and west of the Aberfeldy–Crieff road, has been monitored by Don MacCaskill and Patrick Stirling-Aird, with the assistance of other local ornithologists. At least 42 post-1950 territories are known, though their history of use up to 1980 is rather patchy. A survey of 23 territories in 1981 showed that 18 were occupied, with 12 probable nests. Fuller surveys during 1986–94 showed that at least 19 territories were regularly occupied. Another 11 appeared to be irregularly used, though it is possible that not all their alternative cliffs have yet been located. A further eight territories were visited too infrequently for their occupancy to be known and the remaining three have not been used since 1974, 1976 and c. 1978. At least 20 pairs bred in 1992, probably 18 of them successfully. During 1992–94, five additional territories had become tenanted by breeding pairs, involving the occupation of former alternative crags or reoccupation of long-deserted nesting places. In 1995, 22 territories were examined and 19 were found to be occupied. Overall breeding population during these last three years is estimated at 30–32 pairs. Only a single tree-nest has been known in this district during the post-war period.

Central and eastern Highlands

In Perthshire, north of the previous district but south and west of the A9 road, including the Breadalbane range, Ravens have been less closely studied. At least 15 post-1950 territories are known and 11 pairs attempted to nest in 1986 (A. G. Payne). In 1994 the Tayside Raptor Study Group checked 10 of these territories, recording eight pairs and two single birds. A single tree-nesting site is known in this area. The remaining, most northerly part of Perthshire, north and east of the A9 road, has few Ravens, though at least seven post-1950 territories are known. In 1986, only one certain nesting was known to Sandy Payne.

In the eastern Highlands, the Raven is an extremely sparse breeder. Seton Gordon (1912, 1915) remarked on its virtual absence from the Cairngorms and this remained the position for many years. A few pairs were established by the 1960, and Nethersole-Thompson and Watson (1981) mention three broods reared on Deeside in 1970–71, but the species declined later (Thompson *et al.*, 1995) and is again almost absent from the area. Baxter and Rintoul (1953) said that Ravens bred fairly commonly in Aberdeen and Kincardine up to *c.* 1850 but were extinct there by 1903. Buckland, Bell and Picozzi (1990) describe the Raven as 'uncommon' in the counties of Aberdeen, Banff and Kincardine and note that it had become scarcer during the previous 15 years and irregular as a breeding bird. Breeding records by 10 × 10-km squares were as follows:

Date	Confirmed	Probable	Possible	Total
1968–72	3	1	8	12
1981–84	1	1	3	5

Mick Marquiss has, however, noted an increase in non-breeding birds in this district within the last 6 years.

Crighton (1976) said that only a few pairs then bred in Angus. In 1995, four pairs nested in Angus, with another pair possibly attempting to breed. Small numbers (often singles or pairs, but sometimes three to seven birds) have been seen in numerous widely separated localities in recent years, suggesting that the species has the potential to colonise more widely as a breeder (R. Downing, M. Nicholl, D. Whitton). Cook (1992) referred to old accounts that it was 'numerous' in 1775 in Moray and Nairn, but had been reduced by persecution by 1844, and was rare by 1863. He reported that only occasional pairs remained in the early 1960s and that no nesting had occurred since 1974. Thom (1986) mentions post-1970 records for Moray and Nairn.

In that part of Inverness-shire south of the Great Glen, the eastern districts also show a present scarcity of Ravens comparable to that in the adjoining counties mentioned above. In the Spey Valley, Weir (1978) reported a marked recent decrease: from an average of 16–17 breeding pairs in 22 territories in 1964–68 to only 5 pairs in 1977. Within the Cairngorms area, occupying the heart of the eastern Highlands, Boobyer (1995) found that comparison of the two *Atlases of Breeding Birds* shows net losses for Ravens of 30 squares (10 × 10 km) in presence records and 18 for confirmed breeding, out of a total of 71 squares, between 1968–72 and 1988–91. Farther west, the bird becomes more numerous, though its breeding density is probably no greater than that of the adjoining part of Perthshire. During 1961–63 I saw occupied eyries in nine places and another five unoccupied cliffs with old nests, and think it likely that the total number of breeders in southern Inverness-shire is at least 20 pairs.

Mainland Argyll

Chris Thomas has recently studied Ravens in Argyll and found them to be widespread throughout the mainland district south of the Firth of Lorne. He mentions four recent tree-nesting pairs. The Oban area, northwest of Loch Awe and south of Loch Etive, contains 24 pairs; while the stretch between Loch Awe and upper Loch Fyne holds another 9 pairs. The high mountain country north of the road from Tyndrum to Connel Ferry probably supports at least 30 pairs, the Cowal peninsula another 40 pairs, and Knapdale–Kintyre another 50 pairs. The total for the district is thus an estimated 153 pairs. In the north they are mainly inland breeders, but Kintyre has a good many coastal pairs on its extensive sea-cliffs.

The northern Highlands

Argyll north of the Great Glen covers the mainland districts of Kingairloch, Morvern, Ardgour, Sunart and Ardnamurchan. Breeding Ravens are widely but rather sparsely distributed here, both on rocky coasts and through the deer forests and sheepwalks of the hills. I came across 14 occupied territories in 1962–63, but covered only part of the ground. During 1982–85, Jeff Watson found 13 of these territories occupied and added 15 places where Ravens were either certainly or probably nesting, and another nine where pairs were seen during the breeding season. The total breeding population is thus probably at least 35 pairs, all in cliffs.

North Inverness-shire, between the southern border along Loch Shiel–Loch Eil and the northern march with Ross-shire, is one of the least-known parts of Britain and its Ravens have probably never been counted accurately. During 1961–81 I saw occupied nests in nine localities, plus another seven with only old nests where breeding was evidently sporadic. The species is evidently at low density pretty well throughout the district but, with an average of 1.6 pairs per 100 km² found in adjoining areas of Ross-shire, would have an estimated 50+ pairs.

Raven density is low throughout Ross-shire, but a high proportion of the land surface and coast is suitable habitat. During the 1960s 34 occupied nesting territories were seen during surveys which covered at least half the county (R. Balharry and D. A. Ratcliffe, unpublished data). Nesting on the sea-cliffs of the east coast has evidently decreased since 1972, but continues in a Cromarty locality. The *New Atlas* shows a good many absences from 10 × 10-km squares in Ross-shire during 1988–91, but this may be partly the result of incomplete coverage in remoter areas. The current population of the mainland county is probably at least 50 pairs.

Sutherland shows a similar picture. I came across 25 nesting territories during 1957–75, while other sources have added another 20 (S. Rae, A. Vittery, D. B. A. Thompson, W. Scobie, BTO Nest Records). Breeding density is extremely low inland and some eastern parts of the county have a shortage of suitable cliffs, but there is much more scope for sea-cliff-nesting

12. The deer forests of the north-west Highlands; Golden Eagles are as numerous as Ravens. Suilven, Sutherland (photo: D. A. Ratcliffe)

than in Ross-shire and the *New Atlas* shows an almost continuous distribution on the north and west coast. The estimated population is 70 pairs, of which 30 are on the coast.

Caithness has at least six sea-cliff pairs, numbers having evidently increased here since 1972 (*New Atlas*), and there is a scatter of inland-nesting places, mainly in old quarries, occupied somewhat irregularly. The present population probably does not exceed ten pairs.

SCOTTISH ISLANDS

Inner Hebrides

Elliott (1989) said that, 'In the mid-1980s Islay held about 100 Ravens' which hunted all over the island but spent most time in moorland and coastal areas. He implied that these were breeders but went on to describe communal roosts with similar numbers. The *New Atlas* shows confirmed breeding in all 10 × 10-km squares (13), so that a total of at least 40 pairs seems reasonable; they are largely coastal nesters. Eric Bignal knew of three nests on the more barren adjoining island of Jura in 1981, but an extensive bird survey of Jura in 1985 noted that Ravens were 'conspicuous by their absence in early June, with only eight birds seen (at two sites)'. The distribution of sea-cliffs nevertheless suggests that Jura could have up to ten coastal pairs.

On the smaller islands of Argyll, Coll had an estimated 12 breeding pairs, mainly on sea-cliffs, in 1987 (Stroud, 1989), while another 12 pairs bred on Colonsay in 1994 and six pairs on Tiree in 1993. A pair bred on the Garvellachs in 1989 (Argyll Bird Reports). The chain of islands from Scarba to Lismore could hold at least 10 pairs, probably more. Nearly all these Ravens are coastal breeders.

Mull

Paul Haworth found 57 breeding territories during 1990–96 and believes a few more still remain to be discovered. Total nesting population may thus be put at 50–60 pairs, of which the majority breed on coastal cliffs. North Mull has 20 territories, with 11 of these concentrated in the area around Calgary, while the central to southern part of the island has another 19 territories. The southern peninsula, the Ross of Mull, has 15 pairs, with another two or three on Iona.

Skye

During June 1962, I saw 22 localities where Ravens had nested or adults were present in likely nesting places. Bruce Philp recorded another 12 nesting areas in 1974. Allowing for ground not covered, including the adjoining islands of Raasay, Rona, Scalpay and Soay, a population of 50–60 pairs seems probable. Most breeding Ravens on Skye

13. *The high sea-cliffs of the Hebrides provide some spectacular nesting haunts. Talisker Head, Skye (photo: D. A. Ratcliffe)*

are on the sea-cliffs, and probably no more than ten pairs are found in the unproductive interior.

The Small Isles

Rum regularly has four or five nesting pairs, and Canna two, while Eigg and Muck have an estimated three and one; a total of at least ten.

Outer Hebrides

Macgillivray (1837) said, ' The Raven is astonishingly common in all parts. It roosts and nestles on high rocks on the sea-shore, as well as in the interior'. Cunningham (1983) said that, in Lewis, the bird is especially abundant at all seasons but, apart from mentioning nesting in the Stornoway Castle Woods and at three rock sites, gives no idea of the breeding population size. Paul Haworth has compiled a list of 35–40 Raven nesting places on Harris and Lewis, and the total may well be at least 50 pairs. Bill Neill has estimated breeders on North Uist, Benbecula and South Uist at 35–40 pairs, with another pair or two on Barra. Nesting is reported for the remoter sea-bird islands, including North Rona, the Flannans, the Shiants, Mingulay, Pabbay and the Monach Isles. At least four pairs were breeding on the St Kilda group in 1959 and more recent figures are for five to eight pairs (Harris & Murray, 1978). The *New Atlas* shows a fairly continuous breeding distribution throughout the Outer Hebrides, with confirmed nesting in 51 10 × 10-km squares. During a visit in 1969, my impression was that breeding density was rather low, so that allowing an average of two pairs per 100 km^2 would give a total population of about 100 pairs, which agrees well with the estimate above.

Orkney

The Raven was described as a 'constant inhabitant' in 1848, but Buckley and Harvie-Brown (1891) thought it was much scarcer than formerly. David Lack found it well distributed in 1941.

On Mainland Orkney, Booth (1979) located 27 Raven territories with a mean of 23 nests per year during 1972–77. The higher sea-cliffs were preferred, with 83% of all nests situated here. Inland-nesting sites included both active and disused quarries, stream ravines and disused buildings (three). Two tree sites were used once each, woodland and trees being scarce on Orkney. Numbers on Mainland increased, subsequently reaching 33–35 pairs in 1994–95 (C. J. Booth). Booth, Cuthbert and Reynolds (1984) reported numbers of breeding pairs on the other islands of the group during 1974–82, which suggested a minimum Orkney population of 75 pairs. Chris Booth has recently (1996) given me an up-date, as follows: Mainland, 33; Hoy, 14; Eday, three; Rousay, three; Sanday, four or five; Shapinsay, three; South Ronaldsay, at least six; Stronsay, three; Westray, six to eight; single pairs nest on Auskerry, Burray, the Calf of Eday, Cava, Copinsay, Egilsay, Eynhallow, Flotta, Gairsay, Glims Holm, Helliar Holm, Holm of Papa Westray, Lamb Holm, Linga Holm, North Ronaldsay, Sweyn Holm and Swona. There is occasional nesting on other holms and islands. The total breeding population of the Orkney Isles thus seems likely to be around 90 pairs. Nesting on old buildings has become quite frequent (Booth, 1996).

Shetland

Saxby (1874) described the Raven as resident and very abundant, and added, 'but large numbers which are certainly not natives arrive in autumn, usually about the middle of October, remaining until the breeding season begins'. During a survey of most of the long Shetland coast in 1982 and 1983, no less than 184 Raven breeding territories were located, of which 174 were coastal (Ewins, Dymond & Marquiss, 1986). By combining local information over a longer period with their survey results, the authors concluded that regular breeding occurred in 161 territories and sporadic breeding in the remaining 23. This points to an average breeding population for this island group of around 172 pairs. The island of Fetlar had eight regular nesting pairs during 1977–83 and nine in 1984 and 1987.

During a further study of Shetland Ravens during 1984–93, Ellis *et al.* (1994) found another ten breeding territories, nine of them in inland locations, making a total of 206. For a non-random but widely spread sample of 43 territories during 1987–93, a mean occupation rate of 80% and a mean of 73% with nesting pairs was found. These were equivalent to a total Shetland population of 165 pairs, of which 150 were breeding pairs. Inland-nesting places included streambanks, active or disused quarries, road-cuttings and, in one case only, a tree – there being very little woodland or even scattered trees on Shetland.

Even taking the lower figures, this is one of the largest Raven nesting populations in Britain, and also one of the densest, with an average territory area of 7.35 km^2 per nesting pair (assuming all land to be used) and a mean length of coastline per territory of 7.4 km. Breeding was confirmed in 92% of the 10 × 10-km squares in Shetland (Gibbons, Reid and Chapman, 1993), compared with only 50% during the 1968–72 BTO atlas survey (Sharrock, 1976). The difference is felt to reflect increased observer effort and coverage rather than a marked increase in population. Shetland also held a non-breeding Raven population, estimated at 400 or more flock birds, during the early 1980s, and Ewins, Dymond and Marquiss (1986) believed that the breeding population was close to its maximum, with all tenable coastline occupied in most years.

Fair Isle had up to six pairs until about 1897, then only one or two during the early twentieth century, but three, occasionally four, pairs bred after World War 2 (Williamson, 1965). There were from two to six nesting pairs during 1957–90, with six in 1983–84 (Dymond, 1991).

IRELAND

Information on Raven distribution in Ireland is extremely patchy, and I have had to rely mainly on the two works covering the whole country. Ussher and Warren (1900) said that it had rapidly diminished throughout Ireland during the previous 50 years: 'It has been driven from most of its inland breeding places, such as cliffs over rivers, and lofty trees in desmesnes, like those of Curraghmore and Clonbrook, and the islands of

Lough Erne . . . and it is now a much rarer species than the Peregrine or Chough'. They said the coasts of Kerry, Clare and western Cork were among its most frequented haunts, though their brief geographical scan of distribution suggested that Ravens still bred in most the main mountain ranges and on all the major sections of rocky coast. Kilkenny, Londonderry and Tyrone were regarded as counties where the bird was extinct.

Praeger (1937) said that the Raven 'has multiplied in recent years, and has even nested within a few kilometres of Dublin. It is seldom that you visit any mountain region without hearing its deep croak, and sometimes you may see quite a flock of them'. Hutchinson (1989) records that the species has increased considerably since 1900 throughout much of Ireland. Breeding recommenced in Co. Dublin in 1917, Kilkenny in 1929 and Wexford in 1944, and was restored in all 'lost' counties by the early 1950s. By 1968–72 it was nesting in all counties except Monaghan, Meath, Longford, Kildare and Offaly. In Kildare, where there were none in 1968–72, six pairs bred in 1987. Tree-nesting is once again quite widespread.

The first *Atlas of Breeding Birds*, covering 1968–72, showed a wide distribution on the coast and in the main mountain systems distributed around the periphery of Ireland, but a general absence from many central and lowland areas. Comparison with the *New Atlas* (1988–91) showed an overall increase of 54% in occupation of 10 × 10-km squares, mainly through an expansion from the periphery inwards, though some parts of the central lowlands are still unoccupied (Figs 2 and 3). Many records were for presence only and, when confirmed breeding records in the two maps are compared, the increase is only 20%.

Obtaining more precise population figures is more difficult, except for a few closely studied areas. Gabriel Noonan's long-term study in Wicklow and adjoining counties (presumably Dublin, Kildare, Carlow and Wexford) had identified 80 breeding territories, mainly inland, by 1987. Tree-nests have been used at least once in 44 of these territories since 1940, though rock sites were preferred in many of these (Hutchinson, 1989). In Co. Waterford, Walsh and McGrath (1988) reported breeding during 1976–86 at 20 coastal localities, and at another 18–19 inland localities, mainly in the Comeragh–Monavullagh and Knockmealdown Mountains. This represented an increase on the first *Atlas* records, though improved coverage was partly responsible. Inland sites included six or seven in trees, two on buildings and one in a quarry, while another building was used on the coast. Birds were present in spring in several more likely nesting areas, suggesting a total for the county approaching 50 pairs. Breeding was confirmed in 17 10 × 10-km squares and birds were present in another four.

The many stretches of sea-cliff in Cork and Kerry give ample scope for nesting Ravens but, apart from a record of six pairs breeding on Cape Clear Island in 1986, there are no meaningful figures. The rugged mountains of these counties must also have an abundance of possible nesting places, but the general impression is that Ravens are at only modest densities in these rather barren uplands. A scatter of nests is reported in Limerick, and the coastal cliffs and limestone hills of Clare must hold moderate numbers.

14. The rugged coast of south-west Ireland, with sheep pastures behind, is good Raven country. Near Waterville, Kerry (photo: D. A. Ratcliffe)

In a survey of the district of Connemara, Galway in 1985–86, Paul Haworth found 23 occupied territories (unpublished report to Worldwide Fund for Nature). These are evidently the 23 nests in Connemara referred to by Whilde (1994). Haworth gives 26 pairs as the total Connemara population. Breeding density is evidently low, though higher in the northern hills of the Twelve Bens and Maamturk Mountains than in the lower southern moorlands. Most of the nesting places were on inland crags, the coast of the district being mostly low, with few cliffs suitable for nesting Ravens. Two pairs were nesting in trees, one on an island in a large lake. The extensive sea-cliffs of Mayo evidently have a good many Raven pairs, but the blanket bogs of the interior provide few suitable nest sites and, inland, the species is mainly in the higher mountain ranges of Nephinbeg, Partry, Sheefry and Mweelrea.

There is a good deal of suitable Raven country in the limestone hills of Sligo and Leitrim, but no figures are available. In Donegal, Mac Lochlainn (1984) found six Raven territories in the 9,600 ha of the Glen Veagh National Park, a higher density than in Connemara. Some of the Donegal hills have fewer suitable cliffs for nesting and numbers there are probably lower, but there must be many pairs on the numerous sea-cliffs.

In Northern Ireland, J. H. Wells and his colleagues estimate a current population of about 250 pairs, of which around 200 nest annually. With the

New Atlas confirming breeding in about 84 squares, this suggests an average density of two pairs per 100 km^2. County totals of breeding territories are as follows (estimates of additional locations, mainly with tree nests, are given in brackets; breeding populations average 80% of these figures): Antrim 71(+15), Londonderry 28 (+6), Tyrone 19(+5), Down 33(+10), Fermanagh 25 (+4) and Armagh 22 (+4). Habitats of known nesting territories are 52 inland natural cliffs, 39 sea-cliffs, 85 quarries and 27 trees.

The *New Atlas* shows breeding evidence in 380 Irish 10 × 10-km squares, and presence in another 306, during 1988–91. Coverage is unlikely to have been complete in Ireland during this period, but it is also improbable that breeding occurred in all the other squares where birds were seen. Few parts of Ireland appear to sustain a density of more than two pairs per 100 km^2 over more than limited areas, and this figure probably represents a fair average across the country as a whole. Taking an estimate of breeding in 500 squares would thus give a total breeding population of 1,000 pairs. Adding up the separate known county figures and estimating the rest gives a lower total of 961 pairs. The figure of 1,500 pairs in the first *Atlas* thus seems an overestimate, while that of 3,500 pairs in the *New Atlas* is wildly optimistic.

GRAND TOTALS

Britain and Ireland

The foregoing figures suggest a breeding strength of 777 pairs for England, 1,250 for Wales, and 1,140 for Scotland, giving a grand total of 3,167 pairs for the whole country. This includes some estimated figures which are on the side of caution, so that the total can be justifiably rounded up to 3,300 pairs. The all-Ireland total is rounded up to 1,000 pairs.

The two countries together have an estimated 4,300 pairs, still below the figure given in the first *Atlas of Breeding Birds* (Sharrock, 1976) but less than half that estimated in the *New Atlas of Breeding Birds* (Gibbons, Reid & Chapman, 1993). The latter seems far too high, probably because insufficient allowance was made for the low density of breeding Ravens in Scotland and western Ireland. The *New Atlas* estimate was reached from tetrad counts and it seems probable that the inclusion of non-breeding birds has been a large source of error. The comparative figures given in the *New Atlas* show that, for 'breeding evidence', the decline in Britain from 1,031 to 785 squares was not compensated by the increase in Ireland from 316 to 380 squares, so that there was a net loss of 182 (13.5%) in distribution. It is uncertain how this relates to population size. There has been an increase in density and a slight expansion of range in some more southerly parts of Britain and this may have cancelled out the extensive losses in parts of northern England and Scotland, where the bird was mostly at low density.

CHAPTER 4

Food and feeding habits

The Raven has an extremely varied bill of fare. As Smith (1905) well put it, 'His dietary ranges from a worm to a whale'. Chapter 1 contains some general observations on food, noting that the staple is carrion and that the species is, first and foremost, a scavenger and, in northern Europe, a vulture substitute. It is not at all particular about the kind and source of carrion, taking into account its range beyond these shores, and noting also its historical opportunities. Wherever larger animals live and die, and their carcasses are left to nature's disposal, there the Raven will usually be found.

The Raven's medieval hey-day was when the townspeople provided it with a copious supply of discarded food of all kinds. In Britain, after the species lost its urban connections, it depended especially on land used for the rearing of domestic animals, including horses, cattle, pigs, goats and sheep. On the remoter uplands, the native red deer must always have provided an important food source. By the nineteenth century, in inland districts, it became largely a bird of the open sheepwalks, which supplied it thenceforth with the main item of food. Many writers have noted the Raven's apparent dependence on this mutton fare. Harvie-Brown and Macpherson (1904) said that, 'The large extent to which sheep farming exists on the fell-lands of Cumberland and Westmorland is correlated with the wide dispersal of *C. corax* over our higher grounds, since the Raven depends to a large extent on carrion for its maintenance'.

The hunter-naturalist Abel Chapman, who watched and wrote interestingly about the Ravens of the Cheviot moorlands, spelt out the theme in more detail:

> Sheep, it is true, form a main source of the Raven's subsistence; but for all that, in no proper sense is the Raven a detrimental, since he is not a raptor, but purely a scavenger. The natural death rate among countless herds scattered over wide fell-areas alone suffices to provide many Ravens with their 'daily bread'. Over and above a normal mortality, and the exigencies of the lambing season, sheep are liable to a whole category of casualties. Thus if cast and unable to regain their feet, they lie helpless and defenceless – awaiting death. Similarly, after lambing, a ewe may be too weak to rise; and mortality is increased by foot-rot, louping-ill and other natural causes. Again, sheep get bogged, or 'crag-fast' – in short *horrida mors* attends upon the fleecy flocks as it attends upon Man; and in each and every such case, the watching Raven is there expectant and ready to act as sexton – if not, on occasion, to anticipate that office. (Chapman, 1924)

A scientist-naturalist of a later generation, Ron Murton, also emphasised the close association between the Raven and sheep in this country during more recent times (Murton, 1971). As Chapman noted, sheep ranging free over the uplands are beset by a variety of maladies and mishaps which contribute to a rather high mortality. The effects of various diseases and nutritional disorders or deficiencies are enhanced by the frequent severity of the weather in winter and spring. Among precipitous mountains, a fair number of sheep come to an untimely end on the crags; seeking a nutritious bite, they jump down onto ungrazed ledges from which they cannot retreat and eventually totter into the abyss. Where lambing takes place on the open hill, the losses can be heavy, especially during inclement seasons. Although the law requires dead animals to be removed or buried, it has been normal practice to leave the carcasses of sheep and lambs to the usually rapid attentions of nature's scavengers – which include not only Ravens, but also foxes, Buzzards, Carrion Crows and, more locally, Red Kites and Golden Eagles. During the lambing season, the placentas lying about the hill are an important extra item of food.

Saxby (1874) said that, in Shetland, Ravens fed a good deal on the carcasses of ponies, which in those days were poorly fed and subject to heavy winter mortality. Nowadays, horses and cattle which die in the open are usually quickly disposed of by farmers, so that they are seldom available as carrion. In the Highland deer forests, the native red deer provides the Raven with a natural carrion source, which may be plentiful during or after hard winters. The birds also feed here on the 'grallochs' (discarded entrails) of shot deer, and are sufficiently conditioned in some places to assemble at the report of a rifle. On the grouse-moors, provided that they survive the poison laid out for miscellaneous predators, or other keeper attentions, Ravens pick up incapacitated Red Grouse or those that have succumbed to their wounds after the shooters have left. Speedy (1920) noted that on one Highland grouse-moor, where no Ravens bred, they congregated at the

15. *Hill sheep after winter snow: a time of hardship for both these animals and the Ravens. Scottish Borders (photo: D. A. Ratcliffe)*

onset of the shooting season and worked the ground for dead and wounded grouse. Batten (1923) also watched Ravens systematically quartering a moor for wounded grouse after a drive. On both Orkney and Shetland, Ravens are reported to feed a good deal on road casualties, especially birds and lagomorphs (Hope Jones, 1980; Ewins, Dymond and Marquiss, 1986).

Coastal Ravens eat dead birds washed ashore, along with other edible pickings, such as fish and marine invertebrates of various kinds, especially molluscs. Dead and beached cetaceans and seals provide a bonanza, and Saxby recorded the record number of 800 Ravens assembled to feed on the carcasses of an unspecified number of whales driven ashore at Uyea Sound, Shetland, in 1864. They forage over the farmed grasslands and arable land, and the uncultivated heaths and other rough ground behind or adjoining the cliffs. Some pairs have sea-bird nesting cliffs within their territories, which provide additional scope for scavenging, and, in large colonies, there is usually a plentiful supply of dead or disabled birds. There are also many opportunities for taking eggs and chicks. The sea-cliff Ravens in Cornwall observed by Ryves (1948) nevertheless invariably flew inland to feed while he was watching them, and he never saw them interfere with nesting bird neighbours in any way. In Iceland, Lewis (1938) said that flocks of Ravens

78   *Food and feeding habits*

were associated especially with the coastal fishing towns and villages, subsisting on refuse and fish remains, and scavenging in the streets as they used to do in medieval Britain.

There are fewer opportunities in modern Britain and Ireland for Ravens to find significant amounts of discarded food, but rubbish tips and slaughterhouse yards are much favoured places and often draw good numbers of non-breeders (see Chapter 5 and Appendix 1). Small parties of Ravens frequent the summit of Snowdon for most of the time that the tourist café is open and in Lakeland the bird has taken to hanging about popular picnic places on the fells for discarded eatables.

STUDIES OF FOOD

Although there are numerous direct observations on the feeding habits of Ravens, the most comprehensive information on food comes from the analysis of identifiable remains in their castings – the pellets of undigested material which the birds regurgitate. These are usually about 5–8 cm long by 2–3 cm thick, dark greyish in colour and typically composed of mammal pelage, which varies from the fine-textured fur of small rodents to the coarser wool of sheep and hair of deer and goats. There are frequently bone fragments and the wing-cases of beetles, as well as vegetable material and large pieces of stone and grit. Raven castings superficially resemble the droppings of foxes, but have a different texture and usually a slightly different shape; they also carry the characteristic musty smell of the Corvidae. They accumulate below favourite perches and roosts, and sometimes amount to a considerable number, though decomposition and weathering gradually break them down.

There have been several major food studies based on pellet analysis in Britain and I have summarised the data in Table 2. Marquiss and Booth (1986) note that neither the frequency nor the volume of remains in Raven pellets are directly proportional to the quantities of the various foods eaten. Single large items may appear in a run of pellets from the same bird and, if they are unusual, may give a false impression of importance. Examination of castings may produce an underestimate of the proportion of soft and readily digested food, which includes some invertebrates and some vegetable matter, but is a good method for comparing the broad spectrum of diet in different areas and identifying the most important items. Allowance obviously has to be made for the extremely wide variations in size and weight between the different items when assessing their dietary value to the bird.

While the range of foods eaten is of great interest, from the viewpoint of population ecology it is important to know the items necessary to the Raven's existence in any area, and the degree to which their possible shortage can be offset by alternative foods. The available studies deal with the diet of birds both in communal winter roosts and in breeding haunts, and sometimes with seasonal variations. They lead to the conclusion that the species is, in common with other wide-spectrum feeders, somewhat opportunistic in its diet, exploiting whatever local or seasonal options are available and switching as necessary when change in supply offers a choice or dictates a shift.

The earliest published analysis was by Bolam (1913), who reported on the contents of 433 separate castings collected at random from a mass of those lying beneath a communal roost inland near Llanuwchllyn, in Merioneth, during December (Table 2, column 2). Of these 39% contained sheep's wool, which was by far the most frequent item. The remains of rabbits, rats, mice or voles and moles were quite important, and 28 castings contained the hair or remains of cattle. Although it was an inland roost, 'seashore subjects' figured in 11% of the pellets. Vegetable remains were also frequent, especially in the form of fruits and seeds (notably acorns and beech mast) and tree buds or seed cotyledons.

From other pellet analyses made in the same area but spread through the year, Bolam noted that, although the locality was far from the sea, at all seasons some pellets showed traces of food obtained on the shore. Fragments of mussel shells and crabs were most frequent, with occasional bits of clam and other shellfish, and also fish bones and seaweed fragments. In summer, skulls of mice, rats and especially voles were very numerous. Moles were frequent and almost equally so throughout the year, some being evidently the remains discarded by mole-trappers. Fur and bones of rabbits were plentiful at all seasons, as was sheep's wool, but the hair of cattle and dogs was less so. Eggshells were found only once, and not more than 10% of pellets contained bird remains at any season, though they were most frequent in winter.

Ernest Blezard's unpublished records of his work on the food of birds included analyses of 62 castings from 11 nesting places in northern England, and 63 castings from 8 nesting places in the Southern Uplands. The results are so similar that I have combined them in Table 2, column 3. All 125 castings were collected in the breeding season. Sheep's or lamb's wool was the predominant item and found in 60% of castings, while remains of rabbits and brown hares were about equally frequent (together 24%) and short-tailed field voles formed 20%. At least 53% of castings contained vegetable matter, mainly as grass or moss, and a few were composed almost entirely of grass. Besides stone and grit, soil was a frequent component.

Mick Marquiss analysed 697 pellets from 29 nesting places in the Southern Uplands and Cheviots (Marquiss, Newton & Ratcliffe, 1978). The results (Table 2, column 4) again show an overall preponderance of sheep remains (at 56% frequency), with rabbit and hare (both brown and mountain) together at 27%, field vole at 16% and feral goat at 14%. Eggshell had the highest frequency of any study, at 28%. No less than 93% of pellets contained grass or other vegetable material. Diet varied here in relation to the amount of afforestation in the vicinity of the nest site, and the significance of the variations is discussed under land-use change (see pp. 206 and 227–230).

In central Wales, Newton, Davis and Davis (1982) reported on 789 pellets, 450 from the breeding season (February–June) and 339 from the rest of the year. They were obtained from 30 nest sites and from various roosts and food sources, including a rubbish tip (Table 2, columns 5–6). Pellets from territorial birds showed 92% frequency of sheep remains during the breeding season and 86% from the rest of the year. No other food achieved anywhere near this proportion, and only rabbits, field voles and beetles (outside the nesting season) were of any importance, though a wide range of other items was taken. Ravens feeding at the rubbish dump took discarded meat or fat especially, but even these birds had sheep remains in 30% of their pellets, showing that they were also feeding away from the tip (where there were no sheep carcasses).

Mattingley (1995) has recently examined 700 pellets over three winters at a Raven roost among the hill ranges north of Crieff in Perthshire (Table 2, column 7). The ground is predominantly sheepwalk, but with a secondary interest in red deer and Red Grouse. Rabbits and mountain hares – both abundant in the area – were outstandingly the most important items, with a combined frequency of 94%. Sheep remains had only a 10% frequency, but sheep in the area are nowadays wintered on low ground close to the farms, so that their carrion is not freely available at this season. Vegetable material was present in 69% of pellets, but its well-digested state suggested that it was mostly of secondary origin, from the guts and dung of mammals eaten, especially lagomorphs. Red deer remains were little represented, but would not show if the birds were feeding on 'grallochs', or opened carcasses.

Ewins, Dymond and Marquiss (1986) have provided data on the food of a northern coastal population of Ravens by analysing a sample of 540 castings collected around nesting sites and roosts in territories, plus another 76 from communal flock roosts, on Shetland (Table 2, columns 9 and 10). Sheep remains were again the most common item for the breeders, present in 48% of castings, but lagomorph remains (mainly rabbit but a few mountain hares) were frequent, at 32%. Birds were often present, especially seabirds but also passerines, and eggshells were frequent. There was a strong representation of sea-shore items, and at one coastal site 65–90% of the pellets contained the calcareous seaweed *Corallina officinalis*, often in such amounts as to show clearly that it had been ingested in considerable quantities.

The largest sample of Raven pellets examined in Britain was 945, from communal roosts and 17 nest territories, in Mainland Orkney (Marquiss & Booth, 1986). The results (Table 2, column 8) contrasted with other studies, apart from that in Perthshire, in showing lagomorph (mainly rabbit but including brown hare) remains to be most frequent (at 80%), and large mammals (mainly sheep, but also cow, goat and horse) as much less important (13%). On these relatively fertile islands, where cattle-rearing predominates, sheep carrion was much less freely available than in most other districts. Feathers, with 24% frequency, were mainly from larger birds and probably mostly scavenged from the shore; eggshells, mainly from large birds, were almost as frequent at 21%. Other important items were Orkney voles (7%), rats (6%) and cereals (10%).

Marquiss and Booth found little variation in pellet contents between 1982 and 1983, but significant variation in frequency of the main food items in samples from different seasons and between separate locations. Seasonal variation included less lagomorph but more large mammal, eggshell, feathers and sea-shore items in the breeding season. There was a greater frequency of terrestrial invertebrates in late summer and autumn, cereal in February, berries in late autumn, and rodents in October and December. Major differences between territories were consistent over the 2 years of the study, e.g. in the frequency of lagomorph remains, which ranged from 36 to 97%. Pellets from territories near extensive seaweed-covered shores contained more sea-shore food and bird feathers. Large amounts of eggshell in pellets were associated with presence of coastal heathland and absence of arable land.

Only one published pellet analysis for Ireland is available, from two coastal nesting pairs in Cork, in the far south-west (Berrow, 1992). The results are rather similar to those for breeding territories in Shetland (Table 2, column 11). Mammal remains predominated by volume (36.2%) and were principally from sheep and other large vertebrate carrion, with rabbit rather infrequent. Marine items were almost as frequent and included a wide range of crustaceans, molluscs (gastropod and bivalve), and much coralline seaweed. Birds and eggshells were frequent. Grass was a major component, with 34.9% volume, and stones were also abundant.

The Perthshire and Orkney food studies show convincingly that in some parts of Britain, at least outside the breeding season, Ravens are not dependent on sheep or other large mammal carrion and can manage well on much smaller items, including a fair proportion evidently taken alive. It would appear that some coastal pairs, or even groups, can subsist largely on gleanings from the shoreline and other essentially maritime habitats. Pairs on small offshore islands must depend especially on food associated with the sea, particularly sea-birds. Ernest Blezard examined 24 Raven castings that I collected at a nest on St Kilda in May 1959 and found remains of Puffins in 71%, Soay sheep in 63%, and grass in 79%.

THE RAVEN AS A PREDATOR

In contradiction to the view that the species is purely a scavenger, Craighead and Craighead (1956) opined that, 'ecologically a Raven can be classed as a raptor'. Doubtless they were thinking of the *Buteo*-type raptors, which readily feed on carrion but also take live animal prey. The extent to which the Raven attacks live vertebrates probably depends largely on the constraints of availability of carrion or other alternative food. It is thus likely to vary over time and according to location. Some Ravens may occupy areas where there is seldom if ever a need to act as predator, but some may do so periodically and still others may have adapted to life in otherwise unfavourable environments by making predation a habit.

Mammals

Just how large a live animal a Raven will attack is unclear. The bird has a big and powerful bill with sharp-edged mandibles, the upper curved and almost hooked at the tip, enabling it to both strike heavy blows and tear at flesh. Smith (1905) noted from his tame Ravens that they 'can hardly ever touch the human hand without bringing blood and cutting rather deep'. At one time the Raven had a reputation for killing sheep, but probably few shepherds in more recent years have believed it would attack a healthy adult ewe or ram. As Chapman (1924) noted, a sick or seriously disadvantaged animal is another matter and it appears that the bird will occasionally tackle a dying sheep. Ravens found at the remains of sheep only very recently dead naturally incur the suspicion of responsibility for, or at least complicity in, the loss. Several observers have surmised, but usually on circumstantial grounds, that Ravens could have killed sheep by driving them over the edge of cliffs or from ledges on which they were crag-fast.

Lamb-killing is a different matter and the older literature on the Raven reveals a very widespread belief among sheep-farmers that this bird was a serious nuisance through its predation on young lambs. The same attitude appears to have prevailed throughout much of the world range of the species, though most of the evidence is anecdotal and often coloured by prejudice. After the Protection of Birds Act 1954 came into force, legal protection was withdrawn from the Raven in Argyll and on Skye, on the flimsiest evidence and to mollify sheep-farmers.

In February/March 1953, M. G. Ridpath investigated reports of lamb-killing by Ravens in Pembrokeshire for the Ministry of Agriculture and Fisheries. In an observation period of 100 hours in a coastal area with five breeding pairs of Ravens, he found no evidence of attacks on either ewes or lambs. By contrast, in an inland and upland area (the Prescelly Mountains), during 220 hours' observation he saw two lambs actually killed by Ravens, and nine other attacks on lambs and eight on ewes. The success or failure of the attacks on lambs depended mainly on the vigour with which the mother defended her offspring by rushing at the attackers. Another lamb was seen by the shepherd to be killed during delivery, when the ewe was at its most vulnerable. Attacks on adult ewes were not really pressed home in earnest, but nevertheless appeared to be made initially with serious intent. Some sallies which appeared to be attacks on lambs were made to obtain the placenta, but Ridpath pointed out that this could be damaging if the placenta was still attached to the lamb. This upland area had three breeding pairs of Ravens, but the attacks appeared to come entirely from birds in a flock of non-breeders, varying between 9 and 19 in number, which were living on the sheepwalks in question.

Ridpath emphasised that his limited observations could not be used to draw wider conclusions about the scale of the Raven–sheep damage problem. He had established that lambs are certainly killed under some circumstances but that, on this occasion, non-breeders were responsible. It may never be possible to make generalisations about this issue. Indications of the existence of 'rogue' birds that have developed the habit of killing lambs, or even adult sheep, and about the possible influence of disposing factors, such as lack of alternative food, severity of weather and condition of the animals, can perhaps never be resolved into consistent rules. Incidence of lamb-killing is likely to vary according to all of these factors. A Raven could be driven to killing a lamb through hunger or opportunity but, having learned to do so, might tend to continue. In many breeding areas, there is sufficient sheep and lamb carrion at nesting time to satisfy Raven needs, so that there is no need for the birds to attack living animals. Non-breeding flocks may live in areas that are more marginal in food supply, or create heavy food demands upon their foraging areas.

Ryves (1948) watched two pairs of Ravens on the Cornwall coast feeding on adjoining lambing-fields. When there were no dead lambs or placentas to eat, the birds walked among the sheep, digging in the turf for invertebrates, but one would also occasionally jump onto the back of a ewe with a lamb at heel, to probe into the wool, presumably for ticks. Not a single ewe ever showed the least concern at the bird's presence, or even at this particular attention, and the sheep clearly did not regard it as a source of danger. Two Ravens were once seen sitting on a sheep's back without upsetting the animal.

The reality is that hill shepherds as a whole have taken a pretty relaxed view of Ravens during the past few decades. The impression of general hostility from flock-masters, which the pre-1920 literature certainly gives, is far from the prevailing attitude that I have experienced. Most of the shepherds that I have talked to were quite tolerant of Ravens and at most seemed to

84   *Food and feeding habits*

16. Food for scavengers: the remains of a hill sheep after the attentions of Ravens, Carrion Crows and foxes. Northern Pennines (photo: D. A. Ratcliffe)

regard them as no more than a minor problem. This was matched by the fact that, in observations at several hundred eyries, I saw evidence of shooting of adults only three times and killing of young only twice. If lamb-killing had been a significant problem in recent times – or more widely believed to be so – there would surely have been much more evidence of attempts to deal with the alleged offenders. Some shepherds take a keen interest or even pride in their Ravens and would not have them harmed in any way. In central Wales, during observations in 277 sheepwalk and farmland territories from 1976–79, Davis and Davis (1986) found eight dead Ravens, of which five had been shot; which again suggests a fairly low level of direct persecution by sheep-farmers.

By contrast, most shepherds still regard the Carrion and Hooded Crows as unmitigated pests – despite the fact that they are much smaller than the Raven and potentially less formidable as predators – and I have yet to meet one who had a good word for either. Houston (1977) studied allegations of killing or mutilation of ewes and lambs by Hooded Crows on hill sheep farms near Oban in Argyll. By careful examination of 297 dead lambs he found that the majority (79%) had been born dead, were diseased or had starved. Crows had fed on 48% of the lambs found dead, but examination of wounds showed that only 17% of these were alive when attacked and most had exhausted body fat and were on the verge of starvation. Crows did not select healthy lambs and usually attacked those which would have died anyway. He concluded that about one in 850 lambs were healthy ones which would have survived but for the Crows. Only one attack on a live lamb was actually witnessed and the bird gave up before causing injury. These findings have done nothing to reduce shepherds' hatred of Crows, since dead lambs with signs of their attacks appear to be such obvious evidence of guilt.

The feature of Raven (and Crow) attacks on ailing animals that gives so much offence is the gruesome habit of first pecking out the unfortunate creatures' eyes. This damning behaviour is remarked upon time and time again in the Raven literature and in the testimony of hill shepherds. It even has a Biblical reference, 'The eye that mocketh at his father, and despiseth to obey his mother, the Ravens of the valley shall pick it out' (Proverbs, 30: 17). Dead animals are treated similarly. Many writers have also noted that the tongue is attacked next, before the main carcass receives attention. A simple explanation of this habit, so revolting in human terms, is that these are nutritious and easily taken morsels that the bird could be expected to eat first – but they are not always eaten. The habit is even more notorious in the Carrion/Hooded Crow, and Houston (1977) has suggested that the birds' hunting impulses lead them to attack the heads of helpless animals, as the most vulnerable parts, even when they are not hungry. A further explanation is that this behaviour secures the food supply; an animal which is blinded is as good as dead, and one which can no longer feed will succumb even more quickly. The predator then has a food source that will last it for some time ahead.

This can undoubtedly be a problem on hill sheepwalks when ewes become 'cast' or 'couped' (rolled on their backs and unable to right themselves). Where the shepherding is intensive, cast sheep can be found and righted before they are attacked; animals which have lost only one eye can be treated and saved but those completely blinded have to be destroyed. Where shepherds make their rounds less frequently, a cast animal will usually die within 12 hours anyway, even if unattacked. Many of the remoter sheepwalks, or parts of them, are inspected less frequently than this, so that any economic loss is then attributable to the quality of the shepherding rather than the activities of crows (Houston, 1977). Again, this is a message not readily taken on board by sheep-farmers.

Ravens can undoubtedly kill rabbits and hares when they are cornered or surprised, and the bird was disliked by rabbit-catchers for raiding their traps and snares. In places where the rabbit is abundant, it can be an important item of food, as the Perthshire and Orkney food studies have shown. Smith (1905) recorded the adults at a Dorset nest bringing five rabbits to their young within an hour. Myxomatosis outbreaks must have provided temporary and local bonanzas, though the longer-term effect would have been to deplete this part of the food supply. Healthy adults are less easy to catch and the young of lagomorphs are probably the most frequently taken. On Orkney and Shetland many rabbits are obtained as road-victims, and this may be true of other areas.

Small mammals, such as the various rodents reported in prey lists (including rats and mice), are probably all killed by Ravens, as well as eaten when found dead. Short-tailed field voles are a frequent item, especially when numbers peak markedly, as they often do on recently afforested hill ground. In northern regions, lemmings and voles are taken freely, though according to their abundance, and become important in Raven diet during times of peak abundance. Moles are often caught and also sometimes taken from fences where mole-catchers have left the bodies hanging. Newton (1864–1902) said that the last surviving pair of Ravens in the Breckland fed especially on this animal. It is unclear whether the stoats and weasels that sometimes appear in pellets were killed or picked up dead.

Ravens have on occasion been seen to eat the dung of larger mammals. Freuchen and Salomonsen (1960) say that, in the Arctic, they take caribou dung and even, in winter, the excrement of sled dogs. Their taste for sometimes putrefying flesh also suggests that they are immune to the *Salmonella*-type bacteria which are believed to have adverse effects on some birds.

Birds

The list of birds eaten by Ravens is fairly long, but many of them were taken as carrion and there are rather few direct observations of the species killing other adult birds. Kittiwakes, Rock Doves and Jackdaws are listed among species seen attacked and killed (Cramp & Perrins, 1994). Any bird – except perhaps the largest – found sick or injured is quite likely to be attacked; the taking of shot and incapacitated Red Grouse has been noted above. Pigeons eaten by Ravens are nearly always homers and these must usually be scavenged as the remains of Peregrine kills. Marquiss, Newton and Ratcliffe (1978) found that the frequency of bird remains in pellets decreased with distance from Peregrine eyries, the correlation being significant. Small passerines, such as Starling, Meadow Pipit and Reed Bunting, are reported killed as prey by Ravens, and a wide variety of nestlings and fledglings is evidently taken, though probably mostly in a casual way and only as a minor item of diet. Macgillivray (1837) said that they took the young of Pheasants, Grouse, ducks, geese and domestic fowls. The opportunities for predation are probably greatest in sea-bird colonies, but there is insufficient information to show whether this is ever systematic and serious. Aspden (1928–29) saw adult Puffins killed in north Wales, and Freuchen and Salomonsen (1960) recorded Ravens catching Brünnich's Guillemots that had landed on pack ice and were unable to take flight. Gaston (1985) repeatedly saw Ravens at Digges Island in northern Canada taking chicks of Brünnich's Guillemots.

Eggs

Several authors have given the opinion that the Raven is not nearly as determined and serious an egg-predator as the Carrion or Hooded Crows. Bolam (1913) found traces of eggshell only twice in many hundreds of castings that he examined carefully, and Blezard (unpublished date), Newton, Davis and Davis (1982) and Mattingley (1995) also found eggshells to be infrequent (Table 2, columns 2, 3, 5–7). In Cornwall, Ryves (1948) saw no

evidence of interference with numerous nesting gulls and other birds, and no trace of eggshells in pellets. He just once watched a Raven return with an egg in its beak – evidently one of duck or fowl.

By contrast, the four other detailed food analyses in Table 2 showed frequent presence of eggshells in Raven castings. The finding by Marquiss, Newton and Ratcliffe (1978) of eggshell remains in 28% of 697 castings, involving 26 out of 29 separate sites in the Southern Uplands and Cheviots, suggests that Ravens are more than just casual egg-eaters in this region. Recognisable remains were mainly eggshells of Curlew and Red Grouse. The Shetland food study by Ewins, Dymond and Marquiss (1986) found eggshells in 14% of 616 castings, and species identified included Shag, Guillemot, Curlew, other waders and gulls. Similarly, on Orkney, Marquiss and Booth (1986) found 21% frequency of eggshells, mainly from gulls and other sea-birds, wildfowl and waders. Shag eggshells were also identified in the sample of pellets from St Kilda and, on Ailsa Craig, a Raven was seen flying with a Gannet's egg impaled on its beak (E. Blezard). Writing of the West Country coast, Lloyd (c. 1925) said, 'I have seen many a dozen Guillemot's eggs eaten by Ravens'; and Smith (1905) mentioned eggs of Cormorant, Puffin and gulls being taken in the same region. Ussher and Warren (1900) remarked that, in Ireland, Ravens nesting beside Cormorant colonies often carried off their eggs. Macgillivray (1837) said that in Scotland they took eggs of all kinds.

Des Thompson and his colleagues, studying nesting Dotterel in three areas of the Scottish Grampians, have found that Ravens can be quite serious egg-predators of these uncommon waders; and Sue Holt has supplied the following information. Ravens were variably present on the Dotterel grounds from year to year, but there was no clear relationship between frequency of sightings and loss of Dotterel eggs. In the year of most frequent Raven presence on one area, no nests at all were lost, yet on another area a single pair of Ravens was believed to account for 53% of Dotterel clutches. On an adjoining area, loss of Dotterel clutches was an estimated 48%, and a probable 96% of Ptarmigan clutches was also predated. Despite these high losses, both groups of Dotterel produced more fledged young than were believed to be necessary to sustain their numbers. Hatching and fledging of the remaining nests were high and some clutches were taken early in the season, so that repeats were laid, with greater success. It seemed that some Ravens were specialising in egg predation, at least in some years.

Ravens are perhaps usually opportunist egg-stealers. On the Isle of Man, Cullen and Jennings (1986) recorded that a pair of Ravens nested in the heronry at Ballimoar, where they sucked the eggs from seven of the eight Heron nests. Nethersole-Thompson (1932) watched a pair nesting in a rookery systematically raiding its neighbours' nests.

Egg predation may also be serious in inland districts overseas. Byrkjedal (1980) found that Ravens were responsible for most of the June losses of Golden Plover clutches on the high mountain plateau of Hardangervidda in southern Norway. He estimated nest losses of this species at 78% due to the combined impact of Ravens, Common Gulls and red foxes, while Dotterel in the same area suffered 47% nest loss to the same predators (Byrkjedal, 1987). In Canada, Fox, Yonge and Sealy (1980) recorded that Ravens accounted for 27 out of 33 (82%) identified predator losses of Great Northern Diver nests on Hanson Lake, Saskatchewan. The taking of Sandhill Crane eggs in Oregon has been regarded as a significant factor in conservation of this species (Littlefield, 1986). Gaston (1985) watched Ravens regularly robbing Brünnich's Guillemots of their eggs on Digges Island. In Iceland, Lewis (1938) recorded that Ravens took many eggs from unguarded nests in Eider colonies and the ground was littered with shells sucked by this bird or Great Black-backed Gulls. Bounties for Raven destruction were paid in some parts of Iceland, especially by Eider-farmers, for as far back as 1800. Skarphedinsson et al. (1990) comment that recent studies did not show any significant effect of Raven egg predation on the Eiders' reproductive output, but the Ravens caused disturbances in the colonies. In the USA, Kilham (1989) watched a Raven family eat the eggs of an incubating Mute Swan.

Not all eggshells in castings result from predation. The birds feeding at refuse tips commonly take the shells of domestic fowl eggs from the rubbish. Moreover, the shells of the Ravens' own eggs have been found in pellets from at least six different sites in the Southern Uplands (E. Blezard, M. Marquiss). Some of these instances may have been the birds' eating of their hatched eggshells, but two were at nests, with eggs, that had been deserted, one as a result of disturbance by bird-photographers.

Reptiles and amphibians

Marquiss, Newton and Ratcliffe (1978) found common lizards to be taken in three different territories. Snakes are reported as being eaten in other countries but not in Britain; however, from the frequency with which they occur in Raven nesting haunts, it seems that adders are quite likely to be taken at times by Ravens, as they are by Buzzards. Frogs and toads are also recorded by Marquiss *et al.*, in 21 pellets from eight different localities, and are probably under-represented in the food lists.

Fish

The only freshwater fish recorded as food in Britain is the browntrout (Newton, Davis & Davis, 1982). Batten (1923) also said that Tom Speedy watched a family of Ravens successfully fishing for trout which had become land-locked in a small pool. Marine fish – mainly of the cod family – are evidently commonly eaten in coastal areas, though detailed records are few. Probably most fish items are found dead, though individuals stranded in shallow streams or rock pools would be vulnerable to predation. Ravens are attracted to fish offal in countries where this is freely available, around harbours and fish-processing factories. In Iceland, government-sponsored Raven control programmes began in 1976, largely because of the raiding of stock-fish being dried on open-air racks – an important export item in the country's economy (Skarphedinsson *et al.*, 1990).

OTHER FOODS

Invertebrates

Cramp and Perrins (1994) give an extensive list of invertebrate groups found as Raven food in the western Palaearctic, and the British studies have determined a fair variety. The hard and durable wing-cases of beetles are frequently remarked upon and, in the hills, commonly include large species, such as dor, rove and ground beetles, but also skipjacks, pill beetles and chrysomelids. The larvae of Lepidoptera and Diptera (especially cranefly) are taken, and the intact cocoons of large moths, such as emperor, fox and northern eggar, often appear in pellets, the enclosed pupae having been completely digested away. Adult moths are also taken, especially those species which swarm periodically, such as the antler and tortrix moth. Cramp and Perrins mention that grasshoppers and ants are important items in Spain. Lorenz (1970) notes that a Raven can spot an immobile grasshopper immediately. Engel and Young (1989) found that Raven pellets in Idaho contained 73% frequency of insect remains, but predominantly of grasshoppers during a summer infestation with these creatures. Locusts are taken freely in north Africa. Spiders also figure frequently but, along with the other soft foods, are difficult to detect in castings. One item which appears from direct observation to be important in the uplands of Britain and Ireland is the large black slug, so abundant on many of the upland pastures, though it is difficult to detect in castings. Snails are taken on lime-rich ground and earthworms readily picked up on the richer pastures. In some countries, Ravens follow the plough for earthworms and other invertebrates, in the manner of Rooks and gulls.

On the shoreline, a range of marine invertebrates becomes important to the Raven. They include shore crab, spider crabs, barnacles, limpets, periwinkles, whelk, Venus shell, mussel, horse mussel, cockle, clam, oyster, sea-urchin, lugworm, serpulid tube-worm and brackish water snail (Bolam, 1913; Blezard, unpublished data; Ewins, Dymond & Marquiss, 1986; Marquiss & Booth, 1986; Berrow, 1992). The potential list of intertidal food items is quite lengthy. Smith (1905) commented that Ravens have been seen to drop shellfish from high in the air onto rocks in order to smash them.

Vegetable matter

Small amounts of plant material, especially grass and moss, could be explained by accidental ingestion along with animal remains lying on turf or other vegetation; but the whole castings of such material found by E. Blezard and the high frequency of vegetable matter (Table 2, columns 2–4, 7, 11) suggest deliberate ingestion. Decomposed vegetable remains are evidently from the guts or dung of sheep and lagomorphs (M. Marquiss and W. Mattingley). Bolam's records included an abundance of acorns, beech mast and smaller seeds, as well as the cotyledons of germinating seeds and the buds of trees. Berries included those of rowan, bramble, elder, crowberry, bilberry and cowberry. Cereal grains are probably taken according to their availability. Macgillivray (1837) said Ravens were fond of wheat, barley and oats, and disliked by arable-farmers as a result. Some of these plant foods clearly have a seasonal bias. Soler and Soler (1991) report that 146 autumn and winter pellets in southern Spain contained 86% plant material by volume, of which 73% was barley. Engel and Young (1989) found 97% presence of vegetable matter in 574 pellets in Idaho, with cereal grain forming 69% by weight. The eating of seaweeds, and especially the calcified coralline forms (Bolam, 1913; Ewins, Dymond & Marquiss,1986; Berrow, 1992; Marquiss & Booth, 1986) is interesting because it indicates a particular need for mineral nutrition.

Refuse

The pellets of Raven flocks scavenging on rubbish tips unsurprisingly show a prevalence of actual refuse (Table 2, columns 6 and 8). Some of this is flesh, fat, bacon rind, bones and eggshells, but there is a good deal of paper, plastic (especially butter, margarine and meat wrappings), tin and aluminium foil, nylon line and polystyrene foam. Much of this is no doubt ingested to obtain adherent food. An item which often appears among the wool of springtime pellets is the thick rubber 'docking ring' used on lambs to remove the tails and to castrate the males.

Stomach stones (gastroliths)

Angular fragments of stone and grit often appear in the castings. These are from the larger store of such material which is ingested by the birds and held in the gizzard to assist digestion by their abrasive action. Stomach stones, of appropriate size, are found in most birds, but they are especially noticeable in Raven pellets. Dissection of an adult breeding female showed 160 fragments, mainly sharp-edged quartz, the largest two measuring $22 \times 10 \times 8$ mm and $17 \times 15 \times 3$ mm. Interestingly, the guts of two well-feathered nestlings, discovered dead in the nest, contained 11 and 66 pieces of stone, which could only have been fed to them by the parents (both records E. Blezard, unpubl.). Quartz fragments are noted as favourite material by both Bolam and Blezard, but pieces of glass were found in the analyses by Newton, Davis and Davis, 1982, and Ewins, Dymond and Marquiss, 1986. Soler, Soler and Martinez (1993) found that, in southern Spain, where Raven diet was mainly cereal, stones and grit occurred in 97% of pellets, and they suggested that ingestion of these grinding materials increased with the proportion of vegetable matter in the food.

FEEDING HABITS

Ravens seek their food primarily by sight. As they range over their feeding areas, the birds probably closely observe the activity of the larger animals in view. They have doubtless developed a keen awareness of a sheep or other potential food source in difficulties of the kind that Chapman listed. Whether they systematically patrol the ground is uncertain, but a sick or incapacitated animal is likely to be detected fairly quickly and marked down for further attention. Charles St John (1884) was long ago interested in how carrion-feeding birds located their food and was convinced that they did so by acuity of vision:

> Now Buzzards, Ravens and other birds who feed on dead bodies are in the habit of frequently soaring, for hours together, at an immense height in the air, wheeling round and round in wide circles. I have no

doubt that at these times they are searching with their keen and far-seeing eye for carcasses and other substances fit for food.

Macgillivray (1837) also believed that the bird located its food by sight, and that they watched each other, but said that the 'quiet soarings ... have no reference to prey'.

As well as using aerial surveillance, Ravens probably watch for food opportunities from suitable vantage points commanding wide views of their terrain, including the faces of their nesting cliffs. On the high fjells of Norway, Ingvar Byrkjedal has often watched them hanging by their feet 'like bats' from the sides of large boulders, with their heads just peering over the top, as though surveying the scene with deliberate concealment.

Heinrich (1990), in New England, has indicated the bird's ability to locate the new food supplies which he provided, even in largely forest-clad situations. Sometimes food piles were undetected for months, especially when snow-covered, but eventually 90% of 135 baits over four winters were located and consumed. Although it was widely presumed in earlier times that Ravens could scent carcasses (and therefore death), such notions are discounted nowadays. From his careful observations, Heinrich found no evidence that sense of smell played any part in the location of food by his birds. Even Chapman, who was convinced (as was St John) that wildfowl could detect humans by smell, concluded from hide-watching that such powers were entirely lacking in Ravens.

Heinrich's study of winter feeding behaviour of Ravens in Maine involved both resident pairs and roving, non-territorial bands, mainly of juveniles. He noted that the birds sought for food in this wooded country either singly or in pairs. Heinrich and his colleagues have also studied the idea that communal roosts act as information centres for the whereabouts of food. (These interpretations of Raven inter-relationships are summarised in Chapter 5.) In Wales, Bolam (1913), who also placed sheep carcasses near an observation hide, noted that, in every instance but one, Ravens came to his bait in pairs.

There is still no telemetric evidence of the foraging range of breeding Ravens in Britain or Ireland, but simple observation has led to the view that most of the food is obtained within 1.5 km of the nest in north Wales, where the bird breeds at high density (Dare, 1986a). John Mitchell watched a pair, which were nesting on the heavily afforested Rosneath peninsula in Dunbartonshire, fly at least 3 km across the Gare Loch to find food for their young. Where breeding density is much lower, as in parts of the Scottish Highlands, feeding ranges are presumably greater and, in barren country, the birds may have to cover large areas to find enough food. Thomas (1993) watched Ravens fly in a direct line until out of sight at over 5 km in Argyll, and over 7 km in western Ireland. Conversely, some coastal pairs may find easy pickings within a short distance of their nests (Ryves, 1948). In central and south Wales, Ravens breeding near to rubbish dumps regularly feed there and may well choose conveniently close nest sites (A. V. Cross, A. Dixon).

There is, however, no evidence that Ravens feed only within the limits of a breeding territory, or that they observe any particular feeding limits. On the Cornwall coast, Ryves (1948) spent many days watching two pairs of Ravens with nests barely 1 km apart. They both used a large pasture-field holding a big flock of sheep and their young lambs (with numerous placentas and a few dead lambs) as a common feeding ground. There were never more than two Ravens in the field at the same time and they more often came singly. No hostility was recorded. Ryves felt that, while nesting birds seemed to have fairly well-defined feeding areas, they were not exclusive but often shared, especially when a rich food supply was available. Gilbert and Brook (1931) noted that, on the Pembrokeshire coast, 'Inland there seems to be no jealousy whatever among the Ravens. The country behind their nests [on the sea-cliffs] seems to be treated as a communal feeding ground, and there are never any demonstrations or fights to establish a claim to any particular area'. A land-fill site in south Wales attracted three breeding pairs (A. Dixon). Houston (1977) found that pairs of Hooded Crows in Argyll fed entirely within their territories when they started nesting, but these are smaller birds nesting much closer together than Ravens.

Some observers have commented that Ravens show the same curious quirk as some of the raptors, in evidently declaring a truce with near neighbours over their predatory habits. Macgillivray (1837) observed that the nests of Rock Doves close to Raven eyries were never plundered and Smith (1905) noted that they left the eggs of gulls close to their eyrie unmolested, but carried off those from some distance away.

Outside the breeding season, Ravens may travel longer distances to feed. In Snowdonia, Orton (1948) saw a family party travel from the mountains to coastal sandbanks, 16 km away, in July and early August. The communal roost inland in Merioneth, where Bolam (1913) found numerous remains of sea-shore items in pellets, was 20 km from the nearest point of the Mawddach estuary and 32 km from the open coast. The data given on p. 125 suggest that birds from communal roosts often commute considerable distances to feeding grounds during the day.

When Ravens have discovered a large carrion item, they do not often descend immediately to feed on it. Sometimes days, or even weeks, will elapse before they return. Heinrich (1988, 1990), who observed their feeding behaviour in New England closely, found that, even when Ravens have decided to feed on a new and large meat-bait, they approach it gradually and timidly, as though fearful of the result. A bird typically landed 5–10 m away and then walked towards the bait but, when within close range, nearly always made sudden violent vertical leaps, assisted by one or more wing-beats. Finally, after a first peck, it would then invariably fly away, not returning for several hours. These antics were then repeated, but the bird now landed on the carcass and continued to jump up and down in a seeming 'dance' lasting several minutes. By means of experiments with hand-reared Ravens, Heinrich showed that this exaggerated bait-shyness is innate and he speculated that it either reflected selection from a long history of human persecution at baits or was a defensive response to a dangerous animal which could be still alive, sleeping, or even feigning death as a protective strategy.

Kilham (1989), in New Hampshire, also noted how Ravens nervously approached food, jumping back repeatedly at quite small items; one bird could alarm its fellows by this reaction. He wondered if this shyness was a learned response to the wide use of traps in the area, but his hand-reared bird also behaved in just the same way to food placed in its cage, supporting the innate explanation. Macgillivray (1837) commented long before that the bird 'eyes its prey with attention', while Ryves (1948), in Cornwall, had remarked on 'how cautiously it will approach even a carcass, as if afraid it may still be alive'. Bolam (1913) found that Ravens were capricious in coming to his baits, even when these had been discovered. And when the birds had detected his presence in the hide, nothing would induce them to return. When undisturbed, Ravens usually return to large food items until nothing edible is left. Birds sometimes call a good deal in evident alarm and hang around in places where there is no nest, and the only likely reason for such response is a carcass to which they wish to return.

While Ravens easily remove the eyes and tongue of a dead sheep or other large animal – as they habitually do first – entering the main carcass is more difficult. From his hides, Macgillivray noticed that Ravens feeding on sheep next attacked the subcaudal region, perforating the abdomen and dragging out the intestines before eating them. Both he and Bolam (1913) remarked that the hook of the beak (which Bolam believed to be especially marked in old Ravens) gave the bird a formidable weapon, for tearing and pulling as well as stabbing, and that they easily ripped off pieces of meat. Chapman (1924) spoke of 'a great cavernous shaft driven into the ribs' but I was told by a shepherd that they opened sheep at the vent to reach the viscera and then worked through the rest of the body. Heinrich (1990) and Kilham (1989) believed that their Ravens usually waited until some other predator, such as a coyote, had first opened a carcass of a large animal before they attempted to feed. Hewson (1981) also suggested that Ravens had difficulty reaching most of the meat of large carcasses unless other carnivores had first penetrated the skin or dismembered the body. However, on Orkney and Shetland, where there are neither foxes nor Golden Eagles (which might open carcasses first), Ravens seem perfectly well able to feed on the carcasses of large animals.

Attempts to catch other birds in flight appear usually to be unsuccessful (Cramp & Perrins, 1994), but a Rock Dove was seen to be caught. Tinbergen (1953) said that, in winter in east Greenland, he often saw Ravens attack Ptarmigan by stooping at them and, although he never saw them actually kill a bird, the local Eskimos assured him that Ravens often killed Ptarmigan in flight. Birds are often taken when at rest or on the ground. Aspden (1928–29) described a party of Ravens in north Wales sitting by Puffin burrow entrances, catching the unfortunate birds as they emerged and killing them by violent shaking. Roosting and nesting Kittiwakes have been seen killed on their ledges and Gaston (1985) saw many instances on Digges Island where Ravens caught hold of incubating Brünnich's Guillemots by the bill or wing in order to pull them off their eggs or chicks, which they then flew off with. The skulls of rabbits or smaller prey are sometimes crushed. Prey may be carried by feet or bill, and large eggs are often spiked before being carried off.

In seeking smaller fry, especially invertebrates, the Raven does a good deal of walking about on the ground, and a whole family of adults and fledged young will sometimes quarter favourable spots in search of abundant items, such as slugs or larvae. Any nests of eggs or young found during these searches will be raided. Aerial hawking of winged insects has been observed occasionally. On Majorca Ravens were seen following the plough while ground under olive trees was being tilled.

Ravens have been seen to combine efforts in taking food from dangerous adversaries. Zirrer (1945) watched a pair co-operate in stealing a mouse from a domestic cat. In the robbery of a Mute Swan nest by a Raven family (see p. 89), one adult fended off the female swan while the others smashed the eggs and ate the nearly hatched cygnets (Kilham, 1989). Kilham also saw Ravens pulling the tails of Turkey Vultures feeding at carcasses in order to distract them.

The Raven is given to piracy with other birds on occasion. One pair in an afforested area was seen on several occasions to rob Short-eared Owls of field voles (Marquiss, Newton & Ratcliffe, 1978). Cramp and Perrins (1994) list vultures, Peregrine, Golden Eagle and Great Black-backed Gull as other predators and scavengers seen to be robbed of their food by Ravens. Kilham (1989) watched Ravens in New Hampshire repeatedly robbing Crows of suet put out for them, and thought that this might be connected with the Ravens' nervousness at approaching food on the ground.

Food caches are habitually made by Ravens, both by breeding birds (Cramp & Perrins, 1994) and wintering non-breeders (Heinrich, 1990). Fat, fatty meat, eggs, bones, bread, dates and dung are materials which have been seen to be hidden, usually in holes or beneath stones, but sometimes in small excavations dug by the birds themselves. Kilham (1989) made numerous observations of caching and noted that wild Ravens and his tame bird did so when satiated after feeding. He commented that the captives which Gwinner (1965b) watched hiding food when hungry were living together in a cage, which may have altered their behaviour. Caching could be in various places, including on the ground under lumps of earth, in rotten logs, in grass tussocks and under snow. Birds showed a sense of camouflage, placing moss over cavities in mossy logs and more grass over tussock hiding-places. Snow was scraped sideways with the bill to make a depression. Kilham believed a nesting pair cached food in various rock crannies around their cliff eyrie with young. Ravens have well-developed throat pouches in which they store food for short periods, especially in carrying it to their young. A bird flying with bulging pouch in the spring is usually heading for its nest and its mate or fledglings.

Food rations for captive Ravens have been given as 170 g of meat per day. Wild birds could be presumed to have a higher energy requirement than captives and are also faced with a larger amount of indigestible matter in many food sources, in the form of hair, bone, cellulose, etc. The larger male Ravens will, on average, need more than females and a youngster's food intake will probably approximate that of an adult over the whole of the nestling period. A pair which rears three young will thus probably create a food demand for the year of around 310 kg of digestible material. It is difficult to match this against food availability on the ground; Brown and Watson (1964) have discussed the complications of such estimates for Golden Eagles and the same must apply to Ravens. From counts of dead sheep, and known mortality rates of sheep and lambs, Newton, Davis and Davis (1982) estimated that, on sheepwalks of central Wales, an average of 190 kg/km^2 of edible carrion mutton was available to scavengers in one year, mostly in late winter and early spring. Large items, such as sheep carcasses, will often be subject to competitive feeding by other birds and mammals, so that only a proportion is available to the Ravens – but just how much and its possible variations are unknown. There are also unanswered questions about continuity of food supply. Small items may vary somewhat in nutritional value from one kind to another. Large communal flocks must create a locally heavy food demand that is satisfied by a stable and copious source in one place, by daily dispersal from a roost over a wide area, or by mobility of a roost. Some of these aspects are discussed later.

As a large bird, the Raven probably has a certain ability to endure periods of fast, when food is unobtainable, as during heavy falls and deep cover of snow. It is able to survive in some of the harshest environments on earth, and at times when food supply seems especially meagre. No doubt individuals starve or perish from conditions exacerbated by food shortage in times of stress. Houston (1977) found that the body condition of some birds in flocks of Hooded Crows showed they were unable to find enough food in winter, and some – possibly the lowest peck-order individuals – died of starvation. Birds holding territory probably had sufficient food throughout the winter.

By means of radio-tracking, Engel and Young (1992a) found that the amount of time spent feeding by communally roosting Ravens in Idaho varied seasonally. From May to October it averaged only 10.3% of the day, increasing to 14.0% during November to January, and reaching 41.4% from February to April. The proportions of time spent flying (13%) and moving about on the ground (11%) varied little seasonally, the greater part of the day being taken up by resting, except in the early spring.

CHAPTER 5

Social behaviour

In common with some other members of the Crow tribe, the Raven is both a solitary and a social bird. The paired breeders are apparently territorial and nest well apart from their neighbours, though the spacing distance varies geographically. These established nesting pairs also lead a rather exclusive life outside the nesting season, normally remaining together in their territories in order to defend these against potential usurpers. This non-social existence of territorial pairs contrasts with the long-known tendency of other Ravens to form into flocks and roost communally.

TERRITORIALISM

Cramp and Perrins (1994) quote numerous authors describing the Raven as a strongly territorial species in its nesting habits, in various parts of its world range. From his observations in Ireland, Howard (1920) cited it as a territorial bird, although others had made this connection well before he first elaborated the theory of territory in birds. The Roman author Pliny the Elder is said to have noted that Ravens were 'strongly territorial'

(Armstrong, 1958). Macgillivray (1837) said that, 'Ravens, if unmolested, breed in the same spot year after year; but it seems strange that although they have a numerous brood, their number in any particular district does not appear to increase; nor do two pairs ever breed near each other'. Smith (1887) quoted an old saying that, 'There are never two Raven nests on one estate'. Verner (1909) said, 'They keenly resent the intrusion of any of their species and each pair of Ravens establishes itself on some cliff or tree at some distance from others'. In Wales, Bolam (1913) commented that:

> It is in spring, when the nesting stations are once more sought, that jealousy manifests itself. By that time parents and offspring have probably forgotten one another; the young are then as independent as their elders, and one pair (no doubt the original one as long as they survive) resumes possession of the family seat, and seldom allows a rival establishment to be set up anywhere very near it, whether by strangers or blood relations.

The direct evidence for territorial behaviour in the Raven is nevertheless rather limited – as distinct from the inferences about it drawn by many observers from the highly dispersed distribution of nesting pairs. Little has been published on the interactions between neighbouring birds or pairs, though many observers have witnessed exchanges between Ravens that appeared to be territorial behaviour.

I have many times seen chases, usually involving two birds, in which one pursues the other from a close distance behind (*c.* 3–10 m). Both appear to be flying at top speed, often on an undeviating course, though sometimes there are twists and turns. The pursuing bird does not close with the other and both often disappear from sight in this manner, though the hindmost Raven is sometimes seen to break off and make its way at a more leisurely pace in the direction whence they came. The chase is often accompanied by rather shrill 'aark-aark' croaks, but there can also be complete silence. When there are calls, it is impossible to tell whether they are from the front or rear bird, or from both. Since this call is otherwise heard only when a Raven is attacked by a Peregrine, I suspect that it is given by the pursued bird and is an alarm call. I commonly saw such pursuits when living in Raven country, in the mountains of Snowdonia, and at all times of the year.

While these headlong chases *look* like territorial interactions, with one bird 'seeing-off' the other, they could have a different meaning – display between the pair, for instance. They most usually occur during the nesting season, but are sometimes seen during the summer and autumn. On occasion, two birds will chase either a single Raven or a pair. Occasionally, unmistakable fighting is observed. Gordon (1938) described how, one autumn, he saw a pair of soaring Ravens suddenly descend to ground, to fight with a second pair sitting there. Two of the combatants rolled over and over down the steep slope, with wildly flapping wings and locked talons. All four stood recovering for a time, then pursuit continued in the air, with one pair eventually flying off. The writer said, 'I have little doubt that a strange pair had been disputing this desirable territory with the rightful owners, and had after a struggle been driven away'. Tony Cross frequently sees territorial chases in mid-Wales and watched one male talon-grappling with an intruder in flight; both birds spiralled downwards rapidly, breaking off just before hitting the ground. This male had a mate which was unable to fly and had to work extra hard in defending their nest. In the Brecon Beacons, Andrew Dixon found that one or both of a nesting pair usually drove away intruding Ravens, which were more often neighbouring territory-holders than non-breeders.

My own observations record remarkably few interactions close to the nest, though one was revealing. I sat watching a cliff where the nest was under construction and saw a Raven fly down to a ledge near the nest. It called and then flew away rapidly, whereupon a *pair* of Ravens appeared and gave hot pursuit, but then broke away and returned to the crag, where they flushed out a second intruder, which they also drove off. The pair once more went back to the site, of which they were evidently the rightful owners. I was in no doubt that this was real territorial behaviour in defence of the nest site.

Davis and Davis (1986) found that territorial pairs in general behaved aggressively towards wandering individual non-breeders within about 0.5 km of the nest site, especially in the breeding season, but normally ignored other Ravens flying high overhead. Occasionally, a nesting pair was unable to prevent penetration of its territory by a non-breeding flock and, during the ensuing conflict, the nest failed. Once, a nest was completely dismantled by the intruders and the eggs destroyed.

It seems likely that Ravens vigorously defend a core area of their territory immediately around their nest but are less aggressive to intruders farther away from this nucleus. Probably, as in many of the raptors, territorial behaviour mainly takes the form of visual signals, as during display flights over the nesting area, that avoid direct clashes and fighting. It also appears that, as nesting operations advance, territorial activity fades away, its purpose having been to space out the pairs before nesting begins.

Ryves (1948), in Cornwall, regarded Ravens as easy-going birds, not given to bickering among themselves; and believed that they do not 'fight for possession of nesting territories but acquire them by mutual understanding'. In Pembrokeshire, Gilbert and Brook (1931) observed:

> So thick [on the ground] are Ravens that the question of their territory becomes interesting. As far as one can judge, it is only on the cliff itself that any jealousy is shown. The Ravens stake out and fight for a length of cliff, though their claim may be very small. If a cock bird, hanging in the breeze, sees his neighbour pass by closer than he wishes, he gives warning of his anger by a series of 'rattling' threatening barks.

They only once saw invasion of a territory during the height of the nesting season, when watching a pair building; a party of eight Ravens came along the cliff, whereupon the pair dropped their nest material and rushed silently at the trespassers, who rapidly flew off. The pair were so disturbed that they did not return to building for an hour. Yet the same pair were, later in the year, seen sitting within 20 m of the neighbouring pair, though one of them had a nest of large young less than 300 m away.

Ryves (1948) also described what he called 'nest-visiting', when a pair feeding nestlings settled on a crag close to the nest and were joined by two other Ravens, the four remaining together for about 10 minutes on 'apparently perfectly friendly terms'. Nethersole-Thompson and Nethersole-Thompson (1979) also watched a pair of Ravens visit the territory of another pair: 'Croaking quietly, the four Ravens flew around without challenging one another'. I once witnessed a similar incident: a single bird pitched near a new nest in a high Lakeland crag and, very shortly after, a pair of Ravens flew into the crag, making straight for the nest. They settled very briefly near the first Raven, and then all three birds took off, soaring in front of the rock for a short while before dividing and the pair going away into the distance while the single bird returned to the nest. The meeting appeared to be perfectly amicable. At two other nesting places in this district I have seen pairs of Ravens, with eggs, joined by a third bird in demonstrating at my intrusion. The simplest explanation is that these extra birds or visitors are offspring of the occupying pair from a previous year, though this is mere conjecture.

This last idea assumes that Ravens will be more tolerant to the presence of their own offspring within the territory than to unrelated birds, but the behavioural relationships between kin are little understood. There has been much disagreement over the manner of eventual separation of fledged young Ravens from their parents. Smith (1905), from observations in Dorset, believed that the young were driven away by the adults. Chapman (1924) asserted that, in the Cheviots, even as early as June, the old birds drove away their offspring, and mentioned an instance where a persecuted youngster had taken refuge in a shepherd's garden. He interpreted this as territorial behaviour, forcing the progeny to find a place of their own instead of staying on the home ground, where their parents needed to maintain exclusive possession. Bolam (1913), in north Wales, believed exactly the reverse, that the whole family stayed together until the autumn and broke up through the urge of the wandering instinct that develops at that time. He believed that sometimes the adults were the first to desert the nesting haunts after rearing young, though this is not the experience of other observers. Ryves (1948) found that, in Cornwall, 'Parent Ravens remain with their young, travelling about as a family party, till late August or early September. After that, the young birds are left to their own devices and the parents resume their old life of close companionship, settling down peacefully in their nesting territory'.

Orton (1948) found similarly, in Snowdonia, that the Raven family groups stayed intact for many weeks after fledging, moving from the nesting quarters to higher and remoter parts of the mountains. He saw them often around the high summits and, on a Highland top, came on a group of 11 birds which divided into two families on taking wing. Orton noted that, as summer went on, the old birds left their young for longer and longer periods, wandering off to neighbouring mountains or even visiting the coast. He believed that, towards winter, many young Ravens dispersed to the lowlands but that, in some districts, they united with other broods during late summer and autumn to form flocks.

In New England, Heinrich (1990) suggested that the parent–offspring bonds are easily dissolved in Ravens. He found a great deal of aggression by adult territory-holders towards roving non-breeders, and the assertive dominance of the former was a crucial factor in his interpretation of the complex social relationships within the population. These interactions were, however, mainly over their feeding on carcasses and food dumps during winter, and not over territory *per se*.

It is possible that instances of undoubted aggression of paired adult Ravens towards juveniles have involved youngsters other than their own offspring, which had wandered into a neighbouring territory beyond that of their parents.

FLOCKING AND COMMUNAL ROOSTS

Raven flocks have been recorded in many parts of their breeding range, typically roosting communally on cliffs or in woodlands. Many observers have noted that they are especially conspicuous during autumn and winter, but thin out markedly towards or during the following spring. Saxby (1874) said that, in Shetland, large numbers of the species usually arrived about mid-October and remained until the onset of the next breeding season. Bolam (1913) noticed that, in central Merioneth, flocking was a feature of winter and that the gatherings had generally all broken up by the middle of January, leaving the regular communal roosts untenanted for the next 6 or 7 months, except for a few birds which were paired but showed no inclination to nest. The seasonally varying numbers at a modest-sized northern Pennine roost of up to 57 Ravens are catalogued in Appendix 1. While there was a marked tendency for numbers to be much smaller – or even zero – between February and July, flocks of 35 on 2 March 1913, 19 on 25 April 1909, 21 on 7 May 1909 and 18 on 14 May 1912 were recorded. This suggests that, in at least some years, the flocks became reduced rather than dispersed completely. Recent records of Perthshire roosts by W. Mattingley show comparable seasonal trends and, in Argyll, Hancox (1985) found that flocking was a feature of the non-breeding season. E. B. Dunlop suggested that Pennine groups seen in May were probably pairs that had lost their eggs, but this was a conjecture, unlikely to be true. Flocks, mostly in pairs, were also seen here outside the breeding season.

Heinrich (1990) has questioned the evidence that Ravens flock, in the sense of establishing socially cohesive groups, as distinct from assembling into gatherings for a temporary purpose, such as feeding. By marking birds, he found that groups feeding in winter in the New England forests were extremely fluid in membership, sometimes showing a complete turnover of individuals from day to day, and changes even from hour to hour. He also watched Ravens assembling to food from four quite different directions. In Britain, there is little evidence to show whether or not flocks remain composed of the same birds. In some instances, the constancy of numbers day by day suggests that the same individuals are involved but, in others, wide fluctuations over a short period implies some degree of interchange between different roosts. Cohesiveness of groups also varies greatly during a 24-hour period. While some communally roosting flocks undoubtedly disperse during the day and regather in the evening, other flocks, equally clearly, retain much of their identity during daytime.

Coombes (1948), in his paper on Raven flocks in the Lake District, defined a flock as 'any larger number than eight to ten Ravens associating together'. He not only saw these Lakeland flocks together during the day, but also believed that they existed all the year round, and that they consisted of local stock. All sightings were of flocks either flying into or leaving a roosting place, in evening or early morning, or flying about the hills during the daytime, especially hills where there was no regular communal roosting place. From the observation that they wandered through the breeding territories of the settled pairs when these had eggs or young, Coombes deduced that flocks were composed of non-breeding birds and surmised that prolonged adolescence may be a factor in such flocking. He concluded that, on any given date, the Raven population is divided into:

1. Breeding pairs, mated for life, holding the same territories year in year out.
2. Birds that for a period of years do not breed, although a proportion are paired. These birds wander about the hills in flocks and make use of communal roosting places.

This Lakeland interpretation appears to hold good for other parts of Britain. Flocks of Ravens during the breeding season are composed of non-breeding birds, but it seems likely also that the larger gatherings in autumn and winter consist mainly or completely of birds that have not bred. Some observers have claimed that the winter flocks contain birds from the breeding stock, but the available evidence points to the opposite conclusion. The established breeders can be observed going to roost in their separate pairs, usually on their accustomed nesting crags, on autumn and winter days when the communal roosts are assembling elsewhere. In most Raven breeding areas in Britain, the territory-holding pairs remain apart from each other, in solitary possession of their breeding haunts throughout the winter. Coombes (1948) thought it possible that some paired breeders might join a nomad flock or form a flock for short periods, but said his only evidence was of birds appearing to be the local residents attracted to a deer 'gralloch' one October. This may be compared with Heinrich's finding that territorial pairs and non-breeders often used the same food stores but, far from mixing, remained as virtual adversaries.

The growth of autumn and winter flocks probably results from the young birds of the year, now detached from their parents, joining up with each other and with residual groups of subadult or adult Ravens which have not been able to acquire nesting territories. Given the presumed advantages for these birds to flock at this time of year, there would be greater benefit in the large assemblies dissolving into smaller groups or pairs and dispersing more widely at the end of winter. They would then be better placed to fill any gaps which became available within the established breeding population, or to reach areas where they could establish new territories of their own. Many observers have agreed with Coombes in reporting that the Ravens in flocks mostly appear to be paired, and some of them evidently represent a population surplus awaiting their opportunity to nest. Mortality no doubt reduces flock numbers through the winter, but losses among the solitary breeding pairs will also siphon off birds from the flocks to fill the gaps. Flocks of up to at least 100 birds nevertheless exist in some districts during the summer.

Bolam (1913) stated that a gamekeeper of his acquaintance, who had shot some of the non-breeders at nesting time, found that they consisted of both the previous year's and older birds. Colour-ringing established the presence of first-, second- and third-year birds, and once a fifth-year bird, in a winter flock in central Wales (Davis & Davis, 1986). Two Ravens from a spring flock of about 20 in the Westmorland Pennines were dissected by E. Blezard, who found that, although both were more than 1 year old, the condition of the reproductive organs showed that neither had ever bred. Heinrich (1990) states that Ravens do not breed until at least 3 years old; this being the case, only the older birds in the non-breeding flocks can act as immediately fertile recruits to the breeding population. It is, however, possible that a bereaved adult might pair with an immature bird, or that immature pairs can take up territory and then wait for the necessary time before breeding.

Given that, in certain districts, there is at any time a reservoir of surplus Ravens which can maintain or boost the breeding population, it is generally in the interests of these birds to stay in flocks. The stock explanation of the advantages of flocking in birds is that it facilitates feeding and reduces predation. Feeding advantage is especially strong when food sources are patchily distributed both spatially and in time – as often applies to the Raven. Hurrell (1956) suggested that, since much of Raven country is shared out between territorial breeders, if non-breeding Ravens fed and roosted singly, they might be routed by territory-holders and repeatedly forced to move on. If a mob of birds arrives, however, the territorial pairs cannot evict them so easily, so that there is an advantage to the non-breeders in feeding and roosting in numbers. This is close to the explanation later developed for the feeding behaviour of flock birds by Heinrich (1990), who suggested that there may also be social benefits connected with mate selection and dominance relationships. Most of the British corvids are given to flocking outside the breeding season – Carrion/Hooded Crow, Rook, Jackdaw, Chough and Magpie, and the colonial nesters, Rook and Jackdaw, often associate with each other in mixed flocks and roosts.

Coombes (1948) made a point of disagreeing with the view that an unusually abundant or attractive food source causes Ravens to flock, and noted that the daytime assemblages mostly spent the time wandering slowly ('idling', as he put it) in loose groups along high ridges and summits. While his conclusion about the composition of the flocks appears to be quite correct, he did not really offer any alternative explanation for flocking of the non-breeders. The birds must have fed at some time, and evidently when Coombes was not watching them. Many observers have, by contrast, noted an obvious association between Raven flocking and abundant food supply.

The regular assembling of Ravens at rubbish tips and knackers' yards is good evidence of communal feeding at sites with plentiful food. Ryves (1948) made this connection in Cornwall, where a roost of 150 Ravens mostly fed in a field where butcher's offal was regularly spread. Hancox (1985) found that 30 birds assembled to feed on the carcasses of some 25 lambs which had died in preautumnal sales gathering-parks in Argyll and 25 scavenged a red-deer carcass during severe winter snowfalls. Macgillivray (1837) said that, when a horse died on South Uist, 30–50 Ravens soon gathered about the carcass, and continued to make daily visits until the bones were picked. He also remarked that, 'When grampuses or other large cetaceous animals are stranded, it is astonishing to see the numbers which congregate from all parts', and he noted 200 or more Ravens so assembled on Pabbay in the Hebrides. Around 1900, B. N. Peach saw a gathering of 500 or more Ravens at five Killer Whales beached in Weisdale Voe, Shetland. The most extreme example was also in Shetland, where Saxby (1874) chronicled the large number of 800 Ravens gathered on the carcasses of stranded and flensed whales in Uyea Sound (see also p. 77).

Such a congregation must have represented an influx from a considerable area and it is likely that, under such circumstances, feeding groups contain established breeding territory-holders. Heinrich (1990) showed that both established breeding pairs and vagrant assemblies of juveniles and subadults came to his artificially created 'food bonanzas' in the New England forests. In a fascinating field experiment, involving the laborious placing of 8 tonnes of meat in 135 baits over four winters, he painstakingly observed the relationships between these two elements of the Raven population, and gradually deduced their significance. Both resident and flock birds searched the same foraging area for food, usually singly or in pairs. Sometimes, the established pairs found one of his meat piles first; if so, they remained silent, since it was not in their interests to attract other Ravens. But if other birds chanced upon the food also, the residents' dominance allowed them to deny access to one or a few flock birds. If the flock birds found the food first, they stayed silent and proceeded to feed on their own. If, however, the flock birds came on resident pairs which had found the food first, or if residents appeared, they soon gave a peculiar call ('yell'), which had the effect of summoning other members of the flock. Once the numbers of flock birds assembling to the food exceeded a certain level, the dominant residents gave up trying to exclude them, so that all were able to share the food, though flock birds sometimes then fought among themselves, especially as any one source became depleted. Low-ranking birds may have been the first to be displaced to seek new supplies elsewhere.

Heinrich concluded that what had at first sight seemed to be a mutually beneficial reciprocal altruism (or 'hopeful' reciprocity) between flock Ravens was more likely to be a device for gaining access to otherwise unavailable food. He also suggested that the recruitment behaviour among juveniles and subadults could be related to sexual selection, whereby an unmated Raven finding food draws potential partners, thereby not only ensuring access to the food, but perhaps also enhancing its status and showing its fitness as a future provider for offspring – attributes likely to increase its attractiveness as a mate.

Feeding and roosting assemblies may thus be temporary gatherings which disperse to some extent during the day. As the most important food items are large carcasses that tend to be rather few and scattered at any one time, there would be an advantage in Ravens spreading out to search for these. In open country, birds searching far apart could more readily watch one another's reactions to food than they could in the extensive forests where Heinrich made his study. Macgillivray (1837) deduced that, in the treeless Outer Hebrides, each pair watched its neighbours, taking its cue to follow when the flight of any one bird or pair indicated food. They may thus resemble vultures in their food-finding technique, as well as in their fondness for carrion. Bannerman (1953) made this analogy and described birds assembling in a vulturine style on Grand Canary when a carcass was discovered. Ravens have also been described as searching in flocks for invertebrate food by walking over the ground in fairly close company. While such a food source is less likely to be highly concentrated than in the case of large carcasses, good areas may still be somewhat localised in occurrence.

Heinrich has pointed out that the nature of Raven roosts/assemblies depends on the nature of the food supply. In the eastern USA, the roosts are highly erratic in location and ephemeral in duration, for that is the pattern of occurrence of large carcasses on which they feed. In the west, food supply is more varied and abundant, but depends on the local productivity of the land and the timing of its production, so that Raven gatherings tend to change in numbers and distribution more gradually.

Various observers long ago noted that flocking of non-breeders serves a social function as well, with the birds frequently indulging in joint soaring aerobatic displays, often in pairs. Gilbert (1946), Lockley (1953) and Hancox (1985) all suggest that this semi-nomadic association of non-territorial Ravens may be the basis of pair formation, and the dominance relations discussed by Heinrich could thereby come into play. Lockley described how a circling flock of 58 birds near Tenby, Pembrokeshire, consisted of pairs, each with one chasing or at least following the other, and singletons which wove to and fro across the track of the others, as though seeking partners.

Flocks which have dispersed during the day usually reassemble at their roosts in the evening, and singles, pairs or small groups may appear from similar or different directions. Sometimes the birds gather up first in some different spot not far away and then make their way to the roost. Flock displays over the roost may occur. Hurrell (1956) noted that, at a roost in a Devon woodland, the birds first assembled on high moorland behind, sometimes sitting quietly on the ground. Some birds often flew, performing 'remarkable aerobatics', chasing each other and at times making vicious dives in 'a recognised kind of horseplay'. Wendy Mattingley described a display over a Perthshire woodland roost on 3 December 1995, as the light faded after a brilliant, sunny afternoon:

The Ravens came to roost late and numbers were low at first. Eventually there was a great deal of activity with 40 then 80+ birds soaring in a huge mass very high over the roost in a typical corvid spiral. The high soaring continued for at least half an hour with some birds disappearing into the roost, but the flock still remaining in high numbers as new birds were drawn in on approaching the roost.

Where the roosts are on cliffs, the rocks usually become heavily 'whitewashed' with the birds' droppings. Quite possibly – though I have no evidence – each bird resorts to the same stance every night. Birds disturbed at a roost by humans will often fly away with mild remonstrations of alarm.

In the pine forests of western Maine, USA, where Raven roosts are much more transient and mobile than in Britain and Ireland, Marzluff, Heinrich and Marzluff (1996) have found support for the idea, first put forward by A. Zahavi in 1971, that communal bird roosts can serve as 'information centres' for food-finding. By means of Ravens held captive for varying periods, then marked and released, and the placing of carcasses at chosen sites, they found that birds ignorant of food location followed knowledgeable roostmates to discovered food sources. Conspicuous social soaring displays in the evening often preceded shift of roost to a location nearer to a major food source. They appeared to be initiated by knowledgeable birds and may alert others to the discovery of a new feeding site. No obvious signals from such birds were detected but, just before the flock departed from a roost in the morning, there was a notable 'honking' by a few individuals, which rose to a crescendo as the others joined in. Marzluff *et al.* identify three features which are important to the roost-as-information-centre idea in Maine Ravens:

1. The dead animals they feed on are patchily distributed and ephemeral.

2. Their communal roosts are composed of vagrant non-breeders unfamiliar with the local food situation and therefore in particular need of information about it.

3. Vagrants derive benefit in arriving in a group at food, in order to overcome the dominance of local territory-holders (as Heinrich explained previously).

In Britain and Ireland, the same roosts are often occupied annually but tend to have a transient character over a period of years. Some are moved to new or alternative sites, while others appear to fade out completely, though this may be because new locations remain undiscovered. The roost detailed on p. 259 was moved between four adjacent dales in the Westmorland Pennines. From 1909–13 it was in Rundale, but by 1925–27 it had moved to Knock Ore Gill to the north-east. In 1931 the roost was in High Cup Dale to the south-east, but by 1957–58 it was located in Scordale still farther south-east. After a last record in High Cup in 1964 this roost subsequently waned, apparently to nothing. Similarly, in mid-Perthshire, Wendy Mattingley has found movements between five different roosts during 1990–94, sometimes within a few weeks (see p. 261). Desertion of the previously favourite roost in Glen Quaich and subsequent shifts of location may have been influenced by shooting of birds in the roosting woodlands. Rapid fluctuations in numbers suggest that some birds are going off to alternative but undetected roosts, and some movements are caused by other forms of disturbance. Exhaustion of food sources and discovery of new ones are likely to be the main cause of shifts in roosts, as Heinrich and his colleagues found in Maine.

Many roosts are in places well away from the nearest extant Raven nesting haunts, but nesting cliffs of a well-established pair are occasionally used during autumn and winter. It is seldom, if ever, that non-breeding flocks are allowed to roost on or near currently occupied nesting cliffs during the breeding season. Small parties are quite often to be seen in vacant ground adjoining occupied breeding territories, sometimes in areas where it seems that only the lack of suitable cliffs prevents nesting. Intrusion by non-breeders has been known to cause failure of one nesting pair, whose eyrie was directly attacked (Davis & Davis, 1986). Yet Tony Cross has known a Raven pair in Shropshire to nest successfully only 100 m from the edge of a rubbish tip where non-breeders were constantly feeding.

There is no consistent relationship between the sizes of Raven breeding populations and of autumn/winter flocks in the same districts. The Lake District has regularly held at least modest-sized gatherings but, in Snowdonia, I saw only small parties during 1951–61 and, even after the large recent increase in breeding population, Dare (1986a) saw only small groups. A huge roost has, however, recently been reported not far away, on Anglesey, with 1384 birds in November 1996 (Appendix 1, p. 258). Few flocks of any size have ever been reported from the Southern Uplands, although there was formerly a fair breeding strength from the Moffat Hills to Galloway. The rather low breeding numbers in the Cheviots were matched by occurrence of only small parties. The northern Pennines, with a tiny resident population, had the above-mentioned roost of up to 50 or so birds, but they were evidently mainly from the adjoining Lake District. In 1959, I saw a late summer flock of about 100 birds on moorland north of Carrbridge, Inverness-shire – an area almost devoid of nesting Ravens – and these must have come from other parts of the Highlands. In northern Perthshire, W. Mattingley has recently found a roost of over 300 Ravens in an area with a sparse breeding population. Yet quite large roosts are, or have been, known in areas with major breeding populations, such as Devon, Cornwall, Pembrokeshire, Islay, Skye and Shetland.

Appendix 1 provides details of the distribution and size of Raven flocks and communal roosts in different parts of Britain and Ireland.

SOCIAL BEHAVIOUR IN DISPLAY

Flight display

The most familiar Raven display is that given in flight, when the bird rolls over sideways. Not only is the significance of this display still somewhat obscure, but the varied descriptions of the actual performance cause some confusion. Usually, the bird appears to roll over onto its back with wings drawn in and, after staying thus for 2 or 3 seconds, reverses the movement by rolling back the opposite way and spreading its wings again, so that its normal flight is resumed. Forward momentum and lift are reduced, so that the bird drops slightly during this manoeuvre, leading to the description of 'tumbling'. Ryves (1948) describes it well: 'Hold out your right hand in front of you with the back upwards and then suddenly turn it so that the palm is upwards. Keep this position for a second and then resume the first position'. Several rolls may be given, one after another. Some observers find that the bird completes a 360° roll in righting itself, e.g. Kilham (1989) said that his hand-reared Raven 'made complete revolutions like a propeller when flying free'. Others agree that a half-roll is more usual. The speed of the movement makes it difficult to be sure exactly what happens, though slow-motion film would doubtless resolve the matter.

The description of this flight-roll as a somersault, made by some observers, is not really appropriate, as this implies a turning head-over-heels, which I have never seen. In a half-hearted version of the usual roll, the bird merely furls its wings momentarily and dives for an instant while staying upright. It may do this several times, checking and levelling-out between dives, but sometimes long stoops are seen in which the bird loses a great deal of height. A double utterance of the ordinary croak ('crok' or 'pruk' call), but sometimes sounding softer, is usually given during the roll.

The roll display is given at all times of the year, though some watchers believe it is more frequent during the breeding season. It is often performed by both of a pair, but sometimes by only one, or by a bird on its own. It may have territorial meaning, but is also sometimes given by birds in flocks, which are often paired, and may be more of a signal from one bird to its partner. Ryves regarded it as no more than an expression of *joie de vivre* and Van Vuren (1984), in California, also believed it to be a form of play, but with what functional significance – as behaviourists would wish to know – is pretty unclear. Van Vuren found that the majority of rolls were half-rolls, which could be to either side, and sometimes alternately so. He also saw a low frequency of full-rolls, in which the wings remained extended, and these were sometimes repeated as a double-roll.

An aerial display commonly seen early in the breeding season appears to have clearer territorial and/or sexual significance; it was first noted by Harlow (1922) and is well described by Geoff Horne as the 'unison flight'. The pair drift around very close together in synchronised flight, usually high up, almost touching wings at times. They will keep this up for 15 minutes or more, one or the other now and then rolling, but joining up in close proximity with the partner again. The courtship or pair-bonding function seems clear.

Ryves (1948) described an unusual display between two Raven pairs over the Cornwall sea-cliffs in early January. First two birds were spotted sitting side by side on a favourite crag:

> They silently rose and, flying into a stiff breeze, simultaneously made a headlong dive with wings almost closed. Checking the dive and

catching the wind, they planed steeply upwards, regaining with effortless grace the height they had lost. Then followed, in perfect unison, tumbles, twists, dives and climbs, now close over the sea and now far above it. Suddenly a second pair appeared and, without a sound, began a similar dance about a hundred yards away. Soon, we noticed the two pairs converging rapidly inwards until they met. Then followed precisely the same acrobatics as before only by four birds instead of two, lasting for about a minute, after which, as if by some subtle prearrangement, the pairs separated to continue their dance in distinct couples. Notwithstanding the pace and the very close proximity of each bird to another, each pair remained intact and distinct from the other throughout every twist and turn. Finally, after about half a dozen turns in distinct couples and half a dozen combined pairs, the home pair soared away to the north while the visiting pair flapped strongly southward towards their own cliff three miles away, where they later reared a brood. The whole performance had been conducted in absolute silence and solemnity, and ended as suddenly as it began.

The display had lasted 10 minutes.

While it sounds as though there was territorial meaning in this exchange, who knows what its full significance might be? Ravens are much given to soaring at times, rising high into the sky on thermals with scarcely moving wings when conditions are favourable. The soaring displays by flocks over communal roosts are mentioned above. Soaring over the nesting haunts could well be a form of territorial advertisement, conveying distant visual signals to neighbours or would-be intruders. Accompanying croaks provide auditory reinforcement. Hurrell (1951) described Ravens on Cyprus using thermals for soaring when neither territorial display nor roosting assembly was involved. He watched a flock of 61 birds soaring in pairs to a height of 600 m; some would peel off in a corkscrew dive and then reascend the air current, while others continued spiralling and then flapped off in various directions.

Perched or ground display

Descriptions of Raven displays when perched or on the ground are given by Coombs (1978), Goodwin (1986) and Cramp and Perrins (1994); they are based largely on the original observations on captive birds by Lorenz (1971) and Gwinner (1964). Heinrich (1990) has discussed the subtleties of their meaning at length and made new interpretations of some aspects. I have drawn upon these texts and refer readers to them for fuller details.

Much display is about the establishment of dominance relationships or pair formation, which Lorenz (1940, 1971) regarded as closely connected. He pointed out that Ravens belong to the large group of birds which show sexual ambivalence and which may adopt male or female behaviour according to the particular circumstances or individuals. Pairing involves the establishment of male dominance over the female, with the suppression of any possible masculine behaviour by the female, but, once this is achieved, the two birds live in mutual harmony.

(a) Ear tuft intimidation (after Hermann Kacher in Lorenz (1970))

(b) Thick head intimidation (after Hermann Kacher in Lorenz (1970))

(c) Aggressive bowing display (after Hermann Kacher in Lorenz (1970))

Social behaviour   111

(d) Defensive threat display
(after Hermann Kacher in
Lorenz (1970))

(e) Male choking ceremony in
courtship (after Hermann Kacher in
Lorenz (1970))

(f) Female pre-copulation
display

   One group of displays may be categorised as intimidatory behaviour and regarded as developing stages in a fixed and stereotyped sequence of increasing aggressive intensity. Such signals are given in both aggressive and sexual contexts 'to overawe individuals of the same sex or in the process of attracting and arousing a partner of the other sex, so that there are patterns of posture movements and sound used in various combinations in both rivalry and courtship' (Coombs, 1978). The mildest form of aggressive exchange between two birds is described by Goodwin as 'self-assertive display', but as 'advertising display' by Cramp and Perrins (1994). Coombs, following Gwinner (1964), calls it 'ear-tuft intimidation' and also

(g) Solicitation/mating postures
(after Hermann Kacher in Lorenz
(1970))

(h) Male caress in pair-bonding
(after Hermann Kacher in
Lorenz (1970))

regards it as the first stage in intimidatory behaviour. In the male, the feathers just above the eye are erected to resemble small ears, while the rest of the head feathers are flattened. The flank feathers are then fluffed out to give a 'trouser-like' appearance and the plumage of the throat, neck and breast is erected sufficiently to form a smooth profile with the flanks. The hind neck and mantle also form a smooth line, the overall effect being to give the bird a swollen-necked appearance (see illustration, part (a), p. 110). The wings are also opened slightly at the shoulders so that the carpal joints stand a little away from the body. The bird salivates copiously, which causes frequent swallowing and throat-feather movement. It strides about assertively and gives a soft 'ko' in time with its steps.

From this state, the bird goes into a posture with all head feathers fully erected, giving it a large-headed and shaggy-throated look: 'thick-head intimidation' (Coombs, 1978) (see illustration, part (b), p. 110). At the highest intensity phase the Raven holds itself horizontal and makes forward bowing or retching movements, uttering 'koo' or 'krua', usually with spread tail and with the pale nictitating membrane drawn (see illustration, part (c), p. 110). Two males trying to dominate each other will strut about, circling or walking alongside each other, and the birds may jostle lightly. This may lead to fighting, or one bird may acknowledge its subordinate position by retreating.

The female version of these stages is slightly different. The 'feather ears' are not as long and are less obvious because the other head feathers are more erected; the bird stays silent during the initial and thick-headed phases. In the most intense bowing and forward-reaching phase, the female bows lower still, with tail spread and wings lifted, and makes squeaking, clicking or clappering noises, along with upward movements of the previously downward-pointed bill. The tail is inclined upwards and, with the raised head, gives the body a U-shape, in contrast to the straight profile of the male.

These displays are shown either towards rivals of the same sex, when the motivation is a combination of aggression and escape (fear and anger), or towards a mate or potential mate, when the impulse is sexual. In more intensely aggressive situations, it develops into a full threat display, in which the bird sleeks all its plumage except the 'ears', stands upright and moves with quick, light steps and bounds. Head-forward threat postures are very variable, with open, often snapping bill and either lifted or drooping wings. An inclining of the bill towards a lower-peck-order bird is often sufficient to keep it away or displace it from food. Intense frontal threat is used only in nest defence and involves erect plumage (especially on the head), opened bill, wings partly opened and drooped, and fanned tail (see illustration, part (c), p. 110); the bird stands upright, with tarsal joints bent inwards and the bill at times pointing vertically down, and makes threatening sounds. Lorenz regarded the ear-tuft display as being a more intense form of intimidation than the thick-headed kind; Heinrich (1990) suggests that, instead of an escalation of self-assertion, the changed postures from erect-ear to thick-head, and then to bowing displays, represent a *change of motivation* from aggression to sexual interest.

This behaviour is distinguished from defensive threat display in which the back and mantle plumage is fully erected into a ragged or ruffled profile, and a hump-backed posture often adopted, with variable bill snapping and calling (see illustration, part (d), p. 111). In the full defence posture all the plumage is fluffed out so that the feathers separate and the bird appears very large. The bill is wide open, and raised or lowered, and the calls become louder.

Fighting may follow any of these threats but, though it may appear serious, with birds thrashing about and pecking at each other while gripped together with locked claws, real damage is seldom done. A few feathers may be knocked or pulled out, but not much more. Lorenz (1952) first pointed out that the birds always avoid pecking at each other's eyes, which would be the obvious target if they were truly intent on causing harm. In the artificial conditions of captivity, where subordinate birds cannot escape, weak or injured Ravens are sometimes killed but, here again, the eyes are avoided. Despite their propensity for pecking out the eyes of dying or dead animals, on which they intend to feed, Ravens have developed a strong social inhibition that keeps them from running this risk with each other. Such modern insights appear to be long pre-dated by an old Scottish proverb – 'ae Corbie will no pyke out anither's een [eyes]' – a saying which has English and German equivalents. There are occasional lapses from the rule, however, for Smith (1905) relates an instance of a fight witnessed between two wild Ravens in which 'one of the birds was pecked in the eye and killed on the spot'.

Threat display may also be followed by submissive behaviour on the part of subordinate birds. Submissive Ravens tend to retract the neck, hunching the head between the shoulders and sleeking the plumage, making themselves appear small and thin, while the tarsal joints are bent inwards, giving a knock-kneed look. The sleekness about the head in newly fledged young is characteristic and could be meaningful in this context. In appeasement the bill is raised towards, but held lower than, the bill of a dominant bird. The bird crouches and sometimes inclines its head to a vertical upward position, but often the head is turned away and the bird may look around. In another variation the head is lowered into a vertical downward (staring-at-the-feet) position and there may be pecks at the ground or preening of the breast feathers. A tail-quivering display is given only in apparent gestures to mates or trusted birds.

Both sexes also signal appeasement by juvenile calls and food-begging actions (either those of juveniles or females soliciting feeding). Begging is often associated with submissive postures by the female, as a reaction to male demands and with intent to appease. The bird crouches low, flutters it wings and often makes noises recognisable as variations on the food-begging calls of the young (Coombs, 1978). Both sexes can show food-begging responses, depending on which partner has food to offer. The display can also be a social signal, apart from its mutual feeding context.

In a sexual context, the advertising display may develop into precopulatory display. In the male this begins with the neck stretched forward, throat feathers extended, tail raised, wings partly stretched and spread back, third eyelid drawn and nasal 'krooyooyoo' calls. The wings then become half-spread and drooping, well away from the body. The tail is spread and raised, and all the contour feathers except the flank 'trousers' are sleeked down, while the head is raised on the stretched neck, with the bill horizontal or pointed down, and the nictitating membrane is held over the eyes for several seconds. Lorenz describes the bird as making peculiar choking movements (see illustration, part (e), p. 111). The female solicits by crouching, variably opening and extending or drooping its wings and shaking or quivering the slightly raised tail. The bill is tilted upwards at nearly 45° and the posture is one of extreme submissiveness (see illustration, part (f), p. 111). Postures immediately before mating are shown in the illustration on p. 112 (part g)). The male was later seen by Lorenz to caress his mate's nape feathers (see illustration, part (h), p. 112). If copulation does not follow, the male goes into the female precopulation posture. This display is given also in submissive and appeasing contexts.

In another account of mating behaviour, Wilmore (1977) described the male standing by the female and stretching his neck over hers, ruffling his head, neck and throat feathers, and then bowing to her and sinking down on his belly. His performance was accompanied by 'curious noises such as mumblings, bubblings and poppings like corks being extracted from bottles'. The pair next engaged in mutual preening and the male jumped in the air several times. He then caressed his mate's bill with his own, tickling her under the chin and 'kissing' her. Eventually, the female signalled her desire to mate by jumping up and down. After copulation the pair took off to indulge in aerobatics, and then landed again, with the male giving a musical, vibrating call.

Allo-preening occurs between friendly birds as well as mates and mutual activity is especially frequent in captives. Sick captives with fluffed heads were often preened by other Ravens. The preener often lifts the long throat feathers and looks within, evidently for parasites. This begins when birds are very young and then evidently lacks sexual significance, but juvenile paired birds preen each other most frequently, and Coombs believes it may facilitate pair formation, overcoming individual distance. An individual distance is nevertheless maintained by allo-preening birds, even when paired. Lorenz (1970) noticed that Ravens would defend their fellows vigorously when they were captured, but only when these were known companions. Such defence was accompanied by a harsh 'attack call'.

Goodwin (1986) notes that, in the Raven, the eyes are very much an expression of mood and even lay-people can tell whether a captive bird is in a hostile, appeasing or loving mood. In hostile contexts, the eyelids tend to be widely opened and the pupils dilated, whereas in friendly ones the eyelids are slightly closed and the pupils contracted. Captive birds also respond appropriately to friendly, intimidating or aggressive looks from humans (Gwinner, 1964; Goodwin, 1986). This giving and receiving of signals that humans can recognise in terms of their own emotions was, no doubt, another factor in the special place accorded to Ravens in the early associations between animals and people.

Play

Play, especially in the form of 'snow-romping' is reported by various observers. Smith (1905), speaking of his tame Raven, said that, 'When the snow was deep on the ground, he would play in it or roll over in it like a dog'. Moffett (1984) published photographs of a Raven pair sliding on their backs for 3 m down a snow-slope, after first bathing in the snow. One bird repeated the manoeuvre twice and then its mate followed suit. Tony Cross has seen Ravens in Wales sliding down snowbanks on their backs in the same way. Heinrich (1990) also described Ravens in Maine snow-bathing, kicking snow, sliding on their breasts and rolling on their backs in it, apparently for fun, 'like happy dogs'. Gwinner had captive birds that repeatedly slid down an inclined, shiny piece of board, and Kilham (1989) kept a tame Raven that enjoyed playing in snow, somersaulting on a bank and sliding down on its side, taking a roll.

Tony Cross has seen birds bouncing up and down on the very end of a long, dead branch, by holding on with their beaks and with their wings closed tight against their bodies. He has on numerous occasions watched Ravens taking sticks, turf and sheep droppings aloft, dropping them and trying to catch them again before they hit the ground. One bird trailed a large sheep's skull in one foot, proceeding to drop and retrieve it many times before flying off. Wilmore (1977) describes Ravens in a party pulling each other's tails and playing with each other's bills; and also passing stones to one another beak to beak, pounding and grabbing at them.

Several writers have remarked on how Ravens appear almost to enjoy flying in a gale, making steady progress or shifting this way and that, but always under full control, however violent the turbulence. Chapman (1924) expressed it in his felicitous way:

> On such blusterous occasions one enjoys to perfection watching the wondrous wing-power of the Ravens – literally revelling in it, toying with the tempest, poising and stooping like falcons; anon tumbling sideways or headlong, or throwing double aerial somersaults amid blasts that threaten to sweep a strong man off his foothold.

The possible element of play in the tumbling/rolling display has been noted above. Jaeger (1963), in the USA, described Ravens flying in and out of water columns shot into the air by an irrigation system, soaring around between times. The soaring in thermals described by Hurrell on p. 109 seemed also to be largely a matter of enjoyment.

Gwinner also recorded his captive Ravens indulging over 300 times in 'upside down hanging', a curious behavioural trick known also in other corvids, but for which no functional explanation is yet forthcoming. It is performed especially by young birds up to 2 years old. The birds hang head down by their feet from a thin branch or the netting roof of a cage, but start by rocking backwards and forwards, then swinging into a pendent position. They may stay thus for up to a minute, looking around and sometimes swinging to and fro, with wings partly open. For want of a better explanation, it may be included as a part of play activity.

One of the most oft-remarked quirks of tame Ravens is their playing of tricks on other animals, especially cats and dogs. A favourite ruse is to sidle up to a sleeping animal and tweak its tail, quickly jumping back or flapping up to a higher perch before the enraged animal can retaliate. Seton Gordon (1938) described one bird provoking a dog into chasing it across a field, landing ahead and then taking off before the pursuer could reach it. These seemingly mischievous but risky actions may well be related to the stealing of food from other predators (see p. 95), especially when a pair co-ordinates efforts; one bird will decoy and distract the animal while the other nips in, seizes the food and flies off with it. Yet some of these attacks seem more in the nature of devilry and Kilham (1989) mentioned

White-necked Ravens in Africa stirring up a party of gorillas, who became both scared and angry.

Heinrich (1990) gives other examples of risk-taking by Ravens, especially in landing at carcasses where dangerous animals, such as wolves and coyotes, are feeding, and contrasts this with the bird's nervousness in other situations where potential danger lurks. He has another explanation for this seeming foolhardiness and suggests that it is part of an effort to gain status and dominance with its fellows by demonstrating superiority at food-winning. This would particularly enhance the standing of suitors with females and increase the chances of pairing and reproducing for the males which belonged to the non-breeding sector.

CHAPTER 6

Raven movements

GENERAL MOVEMENTS

Ordinary observation suggests that the Raven is one of our more sedentary birds. Local bird reports have few records of Ravens seen away from breeding areas and there are large areas of lowland country where the bird is never seen nowadays. The *Atlas of Wintering Birds* (Lack, 1986) well illustrates the point, with only a handful of sightings in lowland England to the east of their main occurrences in upland regions. Of Lancashire, Mitchell (1892) noted that, 'It used to be not uncommonly seen on the sea-shore, but it is now exceedingly rare' and, more than half a century later, Oakes (1953) said that single Ravens or pairs were sometimes seen flying over the mosslands at the head of Morecambe Bay, but that it was only very rarely observed away from the hills on the Lancashire lowlands and coast to the south. For Cheshire, Coward (1910) reported the bird as seldom seen, but remarked that Brockholes referred to its former abundance in winter on the Dee marshes, while Bell (1962) mentioned only occasional records of passing birds in the nineteenth century.

This illustrates the widespread experience that Ravens nowadays are little inclined to seek flat shores or lowlands well away from the nearest nesting haunts in England. During a long acquaintance with the Solway plain and shore in Cumberland I never saw a single Raven, even though the nearest Lakeland breeding places were only 25 km distant. Ravens are, however, now quite frequently recorded on the sea-shore in south-west Cumberland, e.g. at Hodbarrow and along the Irt estuary (Cumbrian Raptor Study Group, 1994), where the nearest inland nesting haunts are only a few kilometres away. Orton (1948) noted that, in Snowdonia, the old paired ravens, after separating from their young at the end of summer, visited the lowlands and especially the sea-shore to feed together, but in the evening flew back to roost on their home cliffs in the mountains nearby.

The bird is also little in evidence much beyond the edges of some hill areas which hold good breeding numbers. Around the northern fringes of Lakeland the only Ravens I ever saw away from the hill-nesting areas were a single pair on the outlying moor of Faulds' Brow in 1989. Perhaps they tend to move straight from one hill area to another, without stopping on low ground in between. Several observers have remarked upon occasional flights of small parties or pairs of Ravens high across the Eden Valley, between the northern Lake Fells and the Pennines, moving in both directions. Severe winter conditions – especially heavy snowfall – may cause both territory-holding pairs and non-breeding parties to desert high ground temporarily and move to lower, more hospitable country not far away. I saw small parties, of up to eight birds, on the low hill overlooking the town roofs of Bangor in north Wales during times of deep snow. Cullen and Jennings (1986) state that, on the Isle of Man, 'During winter Ravens do venture much more readily from their typical remote moorland and coastal habitat to the cultivated lowlands'. Such local movements are reversed when milder weather returns.

The general experience in Britain is that the established breeders remain close to their nesting places throughout the year and are extremely sedentary except during spells of severe winter weather. This is reported by Davis and Davis (1986) for central Wales, Dixon (in press) for south Wales, G. Horne (unpubl.) for Lakeland, and Ewins, Dymond and Marquiss (1986) for Shetland. Substantial movements are thus largely confined to the non-territorial Raven population.

For more precise information on movements I have to turn to the ringing data. The BTO's register records a total of 7,625 Ravens ringed and 533 recovered (mainly found dead) up to 1993, the first being in 1924. This represents an overall recovery rate of 7%. All but 13 of the 533 were ringed as nestlings. The recoveries included seven individuals still at large, six of which were identified by wing-tags, and another 20 of birds found in a weak condition but later released again to the wild. Two birds were recovered a second time, one after a sight record and the other after capture and release.

The prodigious efforts involved in visiting and climbing to so many nests to ring the young owe much to Raven enthusiasts, though mostly over different periods. To some extent, and taking the period 1924–93 as a whole, the numbers ringed in different regions have been a reflection of Raven numbers and distribution across Britain and Ireland, but south-west England, south Wales, the Scottish Highlands (other than Orkney and Shetland) and western Ireland are under-represented.

I have assumed that nestlings were ringed on average when they were 25 days old, and this period is added to the time between ringing and recovery in order to give an estimate of age. Recovered birds are then assigned to yearly age-classes. For the 13 Ravens ringed as adults or subadults, the time between ringing and recovery is clearly well short of their actual age and they are omitted from Table 5. The figures for age may be slightly biased upwards, since the time of death could often not be established accurately for the Ravens found dead. The recovery data use the expression 'not fresh' in 44 instances and state that, in another 20, only the ring or ring with leg were found, or that remains were skeletal or long dead. I have excluded the latter group from any analyses involving age, but since 'not fresh' included many yearling birds (some less than 3 weeks after ringing), I have retained this category.

Analysis of the recoveries amply confirms the impression that the Raven is a somewhat sedentary species in Britain and Ireland. Of the 533 recovered birds, 80% were found within 50 km of the places where they were ringed (which were mostly their birthplaces) and only 5% were more than 101 km away (see Table 4 and Fig. 5). There was little movement between the different regions listed in Table 4, even when these were adjacent, and a majority of recoveries were in the same county as that in which they were ringed. Only four recovered birds (all ringed as chicks) had travelled more than 200 km (Table 3).

Although Ravens are powerful fliers, the ringing returns suggest that they are disinclined to cross large expanses of open sea. There are no instances of Ravens ringed in Britain or Ireland being recovered abroad and no recoveries in these islands of birds ringed in other countries. Only two birds had travelled from Ireland to Britain, and only two more in the reverse direction. The individual which flew 229 km across the North Sea to an oil rig was the exception. Twenty-four recoveries of Ravens ringed on the Isle of Man were all on the same island. Of 12 recoveries of Orkney-ringed birds, three were on the adjoining mainland, in Sutherland and Caithness, and the rest in Orkney; while all 18 recoveries of Shetland-ringed birds were refound in this island group, except for the one on the oil rig. Yet Saxby (1874) found that, on Shetland, large numbers of Ravens, which he believed were certainly not natives, arrived in the autumn and stayed until the following breeding season. His observations were supported by both Williamson (1965) and Dymond (1991) on Fair Isle, though they found that flocks of migrant Ravens (40–45 birds) occasionally reached the island during March–May as well as during September–October (25–30 birds). D'Urban and Mathew (1892) recorded a Raven caught when it came on board a ship 800 km west of Ireland in November 1851.

Even the larger flocks may represent no more than the aggregation of non-breeders from a single district, involving only local movements as birds assemble and sometimes move around or later disperse again. Seton

Gordon suggested that some Highland non-breeding flocks were immigrants from Scandinavia. While this is possible – and from Iceland, too – there is not a scrap of evidence to support the idea. They may simply be from other parts of Scotland.

Table 5 does nevertheless show that 98 out of 192 (51%) of first-year birds which survived to more than 75 days were recovered more than 25 km away, so that there is a fairly marked tendency for youngsters to move away from the immediate vicinity of birthplace. Three of the four Ravens which travelled more than 200 km were first-year birds, including the record-holder at 551 km (see Table 3). The similar overall figure of 249 out of 519 (48%) for all birds ringed as nestlings and recovered more than 25 km away suggests that most of the dispersal takes place during the first year. The potential for colonisation of new breeding areas at some distance (over 50 km) certainly exists and the ringing recoveries belie the infrequency of live Raven sightings in many lowland districts. Ravens in northern England (excluding the Isle of Man) and Scotland (excluding Orkney and Shetland) tended to travel farther than those in Wales or Ireland. Recoveries more than 25 km from the ringing site accounted for 69% of those in northern England and 75% in Scotland, compared with 40% in Wales and 41% in Ireland. Reasons for this difference are unclear, other than the possibility of greater hard-weather movements in the colder north.

It is hardly surprising that, of 60 ringed Ravens recovered at or close to the nest site, 52 (87%) were less than 1 year old (Table 5). A few died even before they had left the nest, or immediately afterwards, and 33 birds died within 50 days of ringing, at the stage when their youth and inexperience must make them especially vulnerable. Similarly, of Ravens recovered within 50 days of ringing as nestlings, only three out of 50 had moved more than 10 km from their birthplace. The oldest birds recovered also showed a marked tendency to be fairly close to their birthplace (Table 5). This is unsurprising, too, for the oldest Ravens are likely to have been part of the breeding population, and birds which had soon found a safe nesting place would be unlikely to make further movements that might reduce their survival chances. It is not known how strongly the urge – found in many birds – to return to the area of their birthplace in order to breed operates in the Raven.

Direction of dispersal is not random, but shows an overall bias towards easterly movements. Of 467 Ravens recovered away from the ringing site, 243 had some degree of easterly orientation – between NNE and SSE – compared with 150 that showed a westerly movement – between NNW and SSW (Table 6). Regional data are presented, as angular measurements of direction, in Fig. 6 and show that the tendency is towards easterly movements in Wales, England and the Isle of Man, whereas in Scotland there is an inclination towards the north-east. In Ireland, the bias is distinctly from south to west. Most Ravens ringed in Ireland were in the south-east, so that most of the land area into which they could disperse lies westwards. Conversely, in Britain, the areas where the majority of Ravens have been ringed mostly lie in the west, so that a greater area of land into which birds could spread lies eastwards.

Ringing recoveries may not entirely reflect geographical dispersal

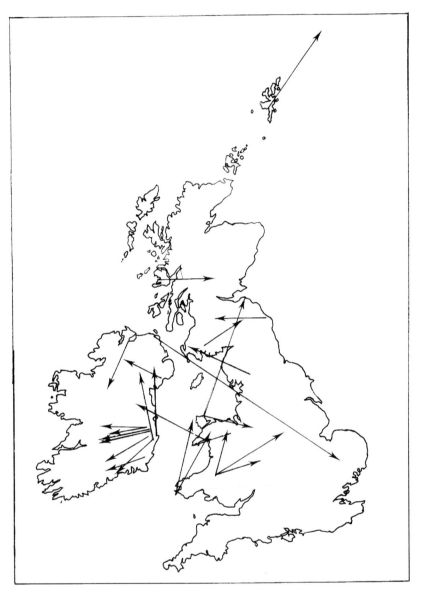

Figure 5. Movements of over 100 km by ringed Ravens.
NOTES: Based on recoveries recorded in the Ringing Scheme of the British Trust for Ornithology.

pattern for the Raven population, since this would be distorted by any regional differences in factors determining ring-recovery rate, i.e. in mortality, and the finding and reporting of rings. Eastward movements in Britain will tend to bring Ravens into contact either with grouse-moors or lowlands, where their chances of survival will be less than if they stay in uplands or on rocky coasts, i.e. eastward movements may well be associated

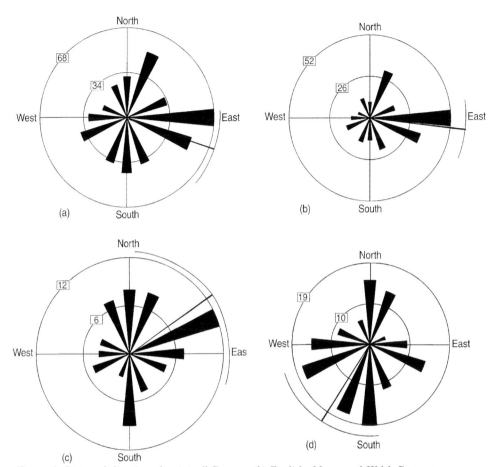

Figure 6. Dispersal directions for: (a) all Ravens; (b) English, Manx and Welsh Ravens; (c) Scottish Ravens; (d) Irish Ravens. (Prepared by Alan Fielding.)
NOTES: Based on recoveries recorded in the Ringing Scheme of the British Trust for Ornithology. Narrow lines to outer arcs show mean angles of dispersal. Scale varies between diagrams as indicated by the numbers.

with heavier mortality and thus increased chances of ring recovery. Of 71 recoveries of birds ringed in Lakeland, 40 were found on the Pennines to the east or south-east and another six on the Shap Fells in between – all notable grouse-moor areas (Fig. 7).

Colour-marking of young Ravens in central Wales has so far proved an unrewarding task. Davis and Davis (1986) ringed 231 young birds with BTO rings and all but four of them also with Darvic colour-rings. The somewhat negative results suggested that the great majority of juveniles left the study area and the surrounding country in their first autumn and did not return. Only two colour-ringed birds were ever detected among 40–100 Ravens at a refuse tip or an abattoir close to the edge of the study area during any winter period. At three Raven-haunted tips between 17 and 24 km from the

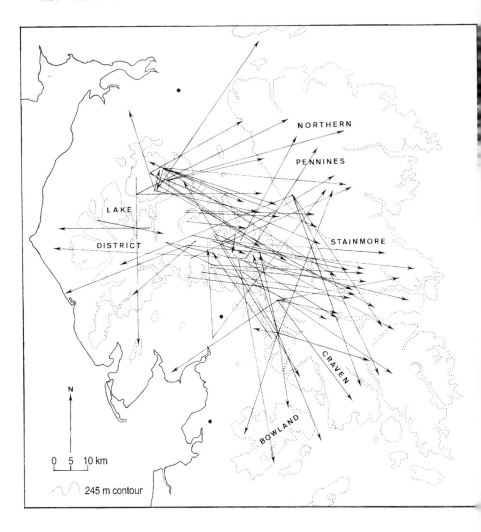

Figure 7. Recovery pattern of Ravens ringed in Cumbria.
NOTES: Based on recoveries recorded in the Ringing Scheme of the British Trust for Ornithology.

study area, only two marked birds were ever seen at once. Tony Cross has ringed the staggering total of 2,016 nestling Ravens in Shropshire and mid-Wales during 1982–96 and over half of these were colour-marked with either Darvic rings or patagial wing-tags. Many marked birds have been observed paired, some to other marked Ravens, but most to unmarked birds. Although two 3-year-old birds held territory in 1995, it was not until 1996 that Cross found any marked birds nesting. This year, in Shropshire, he found two 2-year-old males paired with females of unknown age: one pair evidently did not produce eggs, but the other bred late and reared a single chick. The successful male was ringed only 1.6 km away from its nesting place. In a third instance, in Wales, a 4-year-old female (also ringed only 1.6 km away, but not known to breed previously in the locality) reared two young. While many wing-tags were lost within 2 or 3 years, and these are clearly an unreliable means of identification, only three BTO-ringed birds have been recorded nesting by Cross. The implication is that recruitment to the breeding population is very low, that birds are nesting well away from their natal area, or that mortality of potential recruits is high. The low recovery rate of metal rings (<2%) suggests that many illegally killed Ravens are not being reported.

Cramp and Perrins (1994) discuss the movements of other *Corvus corax* populations overseas, noting that those south of *c.* 60°N are essentially sedentary, though their juveniles may make extensive movements. Altitudinal movements occur in alpine regions of Europe and the Himalayas and, in general, there is a tendency for the bird to become more widespread in winter. In northern regions, movements southward tend to increase with severity of winter conditions. In Fennoscandia, the farthest movements (up to 350–510 km) are mainly by birds from northern areas, towards the south or south–south-east. Gothe (1961) reported having seen several flocks moving in a south-westerly direction, evidently from Scandinavia to Germany. Some populations (or segments of them) appear to winter far north, within the high-arctic breeding range (e.g. on Melville Island, northern Baffin Island and northern Greenland), but there is some evidence for partial migration southward from these frigid regions. Ravens banded in northern Greenland have been recovered up to 1,120 km to the south (Freuchen & Salomonsen, 1960).

DAILY MOVEMENTS

Besides the seasonal or permanent disperal, there are the local movements that Ravens make during their day-by-day activities, mainly in search of food but also in territorial defence or, on occasion, quest for a mate. Ordinary observation suggests that most settled pairs of Ravens feed within 2–4 km of their nest while it contains eggs or young. On Mull, Paul Haworth and Alan Fielding have found that Ravens appear to make use of the entire landscape, and sightings showed little evidence of clustering around nest locations. As noted in Chapter 10, once the young have fledged, the whole family may drift farther and farther away from the nest location in the ensuing weeks, their earlier territorial constraints having evidently faded away. It is a common experience to visit Raven nesting haunts during late summer and to have neither sight nor sound of the birds for a whole day. They have moved elsewhere and, although the distance they cover each day may be quite small, they may be 10 km or more from the nest site.

Birds in communal autumn and winter roosts often travel much greater distances during each day in quest of food, although most of the concrete evidence comes from North America. Heinrich (1990) mentions that Richard Stiehl tracked birds that ranged daily up to 43 km north and south of a communal roost in Oregon, and that Skip Ambrose found radio-tagged Ravens which made a regular daily round trip of 128 km commuting between their roost and a land-fill on the outskirts of Fairbanks, Alaska. Marzluff, Heinrich and Marzluff (1996), extending Heinrich's work on winter feeding behaviour of Ravens in the Maine forests, studied a roost 25 km from a large carcass which they had placed as bait but periodically covered over. Over a fortnight the number of birds in the roost fluctuated according to the availability of the carcass. When a large food source was located, it was usual for the Ravens to establish a new roost close by and, during the soaring displays that preceded these shifts, some groups flew lengthy routes around the roost, covering at least 200 km^2. Individual Ravens were attracted from at least 10 km away to those that were soaring. Heinrich (1990) watched birds assemble at dawn to a food-bait from four different directions, evidently representing separate roosts. Radio-tracking of Ravens from their roost sites in Idaho showed an average distance travelled of 6.9 km, with an extreme of 65.2 km. Habitat was preferred according to feeding value rather than availability, and the birds spent an average of 54% of their time on agricultural land (Engel & Young, 1992b).

At communal roosts in Britain, birds commonly gather in the evening from at least several kilometres away and it seems likely that, when flocks disperse during the day, they must in total range over considerable areas from the roost sites. Just how widely they travel during the day is at present unknown. Coombes (1948) watched a loosely strung-out flock drift slowly during the day over a distance of at least 4 km along the Scafell range in Lakeland. In the same district, Geoffrey Fryer has observed a regular evening flight-line of Ravens to a roost west of Windermere; the birds appear to come from the higher fells in a sector from east to north of the town, and with still another 3 km to go.

CHAPTER 7

Associations with other animals

BIRDS

The Raven shares its haunts in Britain and Ireland with some other birds which might seem to compete with it in some degree, either for food or nesting space. I shall look briefly at some relationships with these neighbours.

Golden Eagle

The Golden Eagle also behaves as a scavenger of large animal carcasses, such as sheep and deer, especially in western Scotland, where live wild prey is often in short supply. It is eminently capable of dominating the Raven, if the two compete for the same items of carrion. Macgillivray (1837) noted that Ravens yielded place to eagles (which could have included Sea Eagles) when feeding on carcasses, though Paul Haworth saw a group of up to 11 Ravens prevent a Sea Eagle from feeding at a sheep carcass on Mull. The Golden Eagle is a predator on the Raven, at least on occasion, and one of the few non-human enemies that the big Crow has to face, since the large owls are virtually absent from Britain. Dick Balharry found the heads of young Ravens in four different Golden Eagle eyries in Wester Ross. Jeff Watson has records of six fledged young Raven remains found in five eagle eyries in north mainland Argyll, Mull and Sutherland during 1982–85. He also found the remains of an adult or subadult Raven in a winter pellet, though the bird could possibly have been taken as carrion.

While predation rate on Ravens may be low, the risk of fledglings being taken would seem considerable and is an obvious explanation for the distance at which breeding Ravens usually keep their nests from occupied Golden Eagle eyries. It is unusual for Ravens to lay within 1 km of an active Eagle nest and the distance is often several kilometres. In Scottish districts where both species breed, I found spacing between active Raven nests and the nearest occupied eyries of Golden Eagles as follows:

District	Range	Mean	n
Galloway	1.7 – 4.7 km	2.8 km	7
Argyll and Inverness-shire	0.6 – 4.0 km	2.4 km	6
Ross-shire	1.8 – 6.2 km	3.9 km	13
Sutherland	1.9 – 4.4 km	3 km	11

Thomas (1993) found the shortest distance between occupied nests of the two in central Argyll was 0.6 km and the average for 13 cases was 2.1 km. In the Highlands, old Raven nests are sometimes to be seen on the cliffs where Golden Eagles are nesting, but they invariably date from years when the Eagles were using alternative crags elsewhere.

Even this degree of avoidance probably underestimates the size of areas over which the Golden Eagle exercises dominion. Over large parts of the Highlands, much of the mountain and moorland terrain appears to be shared out between Golden Eagles, and the Ravens are mostly confined to the more fertile edges and valleys of the main massifs. In some areas away from the coast, Ravens are perhaps even the less numerous of the two as breeders. Some Ravens may forage within the home ranges of Golden Eagles, but their numbers are evidently severely curtailed by the presence of their bigger neighbours. The map in Fig. 8 shows that the distribution of the two species in a fairly typical part of the north-west Highlands is complementary, so that they appear to be interchangeable. It is tempting to imagine that, if the Golden Eagles were not there, Ravens would take their place. There is also clear evidence that Golden Eagles are able to dispossess Ravens by taking over their nesting quarters.

The following observations from Galloway and Carrick show displacement of Ravens from once regular nesting territories by Golden Eagles returning to breed in the district.

Associations with other animals 129

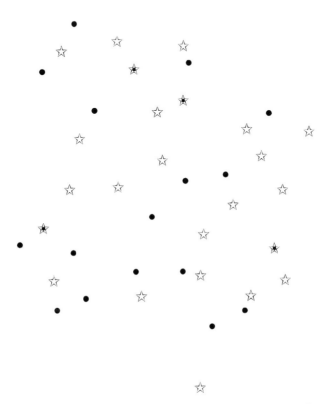

Figure 8. Breeding dispersion of Ravens and Golden Eagles in part of Wester Ross.
NOTES: Not all cliffs are occupied annually by either species. Data from D. A. Ratcliffe, J. D. Lockie and R. Balharry.

1. The first pair of Golden Eagles bred in 1945; the pair of Ravens on the same cliffs at once ceased nesting and none has ever bred since.

2. The second pair of Golden Eagles bred in 1948; a pair of Ravens continued to nest in the vicinity, but not within 1.8 km, and has nested on the Eagle cliffs only when these birds have been using alternative rocks 2.3 km away.

3. The third pair of Golden Eagles bred in 1952; one pair of Ravens breeding on the same cliffs or 3 km away ceased nesting and by 1955 a second pair, 2.3 km away, disappeared.

4. The fourth pair of Golden Eagles appeared in 1963; a pair of Ravens on the same cliffs disappeared that year.

The situation in this district has become complicated by the spread of afforestation, which has caused difficulties for both species. Golden Eagle pairs 1 and 3, which occupied adjoining territories on the same hill, are now reduced to a single non-breeding pair or bird which ranges over the whole area. Ravens bred in 1994 at both the localities where they were displaced by Eagle pair 3.

In Lakeland, the return of Golden Eagles to the Haweswater valley in 1957 resulted in desertion of a Raven haunt on the same crags, even though the Golden Eagles did not lay eggs until 1969. From 1970 onwards, the Golden Eagles moved to another range of cliffs, whereupon a second pair of Ravens in that locality dropped out, while a pair returned to the first place. The closest distance between occupied nests of the two is 2.0 km.

Ravens will not hesitate to launch aerial attack on an Eagle which intrudes to closely into their domain. In 1957 I watched an intrepid bird almost closing with one of the newly returned Lakeland Eagles and Bannerman (1953) mentions an attack in the Balearic Islands in which a Raven pitched on a Golden Eagle's back in flight. Seton Gordon (1938) wrote about a pair of Golden Eagles on Skye 'whose lives must be made miserable by the continued attacks of the Ravens in the district'. The sight of one of these Eagles rising above its nesting cliff often brought one of the Ravens hurrying noisily from its own nest, nearly 2 km away, to do battle by stooping close to its enemy. During these exchanges, the Golden Eagle rarely showed resentment or attempted even half-hearted retaliation. Hebridean crofters were also said to approve of Ravens because they were seen to drive away Eagles. Ravens nevertheless have to be careful, since their powerful adversaries are surprisingly agile and the tables could quickly be turned. The Raven becomes vulnerable when on the ground. Probably the instances of open warfare between the two that are witnessed are misleading because they suggest that the Raven is the aggressor and may even have the upper hand. Ravens respond more to the latent threat by avoidance reactions in choice of breeding place, so that the Eagle is really the winner in this less obvious contest.

In Idaho, USA, where Ravens and large raptors nested on power-line towers, nests of Raven and Golden Eagle were never closer than 1.07 km and usually much farther apart, whereas those of Red-tailed Hawks were as close as 0.36 km, and of Ferruginous Hawks only 0.10 km away (Kochert *et al.*, 1984).

Buzzard

Ravens typically share their breeding grounds with Buzzards, which also feed on sheep carrion but appear to be less dependent on this item than the Raven. In central Wales, where the two species also often share their range with Red Kites, Newton, Davis and Davis (1982) found no evidence that either species was limited in numbers or distribution by the other.

Where the two nest in crags, the Buzzards often use easily accessible ledges, on smaller or more broken cliffs, at 500m or more away from the Ravens. Tree-nesters are frequently closer together and, in central Wales, the two species commonly breed in the same small woodlands. Buzzards sometimes take over and renovate an old Raven nest, both in rocks and trees, and the reverse happens also. On Mull, Paul Haworth watched Ravens take over a favourite Buzzard site with a large old nest, which they moved 4 m along the crag, to rebuild under an overhang. The Buzzards returned the following year and built a new nest on their old site, but were then dispossessed by the Ravens, which removed their neighbours' nest completely and then reused their own around the corner. Ravens have been recorded as predating Buzzard eggs, but this is probably rare. The two species frequently spar with each other in flight during nesting time, but such exchanges seldom appear to have a serious intent.

Hewson (1981) found that Buzzards were dominant over two different pairs of Ravens when feeding at sheep carcasses on the Argyll shore, but Bolam (1913) in Wales noted that 'From such a meal, the Buzzard will generally suffer itself to be driven by a Raven'.

Red Kite

Kites and Ravens select similar tree sites and often nest in close proximity in central Wales. While there is usually a mutual tolerance, serious friction occasionally erupts. Peter Davis recorded that a pair of Ravens killed an adult male Kite which was nesting in an adjoining tree when their fledged young moved towards the Kite's nest. Conversely, Tony Cross found the remains of a recently fledged young Raven in a Kite's nest with a large chick, though it was uncertain whether the Kites had killed the fledgling. Tony Cross also notes that Ravens occasionally take over and rebuild an old Kite nest for their eyrie, while Kites sometimes adopt Raven structures, including those used earlier (though unsuccessfully) the same season.

Peregrine

The Peregrine is the Raven's most frequently remarked neighbour in its cliff-nesting haunts, both inland and by the sea. Their customary sharing of both nesting cliffs and, in different years, the actual eyrie ledges has been described elsewhere (Ratcliffe, 1993). Ravens are absent from some Peregrine areas, such as the grouse-moors, eastern coastal districts, and lowland areas where falcons nest in quarries or on buildings. And, since Ravens are the more numerous species, there are many cliff-nesting places where the Peregrine is absent. The association between the two is nevertheless a notable feature. Not only do they have a similar choice of nest sites, but they commonly breed close together, often within 200–300 m and occasionally within 30 m. Nethersole-Thompson (1932) twice saw occupied nests of the two species only 5 m apart.

In some districts, Peregrines frequently lay their eggs in vacant Raven nests, which vary from completely new structures (usually robbed within the previous month) to ancient weathered-down remnants. My own records showed that, in southern Scotland, 60 of 164 Peregrine clutches were laid in old Raven nests; 21 of 77 layings in northern England were so placed, 18 of 76 in the Highlands, and only 3 of 31 in north Wales. In both the localities where tree-nesting by Peregrines has recently been reported, old Raven nests were used. Ravens also sometimes build on bare ledges previously used by Peregrines.

For all their customary closeness in nesting arrangements, these are famously hostile neighbours. As Blezard (1943) put it, 'The two are forever in conflict, and when either rises from the crags, it is usually the signal for both to display their superb powers of flight in aerial skirmish enlivened by deeply contrasting battle cries'. The Peregrine usually begins hostilities and, when Ravens make opening sorties, the effect can be to arouse the falcons to greater aggression. Some Peregrines are extremely aggressive, stooping down repeatedly, often one after the other, at their adversaries. Blows are not often struck and the Ravens roll over on their backs, extending bill and feet, as their tormentors flash past. They also give an angry snarling call – aark-aark-aark, made in quick succession on a slightly ascending scale – which is otherwise only heard during the pursuit of one Raven by another in apparent territorial interactions. Sometimes this call is given weakly in anticipation of attack, and it may serve as the first warning that a

Peregrine is around. The big Crows are not uncommonly driven to ground, where they stay until their attackers lose interest, or make short and furtive flights to new stances.

In his journal entry for 16 March 1924, Ernest Blezard gave a vivid description of such an encounter in the Scottish Borders:

> Both Ravens came slowly flapping along the hill above us. From a great height the Falcon had spotted them, and making a magnificent stoop, drove them close to earth. Time after time she hurtled down on to them till without attempting to retaliate they sought refuge by settling among loose rocks. Rising to proceed on their way they had no sooner left ground than the Falcon 'waiting on' came hurtling down again screaming loudly. One Raven whose head appeared to be in imminent danger turned over backwards to repel the attacker with its bill. This apparently added to the Falcon's ire, for swinging aside she rapidly rose and with harsh screams of rage stooped with such deadly intent that both Ravens literally fell to earth and took cover under a large boulder. The Falcon now climbed high in the air and circled above the retreat awaiting their reappearance. The sable ones had had quite enough and did not emerge until the Peregrines vanished over the hills; then they made off hugging close to the hillside as if fearful of their tormentor's return. The tiercel took no active part in the attack, merely acting as spectator.

While I have witnessed a large number of these aerial conflicts, I have never seen so much as a feather dislodged. Other observers have occasionally seen fatal attacks on Ravens. In Northumberland, Bolam (1912) recorded that one of a nesting pair was killed by a Peregrine. Aggression by falcons may be heightened or even provoked by human intrusion into the nesting places, so that these become redirected attacks. On the coast of south-west England, D. Nethersole-Thompson and G. Garceau flushed a Peregrine from its eyrie and watched it strike first a Buzzard, which was not seriously hurt, and then a Raven, which died. Female Peregrines demonstrating around their eyries have twice been seen to knock down and kill newly fledged young Ravens from nests nearby, by Geoff Horne in Cumberland and David Wilson in Perthshire. John Morgan saw pluckings of an adult Raven at an eyrie in the western Highlands. In Ross-shire, Dick Balharry found three instances of recently fledged young Ravens being treated by Peregrines as food, two of these remains appearing in the same eyrie in different years. Two out of three newly fledged young Ravens were found by Paul Burnham to be killed by falcons in a Pennine nesting haunt. Two of the BTO Raven ringing recoveries were of rings found below Peregrine eyries.

Now and then, the Raven turns the tables on its persecutor, for the literature records at least three instances in which Peregrines were brought down through aerial combat. In one of these the falcon later recovered and flew off, but in another the bird was badly injured and unable to fly; in the third the Peregrine had apparently been killed (Ratcliffe, 1993).

When Peregrines and Ravens nest very close together, their eyries are usually hidden from each other, so that the Ravens can leave and return out of sight from their fierce neighbours. Yet they are occasionally in full view of each other, and it seems almost as though the two species had declared a truce in their domestic arrangements. Some falcons appear to be tolerant and unconcerned about Ravens sharing their nesting quarters and there is a good deal of individual variation in antagonism between the two. Ravens are now and then made to desert their nests with eggs when Peregrines decide to occupy a site within 10 m, and I have seen three such instances, all in inland localities. Peter Dare recorded another case in Snowdonia. At small cliffs, such as those of the Cheviots, Ravens have also been known to be evicted before laying by Peregrines, or prevented from settling, there being insufficient space for both to breed at the same time.

Peregrine hostility is unsurprising, since the Raven is a potential predator on its eggs and young, and may serve as a warning not to attempt such behaviour. Although the two overlap only very slightly in food preferences, Ravens often scavenge from the remains of falcon prey left in the vicinity. There could also be mutual advantage in close nesting, the watchful Ravens giving the alarm at the approach of predators, such as foxes, while the Peregrines are better adapted for attacking them. Or, simply, the more pairs of eyes, the more readily are any intruders detected. Commensal associations are reported from the arctic tundras between nesting Peregrines, Snowy Owls, or other large predatory birds, and potential prey species, such as geese and ducks. These have not been recorded for the Raven.

Kestrel, Merlin and other raptors

Although both these raptors commonly breed within Raven territories, they tend to keep their distance from active eyries in choosing their own nest sites. Both will, on occasion, take over vacant eyries in which to lay, when the Ravens have moved elsewhere. And, in north Wales, I saw a Kestrel incubating in a deep crevice only 5 m from a Raven nest with large young. Neither Kestrels nor Merlins are afraid to attack Ravens which intrude too closely on their nesting places. Chapman (1924) described the antagonism between these two little falcons and Ravens in the Cheviots. When the Kestrels returned to occupy their rock sites, the Ravens, with half-grown young nearby, were the aggressors, stooping at and chasing their smaller neighbours persistently. A month later, the roles were reversed and the Kestrels relentlessly harried the Ravens and their fledged brood, sometimes striking and knocking feathers out. Such exchanges are not without hazard for the falcons, since another observer once saw a Kestrel killed by a Raven. Chapman also remarked that 'neither party learns by experience', since he watched a pair of Merlins persistently attacking the Ravens in a range of crags in Redesdale, which the two species had shared for over 30 years. There appears to be advantage in maintaining this defensive behaviour.

The Raven and the Hobby probably overlap rather little in range nowadays, but Stevenson (1866), speaking of a former eyrie of Ravens at Highgrove, near Gelderstone in Suffolk, said, 'The Hobbies always used it after the young Ravens had flown'. Dixon (in press) has seen demonstration from a pair of Goshawks nesting close to a Raven eyrie. In Sweden, I watched a family of Ravens immediately stoop at a Goshawk which had perched in a Scots pine.

Other corvids

Among their own tribe, Ravens fairly constantly have the company of Carrion or Hooded Crows, both in the nesting season and at other times of year. Crow nests are only occasionally within 400 m of active Raven eyries, but they are often spread throughout the wider feeding range of each pair. Either will attack the other if their nest sites are too closely trespassed upon. The Crows themselves tend to keep away from the immediate vicinity of active Raven nests, but they can soon appear there if the bigger relatives have failed and moved away.

The two species overlap considerably in diet and Crows take a good deal of the carrion on which Ravens depend, so that there is some apparent competition for food. The Raven is dominant if the two seek the same items, driving the other away from carcasses, and the Crow may be able to subsist sufficiently well on food other than carcasses in order to coexist with the larger species. The fact that Crows nest on average a month later than Ravens suggests some degree of ecological separation between the two. Both species sometimes consort out of the nesting season and, in Wales, Bolam (1913) noted parties of the two roosting together on the same rocks, though on different series of ledges. Ravens and Crows also overlapped in a woodland roost in Merioneth during 1937–47 (W. A. Cadman). Mylne (1961) noted that, in April 1959, 50 non-breeding Ravens were feeding at a Pembrokeshire farm midden in the company of four Magpies, two Buzzards and a large number of Carrion Crows. In North America, Kilham (1985) watched Ravens repeatedly robbing Crows of food.

Ravens have been reported several times nesting on the edge of rookeries in small woodlands. In one such instance, in south Wales, the bigger birds largely escaped notice and so may have derived advantage from the association (Heathcote, Griffin & Salmon, 1967). The same is by no means true for the Rooks, for in another instance in south-west England, Nethersole-Thompson (1932) watched the pair systematically helping themselves to the eggs of their smaller relatives, who hovered overhead in a noisy crowd. Morris (1895) also said that Ravens which built near a rookery took the young rooks to feed to their own nestlings. Ravens nesting in another rookery were seen being mobbed by the Rooks but, in a shared nesting place on the Isle of Man, there was evidently no friction between the two species (W. S. Cowin). Blezard (1928) described the apparent panic which the appearance of a pair of Ravens caused among a party of Rooks feeding on a Pennine fell: the birds promptly fled *en masse* with much noise for their fell-foot rookery. In the Southern Uplands I also witnessed such a response from a party of about 30 feeding Rooks when a pair of Ravens that I had disturbed from the nest flew in their direction. Yet I have also watched Ravens feeding among a large flock of Rooks on a hillside in north Wales.

Blezard saw a throng of clamouring Jackdaws in Lakeland suddenly dive for the safety of their cliff nest holes when a pair of Ravens passed high overhead. A large flock of Jackdaws feeding on a Snowdonian hillside flew down to lower ground when a pair of Ravens came over. Jackdaw colonies do not usually occupy the same inland nesting cliffs as Ravens, though there are some places where active Raven nests are right in among good numbers of the 'daws. On the coast Jackdaws often have nests in crannies close to those of their large relatives. They have on occasion been known to predate unguarded Raven eggs (Wilmore, 1977). The nests of Choughs are often no more than 100 m from occupied Raven eyries, and Cullen and Jennings (1986) mention the two species nesting together in the mine wheel-case below Beinn y Phott, on the Isle of Man, in 1938.

In the Nant Ffrancon, in north Wales, I once witnessed a curious incident from the house where I lived. A man with a .22 rifle shot a Carrion Crow feeding in a nearby field, but remained concealed. The bird flapped for a few moments in its death throes and then became still. At once, several other Crows in the area flew around, calling a great deal. Then a mixed flock of Rooks and Jackdaws came overhead, adding their voices to the chorus. And finally, to crown the show, a pair of Ravens circled around, croaking repeatedly. An example of solidarity among the Corvidae in time of grief? Lorenz (1970) found that some of the Crow family give a 'sympathetic' alarm response when other members are threatened by humans, but that Ravens were much more selective in reacting (see p. 115). He also noticed that, when he took his tame Raven with him to the Danube banks, otherwise timid migrating Crows and Jackdaws alighted beside him.

Sea-birds

Many coastal Raven nesting stations are on sea-bird cliffs, though eyries are seldom placed among the densest throngs of breeding auks, Kittiwakes or Herring Gulls. There are sometimes scattered nests of Shag and Herring Gull fairly close at hand, and both Lesser and Great Black-backed Gulls sometimes nest in the immediate vicinity of Raven eyries. Ravens frequently give chase to the larger gulls that fly along the cliffs which hold their eyries and seldom leave these unguarded. It appears that Ravens prey upon these sea-bird neighbours, or their eggs and chicks, only in a few places, though they doubtless scavenge any corpses that come their way. They have been seen, on occasion, to predate the eggs of Guillemot, Razorbill, Puffin, Shag, Kittiwake and Gannet.

The big gulls have occasionally been seen to despatch Ravens, but these were probably mostly sick or injured birds that could not defend themselves against attack. However, J. F. Peters saw a Great Black-backed Gull kill an apparently healthy Raven at Loch Ewe, Wester Ross, on 14 June 1921. The gull stooped at the flying Raven (evidently a recently fledged youngster) and both fell into the sea. The Raven climbed out on to some rocks, whereupon the Gull again knocked it into the sea and the lifeless bird floated away. Hewson (1981) noted that Great Black-backed Gulls were dominant over Ravens when feeding at sheep carcasses on the Argyll shore, but much may depend on individuals in such interactions. On the Isle of Man, Cullen and Jennings (1986) twice knew Shags to take over new Raven nests under construction, evidently evicting the rightful owners.

Among the sea-bird colonies, it is the Ravens' association with Fulmars which is likely to have the greatest significance. The Fulmar's aggressive habit of ejecting its stomach oil at neighbours is known to affect numerous bird species and is regarded as a serious adverse population factor for even so intimidating a predator as the Peregrine (Ratcliffe, 1993). There is growing evidence that the Raven is among the victims of this combative denizen of the sea-cliffs. Ewins, Dymond and Marquiss (1986) refer to records of contamination of Raven plumage by Fulmar oil on Fair Isle and Shetland, and their observations indicate that it is a major cause of fledgling and adult mortality there. Of 37 nest failures in Shetland, 14 resulted from interference by Fulmars and Fulmar-oiling was blamed for the failure of three out of seven nests on Fetlar in 1982. Active take-over of Shetland Raven nests by Fulmars has been strongly suspected and Fulmars were seen persistently attempting to alight on nests, even while the Ravens were incubating. On Fair Isle, Dymond (1991) refers to 'competition with Fulmars which can sometimes successfully oust Ravens from nest sites and often contaminate nestling or fledgling Ravens with oil'.

On Orkney, Chris Booth has 26 records of Fulmar-oiled Ravens, mainly juveniles, some incidents involving up to three youngsters in the same brood. He finds circumstantial evidence of competition between the two species for nest sites, but thinks it unlikely that a Fulmar could displace a sitting Raven. Booth repeatedly finds Fulmars sitting in nests that a week or two before had held Ravens incubating eggs or brooding small young, but these nests may have failed for other reasons. He notes that Ravens can be very aggressive to Fulmars, dislodging them from ledges by attacks; one bird almost lifted a Fulmar off after grasping its crown feathers in its beak. One Fulmar died through becoming entangled in barbed wire with which a Raven had begun to refurbish an old nest. For the Isle of Man, Cullen and Jennings (1986) noted that, in 1977, four former Raven nests were occupied by Fulmars, but it is unclear whether there was competition between the two for nest sites.

Other birds

Ravens occasionally elect to nest in woodlands with breeding colonies of Herons, perhaps again deriving some advantage of concealment from the presence of the more numerous species. Examples of heronry nesting places are at Pennybont, Radnorshire, for several years up to 1942 (A. W. Bolt); Yealmpton, Devon, in 1945 (O. D. Hunt) and in Ayrshire, in 1946 (G. H. Onslow).

Birds which are subject to egg and chick predation by Ravens sometimes show defensive reactions at their approach. I have seen a feeding Dotterel on a hilltop crouch and remain motionless until a passing Raven was out of sight. Incubating Dotterel have sometimes been seen to fly from their eggs in response to intruding Ravens (Des Thompson). Sitting waders often fly from their eggs in order to mob intruding predators, such as corvids and Hen Harriers, and Nethersole-Thompson and Nethersole-Thompson (1979) have watched Greenshanks with chicks vigorously attack low-flying Ravens, diving at them repeatedly with loud alarm cries. Oystercatchers, Curlews and, even more, Whimbrels will also often try to drive Ravens away from the vicinity of their nesting places, and Bannerman (1961) mentions how a pair of Whimbrels completely ousted a lone Raven. In the USA, Kilham (1989) saw a Killdeer Plover, whose nest was under attack, land on a Raven's back and peck its head. Lorenz (1970) described how a Mallard duck flew after a Raven carrying off a duckling, seized it and brought it down, then vigorously attacked the bird, so that it released the prey and flew away. Smaller birds, such as the general run of passerines, do not obviously show alarm at the presence of Ravens, but perhaps do not seek too close a proximity either.

MAMMALS

A number of larger mammals appear to be potentially either predators upon, or competitors with, Ravens. The vulnerability of Raven nests to the fox, pine marten and polecat is mentioned in Chapter 8 but, while newly fledged young are at risk to ground predators, there is no reason to suspect that older birds or adults are ever more than very occasional victims. In North America, Ravens have on rare occasions been found to be killed by coyotes and wolves, apparently through exposing themselves to attack by landing at carcasses where these predators were feeding. Heinrich (1990), in discussing this risk-taking by Ravens (see p. 116), noted that these predators, and also sled dogs, usually ignored Ravens attempting to feed close-by.

Several writers have observed that there can be something approaching a symbiotic relationship between wolves or coyotes and Ravens. Kilham (1989) quoted evidence that the birds will guide both these predators to food during times of deep snow by croaking in the vicinity of prey such as deer, and then partake of the leavings from kills. Lawrence (1986) said that wolves and Ravens could sometimes be heard howling and croaking in unison: 'A wild concert not infrequently heard in the breeding season of wolves and after the pack has made a kill'. Freuchen and Salomonsen (1960) stated that Eskimos believe Ravens show them the presence of bears and caribou by their flight, and Ravens follow hunting parties – as they do polar bears – in the evident prospect of sharing kills. And, if nothing better is on offer, they will take the dung of various mammals.

Several of those who have kept captive Ravens have described how the birds have formed friendships with dogs and that the attachment was especially strong when the dog was sick or injured. On occasion, such birds were seen to bring food for their canine companions.

CHAPTER 8

Breeding: nest and nest site

The nesting places are the central and fixed point in the Raven's life, once it has joined the breeding population. They become the focus of its territorial urge and are used as roosting locations outside the nesting season, defended year in year out against all-comers. Typically, Raven nesting places are traditional, occupied for as long as anyone locally can remember and sometimes with a history of use going back for well over a century. There are fewer documented records of ancient eyries than for the Peregrine, with its falconry associations, but a much larger number of rocks named, after the bird, 'Raven Crag', whether in English, Welsh or Gaelic. Many of these names evidently denote long-continued occupation, often maintained to this day. On the 1:25,000 Ordnance Survey maps there are at least 33 places in Lakeland named Raven Crag, of which 22 have certainly been used since 1950, though probably only six are still occupied regularly. The last nineteenth-century Peak District nesting place, deserted after 1863, was again the preferred site when Ravens returned to the area in 1996.

Pair formation in Ravens appears to occur mostly within the non-breeding flocks, so that courtship activity between couples in nesting territories is more in the nature of pair-bond renewal. When one of a breeding pair dies, the survivor evidently seeks a new mate from the non-breeding sector, though it would perhaps be possible for a new and dominant pair immediately on hand to evict a lone widowed bird. This is a part of the Raven's life history about which little is known.

As breeding time approaches, each resident pair is to be seen more constantly in the vicinity of its nesting haunts and aerial display by both birds together becomes evident. In particular, there are frequent unison flights in which the pair stay extremely close together and engage in the same evolutions: soaring, diving and rolling as one. The male also indulges in frequent flight-rolling and tumbling, and often dives headlong over a considerable height. Courtship feeding of the female by the male, involving regurgitation from the throat pouch, also takes place. This is followed by the onset of nest-building.

NEST CONSTRUCTION

Ravens nest in trees, on cliffs and on buildings. In common with others of the Crow tribe, they build their own nests – substantial constructions which allow them a certain latitude in choice of the actual site. Coombs (1978) details nest construction in four separate layers. First is the outerwork: a basket of sticks and large twigs, 25–150 cm long and 0.5–3.0 cm thick. These are skilfully woven together and intertwined so as to confer a considerable rigidity to the structure, which has a neater appearance than most large-raptor stick-nests. This outer wall is about 20 cm thick. There may be an inner layer of thinner (0.4-cm) twigs spreading out over the upper part of the outer basket, but this is not always present. In some areas, there is then a third layer of clay or earth and rootlets, or even dung, forming a coarse lining to the nest cup. Ryves (1948) watched the process of mud-plastering of the inside of the stick-basket, conferring extra strength. He described the dexterous weaving of the outer structure and the complex gyrations by which the bird controlled the diameter and depth of the symmetrical nest cup. The deep bowl is then lined first with rather coarse lumps of soily turf, tufts of grass, moss, dead leaves or other plant litter, strips of bark, paper, string and cloth, and finally finished off with a thick pad of soft, heat-retaining material, usually animal hair. The completed nest retains the deep bowl-shaped interior which holds the eggs securely and provides a snug cradle for the young Ravens.

The eyries vary a good deal in external dimensions, especially in rock sites. Occasional crag-nests are rudimentary affairs, with little stick-work beyond a fringe to the nest bowl, or only on the outer side of this. These shallow structures are sometimes difficult to see from below. Andrew Dixon has found that Ravens nesting on power-line pylons in south Wales sometimes use the open box-like structures at the ends of cross-arms on some pylons. They then use few or no sticks in the construction and simply line

Breeding: nest and nest site 141

17. *One of the many Lakeland Raven Crags, annually tenanted by Ravens to this day. Westmorland (photo: D. A. Ratcliffe)*

the box with wool, so that the nest is invisible from below. Most nests are 0.5 m tall and from 0.5 to 1 m across the base. A few are massive piles, usually representing several years accumulation, but tall nests are occasionally run up in a single season. Chapman (1924) described one nest, in the chimney of a Southern Upland crag, which reached a height of up to 3 m before collapsing. I saw another 3-m nest in the Tweedsmuir hills; its base was on a ledge below a steep bank and it was built up the slope so as to bring the nest bowl under an overhanging rock above. Yet, after all this hard work, the owners had left the nest unfinished and moved to another site. Abraham (1919) reported a Lakeland nest 2.1 m high and the same width across its base, and I have seen a 1.8-m tall nest in this district. A Galloway nest was lower, at 1.2m, but of equal width throughout, giving a chimney-like appearance. Tree-nests are often bulky and occasionally reach a height of 1.5–1.8 m. Simson (1966) saw one structure of over 1.8 m with eight old nests countable in the strata. The nest bowl varies less in size, being usually about 25–35 cm across by 15–25 cm deep.

The Ravens mostly pick up or break off dead sticks and twigs, but they sometimes wrench off living material with their powerful bills. They tend to take whatever is freely available in the immediate vicinity. Heather sticks are favourite material in the uplands but, on the grassy sheepwalks, the twigs of rowan, birch, oak, ash, beech, sycamore, conifers and hawthorn are used instead. Tree-nests are of a still wider range of stick material, including pine, spruce and larch in plantation areas. The gnarled dead stems of juniper are often used where available. Sea-cliff eyries are of similar construction but gorse, blackthorn and bramble often figure in the stick-basket. There is occasional recourse to unusual items, such as a nest in the Westmorland Pennines built largely of bleached sheep's bones (E. Blezard) and another in eastern Lakeland with several small antlers of red deer worked into the outer basket. Lengths of rope, string and wire also appear in some nests; barbed wire is a favourite material on Orkney (C. J. Booth).

The association between the Raven and sheep extends to the nest materials, for the final lining is most usually a thick pad of sheep's wool, packed in densely to form a beautifully insulating bed for the eggs and young. It is absolutely ideal for this purpose. Local variations include the hair of wild goats and red deer. Small amounts of rabbit fur and the feathers of other birds sometimes appear in the lining, and there can be plant material, such as finer-leaved grasses (e.g. sheep's fescue and mat grass) and mosses. Fox fur is occasionally reported. Ryves mentioned a wartime nest near a military hospital that was lined with cotton-wool swabs. Simson (1966) saw cow and horse hair used in Schleswig-Holstein, and nests in arid regions of north Africa are reported to be lined with camel hair.

Ryves (1948) found that both birds foraged for nest materials and carried them to the site, but supported Harlow (1922) in noting that the female alone was the architect. Davis and Davis (1986) observed that one bird seemed to do most of the stick-carrying and nest-building, supervised by the other, and supposed that the labour was by the female. Gwinner (1965a) observed his captive Ravens building; the male made much of the outer structure, but the female added most of layers two to four, with her mate helping. Other observers have found that the sexes share the work, but variably according to individuals and the stage of construction (Cramp & Perrins, 1994). Building takes from 2 to 3 weeks for an average new nest, though repeat eyries after loss of a first clutch often take only a week or even less. Nests from which young have previously been reared need to be completely rebuilt but, when an old nest in decent shape is renovated, the time to completion may also be quite short. Only heavy snow appears to

Breeding: nest and nest site 143

18. *Nest in a traditional site on the Raven Crag shown on p. 141 being examined by Geoff Horne (photo: D. A. Ratcliffe)*

delay operations once they have begun, and the ground is often extensively snow-covered when nest-building begins. Completed first nests in the year are sometimes left empty for several days – even up to a month is reported – but laying usually begins soon after completion.

Ravens are quite adept at building a nest on small or sloping rock ledges or tree limbs, where it would seem to the human eye that there is insufficient lodgement for such a structure. A few are, indeed, insecure and easily toppled by high winds or a careless human visitor. A Raven nest is nevertheless often a remarkably solid and durable structure, and the weathering of the base of an old eyrie helps to solidify the foundations. Nests from which young have fledged tend to disintegrate more quickly if they are not subsequently rebuilt, for they are usually trampled almost flat by the youngsters before they fly, and their droppings tend to promote decomposition of the nest materials. When nests are rebuilt, the older sticks at the base are often green and decaying, in contrast to the fresh material of the current year. A nest in which eggs have not hatched can last for quite a long time – 10 years at least, and perhaps a good many more in dry, sheltered rock sites. Sometimes, however, nests in well-protected sites disappear between the end of summer and the following nesting season, and it appears, from the litter of sticks below, that they have been pulled down. Birds from non-breeding flocks have been seen to destroy nests (Davis & Davis, 1986; Hurrell, 1956) but, in most instances, it is impossible to know whether such dismantling has been done by intruders or the owners themselves.

Raven nests have the reputation of being difficult to reach, but much depends on site availability, whether in rocks or trees. When the choice of situations allows, Ravens will nearly always build in a site that is awkward of access to a human or an animal predator. Yet, where the choice of sites is more limited, the birds will often take their chance with easily accessible nests, rather than forgo the opportunity to breed.

NEST SITES

Tree-nests

The kind of tree depends somewhat on the available choice, but there is an evident preference for conifers, especially Scots pine but also both Norway and Sitka spruce, and larch. Walpole-Bond (1938) said that nearly all the former nest sites inland in Sussex were in Scots pines, and this tree was also preferred in Hampshire (Kelsall & Munn, 1905). Table 8 shows that Scots pine has been the favourite nesting tree in more recent years. Various exotic conifers, such as cedar, Wellingtonia and noble fir, are used and occasional nests in yews and junipers are reported. Of the broad-leaves, oak and beech are used most often, but ash, birch and sycamore are also quite commonly employed. In parkland nesting haunts, elm was sometimes used, and nests in both types of chestnuts have been recorded. Alder, rowan and hawthorns are chosen less frequently, and elder and holly occasionally. Elsewhere in Europe – Poland, Byelo-Russia, Finland and Sweden – Scots pine is also the preferred tree but in Germany and, formerly, southern Sweden, beech is more favoured.

Perhaps because the Raven depends so much on open country for foraging, few nests are deep in large forests – in Britain and Ireland, at least – and the favourite location is a shelter clump or belt of plantation (typically

19. The bleak conditions of early spring at an historic Lakeland Raven haunt; the nest is under the middle one of the three sharp overhangs. Cumberland (photo: D. A. Ratcliffe)

conifer) from half to a couple of hectares in size. There are haunts in remnant oakwoods in the West Country combes and the equivalent Welsh cwms, often in steeply sloping 'hangers' (photograph 6, p. 49). Some pairs nest in the extensive new plantations of Sitka spruce and other conifers, but close to their edges, where these abut still open ground. A group of older trees within young forest is sometimes used. In central Wales a favourite situation is in old clumps or lines of oak, beech, ash or sycamore so typically present around deserted hill cottages, and especially in the often ruined former summer residences (*hafod*) on the upland sheepwalks. One such eyrie, in a house left empty just a few years before, was only half a metre or so from the rear wall and overlooked by a bedroom window. In this district, Davis and Davis (1986) found that conifer nests were slightly but not significantly more successful than those in broad-leaves. Nests in hedgerow trees are frequent in Devon and sometimes completely isolated trees are used. A small rowan on the islet of a Hebridean loch was an unusual site.

20. A Raven nesting site in a shelter clump on upland sheepwalk. Southern Uplands, Kirkcudbrightshire (photo: D. A. Ratcliffe)

Tree-nests in the lowlands are mostly built in tall trees, either broad-leaved or coniferous, and placed high in the crown; though some are at the base of a main limb where it joins the trunk, or towards the extremity of a lateral branch. Typically they are 15–30 m above the ground, in growths either lacking branches for a considerable height or with massive lower trunks and limbs. Many of these Raven trees are not climbable without climbing-irons or ropes. In a famous passage, Gilbert White (1789) enlarged on the impregnability of a Hampshire Raven eyrie:

> In the centre of this grove there stood an oak, which, though shapely and tall on the whole, bulged out into a large excrescence about the middle of the stem. On this a pair of Ravens had fixed their residence for such a series of years, that the oak was distinguished by the title of the Raven Tree. Many were the attempts of the neighbouring youths to

get at this eyry: the difficulty whetted their inclinations, and each was ambitious of surmounting the arduous task. But when they arrived at the swelling, it jutted out so in their way, and was so far beyond their grasp that the most daring lads were awed, and acknowledged the undertaking to be too hazardous.

The observant countryman-poet, John Clare, who lived in Northamptonshire in the early nineteenth century, wrote of an historic Raven tree – 'an hugh old oak' – which invited numerous attempts at nest-storming to take the young. Yet despite being assailed . . .

> with iron clamms and bands adventuring up
> The mealy trunk or else by waggon ropes
> Slung over the hugh grains and so drawn up
> By those at bottom one assends secure
> With foot rope stirruped – still a perrilous way

. . . only once in living memory had the eyrie been reached. Ticehurst (1909) referred to a nesting site used regularly in Lullingstone Park, Kent, up to 1844, always in the same tree with 'a bulbous enlargement halfway up the trunk' and unclimbable without a long ladder. Bolam (1912) also mentioned a traditional Raven tree at Ilderton, Northumberland, in the early nineteenth century – a huge Scots pine, which again offered challenge to the local youths. He quoted a Dr Johnston as saying, 'it was a deed of hardihood to reach the top and harry the nest of the Raven that annually built thereon'. No doubt there were many such that were marked down for annual plunder as a test of strength and skill.

The most vivid tale of assault on a lowland Raven tree is that told by Smith (1905) when, on 24 February 1855, he and a school-mate trudged 10 km through the snow to a traditional nesting site at Badbury Rings in Dorset. This eyrie too had the reputation of inaccessibility and the boys were weighed down with a hammer and bag of 60-odd 250-mm nails as tackle for the job. Sure enough, the sitting Raven flew from the nest in a lofty Scots pine with its first branch some 15 m from the ground. The account of its ascent, by means of a ladder of nails driven into the trunk, is as hair-raising as any description of a great alpine conquest.

> It was a task of delicacy and difficulty, not to say of danger . . . As I climbed higher, the work became more dangerous, for the wind told more; and a slip would now not only have thrown me to the ground, but have torn me to pieces with the nails which thickly studded the trunk below.

Finally the nest was gained as dusk fell and 'to my inexpressible delight', found to contain four eggs, which were carried back in triumph to school at Blandford and shared as the crowning prize between two youthful collectors. Seebohm (1883) also describes the storming of a difficult Raven tree near Earls Colne in Essex, by two youths who were menaced by the parent bird as they clung to the topmost branches.

Yet in situations where big trees are lacking, Ravens will use quite small

Breeding: nest and nest site 149

growth in order to nest, and here their nests may be only 3–10 m above ground. Streamside alders seldom present climbing problems and, in secluded gills deep in the hills, nests in the sparse fringing birches and rowans are usually quite easy to reach. The simplest I have seen is shown below. I could touch the bottom of the nest when standing at the foot of the

21. *Raven tree-nest in a rowan on the bank of a secluded small gill; the sitting bird was shot by a keeper. Langholm Hills, Dumfriesshire (photo: D. A. Ratcliffe)*

rowan. Nests are occasionally built in small trees, usually birch or rowan, growing on cliffs where good rock sites are few. Large bushes of hawthorn and elder in quite accessible places are also used in central Wales.

Davis and Davis (1986) found that success of tree-nests increased with height above ground; of those above 14 m high, only six out of 34 (18%) failed, compared with 63 out of 145 (43%) below that height. Tree-nests below 6 m generally have a low success rate, mainly because of their accessibility to humans. Tree-nests are not safe from the more agile kinds of mammal predator, but they give the owners an advantage in warding off possible enemies. Tony Cross saw a Raven pluck off an intruding grey squirrel from just below the eyrie and drop it into an adjoining tree.

Rock-nests

Rock-nests are, on the whole, difficult to reach and probably the majority are inaccessible without the use of ropes. Some are a formidable proposition, even for expert modern climbers. In Lakeland, master cragsman Jim Birkett was never beaten by an eyrie and made a point of reaching every one that he came across, some of them in notoriously difficult places. His brilliant climbing technique included the ability to pendulum in under massive overhangs while abseiling, grip the rock and adhere to it. Geoff Horne is also a skilled exponent of this technique. Less accomplished performers trying this manoeuvre find themselves spinning uncontrollably under the overhang, and I have heard of at least three disastrous attempts when wild clutching at the nest itself ended with the whole structure and its contents crashing to the foot of the cliff.

Nests tend to be easier to reach in areas where the crags are mostly small, such as the Southern Uplands and Borders, but even on minor outcrops the Ravens often show a distinct flair for choosing the most awkward possible nest site. Many are readily approachable to within a short distance, but the last 2 or 3 m give real problems. Since the birds make their own nest they can place it for maximum security. The overhang of rock above is nearly always present, giving protection against falling stones, ice and snow, and high winds, as well as predators. Nests in exposed places are occasionally blown down in severe gales, but it is quite usual for the structure to be surprisingly well sheltered, however ferocious the storms raging around the cliff. Where good, overhung ledges are few or lacking, Ravens quite often use a bush or stunted tree growing from the rock face on which to lodge their nest (see p. 162). While these can be classed as rock sites, they are unusable without the woody growth.

Sea-cliff nests tend to be still more difficult to reach than those on inland crags, for they can seldom be approached or even viewed from below, and may be awkward to locate from above. Coastal cliffs often have a good deal of loose and unreliable rock above the level of tidal scour, and their vegetation can be of a particularly treacherous kind to climb upon. Some faces are desperately sheer cut or overhanging. Walpole-Bond declared that some coastal eyries 'could not be stormed with all the tackle in Christendom', but the modern climbing 'tigers' would no doubt respond to this challenge. In sheltered coves, or faces with an undercliff or beach, the nest may be in low cliffs and/or near to the base, but on exposed sea-cliffs the heavy wave and spray action during time of storm compels most eyries to be built at least 20 m above high-tide level. In Shetland, Ewins, Dymond and Marquiss (1986) found that coastal nest sites were usually in the upper third of the cliff face. While, as in inland areas, the biggest cliffs tend to be most favoured, because they offer the greatest nest-site security, Ravens show occasional lapses from habit. The most accessible sea-cliff nest I ever saw was on St Kilda, where one of several pairs had that year chosen a simple situation on Hirta, despite having a choice of some of the most formidable precipices in our islands. A few nesting places are on isolated rock stacks, including the most spectacular pillar of all, the Old Man of Hoy on Orkney (138 m).

In common with most of the cliff-nesting predators, Ravens prefer a nest site with a commanding outlook, but they are more often to be found nesting in shut-in ravines than are Peregrines. Some nests tend to escape notice through being in hidden gullies away from conspicuous frontal crags. Ravine nests are often in picturesque situations, with waterfalls and rock pools, and fragments of native woodland clinging to the sheltered sides. Wooded cliffs are just as acceptable as those devoid of trees. In south Wales, a nest was found below ground level on a limestone face within a swallow hole (A. Dixon). Quarries are much favoured as breeding places, there being at least 25 known examples in south-west England, 57 in Wales, 19 in northern England (including Isle of Man), 19 in Scotland, 25 in northern Ireland and at least seven in southern Ireland. Most of these are abandoned workings, but at least 17 nesting places are in active quarries, where the birds usually find a face no longer worked. Dare (1986b) recorded successful breeding within 200 m of machinery and blasting in an operational quarry. Some of these man-made cliffs are in areas where the lack of natural cliffs would otherwise have prevented the Ravens from nesting, unless they resorted to trees. Old railway and road rock-cuttings have very occasionally been used.

Ravens have bred on a variety of man-made structures and doubtless this was commonplace during the earlier times when the bird was a numerous town-scavenger. Ussher and Warren (1900) said that ruined towers and castles were sometimes used in Ireland, but named only one example (Table 9). The site at Glastonbury Tor gives a nice association with the legend of King Arthur and his knights. Recorded examples of nests in buildings in Britain and Ireland are given in Table 9, grouped into old buildings, modern constructions, old mine buildings, deserted moorland cottages, railway viaducts and bridges, and pylons and masts. Booth (1996) has found such nest sites to be used especially frequently on Orkney, where their success rate is higher (at 74%) than those on the sea-cliffs and other habitats (54%). In Orkney 11 territories have recently held building-nests: five on Mainland and six scattered over the smaller islands. In some other countries, Raven nests on man-made structures are still more frequent: for instance 48% of pairs used electricity transmission towers, empty farm buildings, barns and grass silos in the south-west lowlands of Iceland

(Skarphedinsson et al., 1990). In the USA, Stiehl (1985) recorded that 23% of 87 nests in his Oregon study were on ruined buildings or windpumps, and Steenhof et al. (1993) followed the build-up of a whole population of 81 pairs on power-line towers in Idaho. Oil derricks are also a popular nesting site in parts of the USA. The Australian Raven commonly nests on electricity pylons.

22. *Raven nest on the roof of a derelict cottage on hill sheepwalks; the conifers beyond provide an alternative site. Elan Valley, Radnorshire (photo: D. A. Ratcliffe)*

Rock-nesting Ravens have the same tendency as Peregrines and Golden Eagles to select the tallest and steepest cliffs available, evidently for reasons of nest security. Table 10 gives figures for the vertical height of Raven nesting cliffs in several inland districts. It shows that, in rugged areas, where high cliffs are plentiful, the birds prefer to nest on faces more than 20 m high. Where tall cliffs are scarce, most pairs are to be found breeding on rocks less than 20 m high and some regularly nest on crags less than the lowest normally used in rugged districts (c. 12–13 m). They will nest on small rocks down to only 6 m high if nothing better exists and if the situation is relatively undisturbed by humans. However, some of the smaller rocks, when they are close to roads and houses, are held less tenaciously than big crags. As with Peregrines, fidelity of occupation is related to height and security of the nesting cliffs. Hickey's (1942) principle, that the minimum height of rock face acceptable to the Peregrine varies inversely with the degree of wilderness to which it is exposed, applies equally to the Raven.

Until around 1966, Ravens nested on a number of small gritstone outcrops on the Northumbrian moors, but most of these were in secluded places well away from human habitations. The locality with the most easily accessible nest sites was, up to the 1950s, 11 km from the nearest metalled road. On Dartmoor, Ravens have nested on many of the granite tors, where the rock faces are mostly under 10 m high. Up to the 1940s, several of these were regularly occupied but, even though 11 different tors have held nesting Ravens since 1985, these recent attempts have mostly been sporadic and unsuccessful (W. Robinson, G. Kaczanow, P. Dare). Dare (1986b) noted that, in Denbighshire, nests at two small rock scarps in fields only 200–300 m from farm-houses were seldom successful. A good many breeding places on low cliffs are still occupied in parts of Wales, the Southern Uplands, the Highlands and Ireland, but they are mostly in rather out-of-the-way situations. Similarly, most of the nests in small trees or bushes are in remote places, often in little glens hidden deep in the hills, where few people ever penetrate. Conversely, quarry-nesting haunts are often close to humans, but in abrupt, vertical faces, with sites difficult to reach (see p. 51).

In Britain and Ireland, most Ravens evidently prefer to nest in cliffs rather than trees, when there is a choice, and even when trees might be the safer option. The infrequent use of trees in regions with important rock-nesting populations, such as Snowdonia, Lakeland, the Highlands and most of the Irish uplands, points to this conclusion. When and where trees are used here, their recent and often sporadic pattern of occupation suggests that they are the secondary habitat for an overspill from the predominantly cliff-nesting populations. In a few territories, both cliffs and trees are used, in different years, but in most such instances the cliffs are 'third class' with inferior sites. It might be wondered why rock sites should be preferred when better tree sites are available. The answer seems to be that it is not so much the security of the nest *per se* that is crucial, as the general vulnerability of the birds themselves in the situations where safe trees occur. No tree is high enough to be out of gunshot range and the problem with many tree sites is their proximity to people.

It is noteworthy that in the districts – notably the West Country and Wales – where Ravens have increased markedly in recent decades, as a result of diminished persecution, they have done so largely by occupation of tree sites. Frequent tree-nesting in our islands appears to be a response to relaxation in persecution. In their central Wales study area, Davis and Davis (1986) recorded that, of 294 nests seen, 210 (71%) were in trees, and 84 (29%) in cliffs or banks. There was no real difference in success between tree- and cliff-nests. For inland areas of Devon, Peter Dare estimates that 90–95% of a population of nearly 300 pairs nests in trees.

Ground-nesting in Ravens seems to be exceptional, because of the dangers from ground predators. The fox has the potential to be a serious predator on too readily accessible nests. I have four records of foxes taking the young from Raven nests which were easily reached in small rocks, all in the Southern Uplands. Foxes hard-pressed by hounds have also twice been seen to take refuge in vacant crag-nests of Ravens. Occasionally, eyries are almost on the ground and reachable without danger. Nests in tiny rocks are reported on Tiree, Mull, Orkney and Shetland – all islands without foxes.

Ewins, Dymond and Marquiss (1986) recorded a Shetland nest on the beach at the foot of a small cliff and, on Orkney, Chris Booth found a 'walk-in' site on the sloping woodrush-covered bank of a stream where a pair nested successfully for 2 years running. One predated Raven brood in the west Highlands was not reachable by a fox, but the more nimble pine marten could well have been responsible. Some eyries may be vulnerable to the polecat, mink and wild cat as well. There is sufficient of a threat from these ground mammals for selection to operate actively against adaptation to nesting in 'walk-in' sites, in the manner increasingly shown by Peregrines – which are probably better able to ward off mammal predators. Ground-nesting is reported from Holstein, Germany, in an area lacking both cliffs and trees (Coombs, 1978).

Alternative sites

The Raven shows the same habit as the Peregrine and other raptors of switching between a number of different nesting ledges or even different cliffs, or between different trees, over a period of years. Sometimes, indeed, the two species occupy each others' vacant crag-eyries in successive years. In the Raven – as in the Peregrine also – the amount of movement between different nest sites depends greatly on the available choice. Where there is a wide selection of good nest ledges, there tends to be more movement between different sites than where the choice is limited. Chapman (1924) suggested that such changes of nest were dictated by the necessity of at least one year's cleansing by the elements of the befouled abodes vacated by broods of youngsters. While plausible as an explanation, this idea does not stand up to actual experience, since very often the same eyrie is used for 2 or more years running, after young are reared, even when a good choice of alternatives is available. On small rocks with only a single serviceable ledge, successive occupants will, perforce, use the same nest indefinitely.

The pattern of use of alternatives is so variable, even within the same territory over a long period, as well as between different territories, that it could as well be attributed to caprice on the part of individual birds or pairs as to any more elaborate reason (Blezard, 1928). The sequence of use can change between one decade and the next – perhaps associated with change in occupation by individuals – and the number of known alternative nest sites tends to increase with the observation period. Most territories examined for no more than 5 years show use of only two or three sites. Some Lakeland and Southern Upland nesting places have been observed over a period of 50 years and this has given good opportunity of recording the patterns of switching between sites. Table 11 gives details of site occupation for eight territories in which records were made over periods of at least 20 years. The overall appearance is of considerable variation in pattern of use. Some territories with a good choice of sites have one or two favourites, but in others there are no clear preferences. Eleven successive years is the longest run of use for the same nest, in Territory 2. In Territory 6, only one site was used during 1924–32, but thereafter no one site was preferred. Territory 1 has seen use of at least 14 different nests.

Another possible explanation for the use of alternative sites is that the tendency to move about might have survival value in working against the memory of predators. The young of Ravens are somewhat vulnerable to ground predators (notably foxes) just when they leave the nest and are still unsteady in flight. One might suppose that any such advantage would be over-ridden by, or traded off against, the greater security of some nest sites compared with others, during the lengthy egg and nestling stage. Once again, there is no support from the observations for any pattern of use that would match a theoretical advantage. In Territory 3, site A is much the safest of seven, and has been most often used; yet in adjoining Territory 2, the most easily climbable site B has been used almost as often as the most difficult site D. The pattern of use in relation to the known success of particular sites appears to be variable (Table 12). In records for three regions I found no convincing connection between success or failure and the use of the same or different sites in successive years, whereas, in Shetland, Ewins, Dymond and Marquiss (1986) reported a highly significant correlation. In Oregon, Stiehl (1985) found no relationship between successful breeding and reuse of the same nest. Both Ewins *et al.* and I found a marked tendency for repeat clutches to be laid in different nests from those in which first layings failed, but Davis and Davis (1986) did not find this pattern (Table 12).

Alternative sites may be extremely close – the Lakeland crag shown on p. 141 had at one time a row of four nests within the space of a few metres along a single ledge, and a ruined Hebridean farmhouse had four separate nests (D. J. R. Counsell). More typically they are spaced out at 10–100 m intervals or more. On an extensive inland escarpment, alternatives are often spread over 1 km laterally and 200 m vertically. Separate cliffs are used less often than by Peregrines, largely because the usually greater numbers of Ravens create more competition for suitable cliffs. Alternative Raven cliffs are usually 1–4 km apart, but tend to be more widely spaced in low-density areas. Exceptionally, in 1950, the Upper Teesdale pair moved 10 km, to the Eden Valley scarp, for a repeat nesting; the female of the pair was recognisable by a distinctive gap in her primaries (H. Watson). Two cliffs used as alternatives in Sutherland are 7 km apart. In recent years, it has often happened that different crags, once held as alternatives by a single pair of Ravens, are now tenanted by separate pairs, thus reducing the number of nest sites available to each. This 'doubling' of territories has occurred especially in Snowdonia (P. Dare), but also to a lesser extent in Lakeland (G. Horne). In the Southern Uplands, under decline in Raven numbers, the reverse has sometimes happened, with single pairs expanding their claims to include crags once held by other neighbouring pairs.

Sea-cliff Ravens show similar patterns in the use of alternative sites. On extensive ranges there will usually be at least half-a-dozen different nests spread over distances of a kilometre or more but, on less favourable cliff sections, the birds have to make do with a more restricted choice. On long, continuous lines of sea-cliff, some pairs of Ravens show the same tendency as Peregrines of gradually shifting their favourite nesting places, so that observers operating, say, 30 years apart, draw quite different conclusions about locations, though numbers may remain unchanged.

156 *Breeding: nest and nest site*

23. *Raven nest with eggs on the Raven Crag shown on p. 141; protection by the massive overhang of rock is typical. Westmorland (photo: D. A. Ratcliffe)*

Much the same applies to use of alternative eyries by tree-nesters, which usually have from two to four different nests within a circle of radius about 50 m in the same woodland or clump. Simson (1966) found that nests in different years were up to 100 m apart in north Wales woodlands. Occasionally two nests are built in the same tree. Some pairs move between different woodlands or clumps up to 2 or 3 km apart but, as we have seen (p. 146), there have also been historic Raven trees – usually large oaks – where the same nest was occupied year after year without a break. In one birch-grown gill, where Ravens nested successfully for some years, one after another the nest trees soon began sprouting bracket fungi and then died, evidently as a result of heavy manuring from the young birds. Here, the birds had perforce to move to new sites every few years.

It is a common finding that, at the onset of the breeding season, a pair of Ravens will partly build or renovate a different nest from the one that the female later lays in. Some pairs will work on two or three, or even more, different structures and then leave these for another. Seldom is the construction taken beyond the half-lined stage in these abandoned nests, and some are left as barely formed outer stick-baskets. Even so, the work involved would seem to represent an unnecessary expenditure of energy, to a degree that raises thoughts about its possible significance. Could it be a form of territorial marking, to increase the chances of any intruding Ravens being aware of possession when the owners are away foraging elsewhere in their territory? Fryer (1986) has concluded that the habit of Common Buzzards in bringing green-leaved material to favourite perches, as well as to nests, has this significance. On this premise, the use of alternative nest sites could in itself have such meaning.

The whereabouts of Raven nests on inland cliffs are typically revealed by conspicuous patches of green on the rocks immediately below, which often extend to the older nest foundations as well. These are produced by the downwash of soluble nitrogen and phosphorus compounds from the droppings, especially of the young, which are colonised by microscopic green algae of unicellular and filamentous types. Much used and frequently successful eyries often have green patches several square metres in size and visible at considerable distances. Smaller streaks develop beneath favourite perches and roosts of the adult birds. The green patches can form after the fledging of a single brood, but they also disappear in time if the nutrients are not renewed; one especially good example had disappeared 12 years after the site was last used. Where nests are placed on vegetated faces or ledges, there may simply be a more luxuriant flush of plant growth than usual. These effects are sometimes seen on sea-cliffs, but growths of lichens, such as *Ramalina scopulorum*, rather than algae here mark the position of some nest sites.

Many cliff-nest sites are probably of considerable antiquity. The identical Helvellyn rock ledge where I saw my first Raven nest in 1945 was used in 1993 and 1994, and the Skiddaw nest site to which I climbed in 1945 was again in use in 1993. The nest shown opposite was used in 1949 and was also occupied in 1996 (Table 11, Territory 4, site B). Good sites are evidently recognised by successive generations of birds, and the identical ledges are often chosen by reoccupying pairs after long periods of desertion, during which all traces of previous nests had disappeared. Some of them could well have been used for hundreds, or possibly even thousands, of years, though this obviously depends on the rate of natural erosion. Many cliffs of hard igneous rock have probably changed little since the retreat of peri-glacial conditions, but others are subject to occasional rock-falls that must chance to destroy old nest ledges from time to time.

Altitude and aspect

Ravens breeding on coastal cliffs mostly place their nests within 200 m of sea-level, since loftier precipices are rather few. Some nests are placed quite close to high-tide mark, especially on lower cliffs. The limiting factor is the need to avoid exposure to spray-drenching during storm-surges, so that most sea-cliffs below 25 m in height are unsuitable and the majority of nests are in faces of 30–100 m. Smaller faces are used in the less exposed coastal situations, in sheltered coves and inlets, above permanent beaches and on slopes standing back from the sea. There appears to be no particular preference for nest-site position when the cliffs are high.

Tree-nesting Ravens tend to be at only modest elevations, since the terrain they inhabit belongs mainly to the zone of enclosed farmland or the lower sheepwalks of the marginal land and foothills. An altitudinal range of 30–300 m encompasses most of this segment of the population, though there is a scatter of more elevated haunts where the birds nest in upland woodlands or little tree-lined gills deep in the hills. Most of the dense inland population studied by Davis and Davis (1986) in central Wales nested between 200 and 400 m.

Clearly, a bird which lives and breeds so widely on coasts and in the lowlands has no need for high altitudes. In its mountain haunts it is simply exploiting a relatively secure refuge and one that provides an adequate food supply. There is, however, a trade-off between the security of high-altitude nesting and the increasingly adverse effects of hostile climate, which increase with elevation. Table 13 shows the altitudinal distribution of Raven nesting cliffs in several upland districts where the highest summits rise above 750 m. South of the Highlands, the median elevation is at 300–500 m. In these districts, most of the lofty inland cliffs are situated above 300 m, but few at over 600 m are tenanted regularly, despite the presence of many very suitable cliffs at these higher levels. The hill Raven is thus essentially a submontane species, nesting largely below the climatic tree-limit, though it forages over the highest ground.

In Snowdonia, many of the new nesting places established during the post-1960 increase have been below 300 m (Dare, 1986a and b), so that a downwards expansion through improved food supply has occurred; this has also depended on greater tolerance to the birds from human neighbours. Nesting cliffs tend to be at higher altitudes in the southern Highlands and it is here where the most elevated breeding haunts are located, probably because, in some parts of the main massifs, good cliffs are all at considerable altitudes. The highest eyrie that I have seen or heard of is at 890 m on Ben Lawers, Perthshire, and another has been seen at 850 m on Beinn Dubhcraig, also in Perthshire (D. Orr-Ewing). There is a haunt at 760 m in Lakeland, and at least one at this level in Snowdonia; in south Wales nesting occurs up to 720 m in the Brecon Beacons. Average elevation falls in the north-west Highlands, where inland eyries are mostly along the flanks of the main glens and the high-lying corries of the big hills are usually left to the Golden Eagles. In Ross-shire, inland nesting places average 269 m above sea-level, and in Sutherland only 188 m. This also agrees with the general descent of altitudinal life-zones with distance west and north in Britain.

This is a hardy bird, well adapted to living and breeding under severe conditions, as evidenced by its wide distribution in the Arctic regions. There are limits to its tolerance of cold, nevertheless, and Chapman (1924) pointed out that extremes of frost can prove fatal to the nestlings. And despite their ability – or that of their parents – to keep going through heavy snowfalls, the duration of snow-cover in the montane zone must be inhibitory. It appears that some of the most elevated nesting sites are not used in winters with much late snow. The general increase in severity of conditions with elevation, including rainfall, cloud-cover and wind speed, must also militate against high-altitude nesting. Nesting elevation falls as climate becomes increasingly cool-oceanic, suggesting that prevalence of mist and rain are more adverse to the Raven than low temperature alone. Cold, wet springs can cause frequent breeding failure (see p. 187).

While aspect of the nesting cliff shows some degree of bias, this appears to be more from availability than preference. In regions such as Snowdonia and Lakeland, the tendency for the breeding rocks to face mainly between north and east is evidently because more intense glaciation on these shaded aspects of the mountains has given a predominance of large crags with this orientation. Where good crags face between south and west, Ravens readily use them. Sea-cliff nests may show local bias in aspect according to the general orientation of sections of coast.

CHAPTER 9

Breeding: the egg stage

PRELAYING BEHAVIOUR

Besides nest-building, Ravens engage in mutually stimulating reproductive behaviour as the breeding season advances. Ryves (1948) has described courtship of a Cornish coast pair:

> Both birds circling above the cliff. They alight on the sward close to the top, a foot or two apart. They ruffle their feathers, shake themselves

and preen. Now they stand motionless for a spell. Now the female creeps closer and closer to her mate in sidelong movements until she almost touches him. The male then gently scratches the back of her neck with his massive bill – the next moment he tickles her under her outstretched wing. She moves away and faces him. He opens his bill and croons. Much rubbing of bills together, bowings and croakings. The love-making lasted about 10 minutes, after which they rose silently and flew out of sight and ken.

This exchange did not end in copulation, but Ryves noted that mating followed similar activities on the ground on other occasions.

Wilmore (1977) has given another description of mating activity:

... the male raising the tufts of feathers above and behind his eyes and giving nasal muffled notes and cracking sounds with his bill. He then stands by the female and stretches his neck over hers, ruffling his head, neck and throat feathers he bows to her, sinking down on his belly, with his neck straight out but his bill pointing downwards; this is accompanied by curious noises such as mumblings, bubblings and poppings like corks being extracted from bottles. The pair then indulge in mutual preening. Stimulated, the male jumps in the air several times and caresses his mate's beak with his, tickling her under the chin and then 'kissing' her. Both birds stand like this until the female jumps up and down, a signal for her desire to mate. After copulation the pair celebrate with aerobatics and then land with the male giving a musical, vibrating call.

THE EGG

Raven eggs are the pale blue to green type characteristic of the Crow tribe. Ground-colour variations in the normal range may be typed as blue, blue-green, green, grey-green and olive, all the expression of the biliverdin pigments. Darker, superimposed markings are of varying shades, from grey, olive, pale and dark brown to black, and of variable size, density and distribution. Some eggs are ticked, freckled or finely spotted; others have larger streaks and blotches, sometimes merging into a cap at either pole, usually the larger. Ashy to lilac 'shell-marks' laid down earlier in the egg development are common and, in some eggs, the darker superficial markings tend to a suffused smokiness. Nearly immaculate eggs are quite frequent and occasional whole clutches are almost devoid of darker markings; they are usually of the blue-ground type. Another distinctive clutch type is blue with scattered black and grey spots and small blotches, rather like large eggs of the Jackdaw or even Song Thrush. Very often, there is an odd egg in the clutch, with a blue egg in an otherwise uniform greenish set, or vice versa. Some clutches show a rather wide variation in both ground colour and darker markings; most show at least slight variability and rather few are markedly uniform in egg type.

Individual females tend to lay the same types of eggs throughout their lives and those producing distinctive eggs can be identified by this means. Size and shape are as diagnostic in this respect as colour and marking pattern. The shell has a slightly glossy texture and occasional eggs or clutches with a matt or even rough surface are probably laid by birds not in full health.

'Erythristic' varieties are rare, but occur more frequently than in other British members of the Corvidae. Normal eggs have both blue-green (biliverdin) and reddish brown (protoporphyrin) pigments, but the former are absent in erythristic variants (Harrison, 1966), giving a ground colour varying from pale flesh-pink to almost white or, more rarely, pale biscuit-brown, with red-brown and purplish darker markings, and ashy to lavender shell-marks. These 'pink' or 'red' Raven eggs were much sought by collectors, and were first known in Unst, Shetland, in 1854; the clutch of four was described by Newton (1864–1902). They have occurred most consistently in the Lake District. In the Langdale area of Westmorland they were known for more than 30 years during the first half of the 1900s, though it seems likely that more than one bird was involved (R. J. Birkett, J. Coward, J. Wightman). In Cumberland they have been known in at least three widely separated localities: during 1920–25 (E. S. Steward), the 1940s (R. J. Birkett) and 1971–74 (L. Caris, G. Horne). I knew an erythristic-egg bird in the Southern Uplands from 1948–54, and R. Roxburgh, M. Marquiss and R. Mearns found two in this region in 1976–77. Andrew Dixon tells me that in Wales another was known on the sea-cliffs of Caernarvonshire from the late 1950s to the early 1960s. In the Highlands, P. Meeson found pink eggs in an eyrie on the edge of Rannoch Moor, Argyll in 1938.

It has been supposed that erythrism in normal cyanic eggs results from a simple recessive mutant condition, but the genetics of egg coloration are somewhat obscure. Curiously, in the Black Crow of South Africa, erythrism in egg colour is the normal condition, so that in this species it has presumably been favoured by natural selection, though its adaptive value is unclear.

Raven eggs are small in relation to the size of the bird. With an average fresh weight of $c.$ 30 g, this is only 2.5% of female body weight (average 1,200 g), or 12.5% for an average clutch of five eggs. Shape of the eggs is typically ovate to long-ovate, but occasionally more rounded, elliptical, pointed or even slightly pyriform. Eggs within a clutch are often similar in shape, but sometimes show marked variations. The average dimensions of 90 British clutches (453 eggs) in collections were length 49.1 mm by breadth 33.6 mm. This compares closely with figures of 48.8×33.8 mm given for 744 eggs of the nominate race by Cramp and Perrins (1994), who also give a range of 42.1–68.0 mm for length and 29.0–38.5 mm for breadth. For 100 British eggs, Witherby et al. (1938) gave maximum sizes of 63.0×34.5 mm and 55.0×37.5 mm, and minima of 44.2×33.5 mm and 52.3×29.8 mm. Size of eggs within the clutch tends to vary somewhat, but the mean is fairly constant for the same individual female, including for first and repeat layings. Some clutches have one egg markedly smaller than the rest; it is often supposed that this is the last egg laid, but such is not always the case.

---

24. *Raven eggs in a typical hill-nest made of heather sticks and thickly lined with sheep's wool. Langholm Hills, Dumfriesshire (photo: D. A. Ratcliffe)*

No difference in mean egg size between south-west England, Wales, northern England, southern Scotland and Ireland was detected, but the Highlands gave a mean egg size significantly larger than the rest. Table 14 suggests that a south–north cline in egg size could exist, but more data would be needed to test this possibility. Egg size is usually related to individual female body size, so that such a trend would be consistent with the tendency of northern races of Ravens to follow Bergman's Rule (see p. 245). Nine Irish clutches were not significantly different from those in Britain as a whole.

In a study of the connection between eggshell-thinning and contamination of various bird species by DDT/DDE residues, a comparison of 222 Raven eggs for the period 1851–1946 with 205 specimens for 1947–65 showed a decrease in shell thickness index of only 0.89% (Ratcliffe, 1970). This was not significant statistically and was consistent with the finding of only low levels of DDE contamination in the contents of eight Raven eggs analysed in 1963 (mean of 1.0 parts per million, wet weight; all eight eggs in the sample contained DDE) (Table 27). Yet samples of eggs of Carrion Crow and Rook for 1958–69 both showed highly significant decreases in shell thickness index of 5% ($p < 0.001$), against still lower DDE levels in samples of 25 and 11 eggs analysed in 1963–66. Even more surprisingly, 27 western Scottish Golden Eagle eggs for 1951–65 showed a highly significant mean 10% decrease in shell thickness index, against a DDE level in egg contents of 0.8 ppm during 1963–65 (i.e. similar to that in the Raven).

There is a problem over these comparisons, since the shell thickness measurements were mostly made on specimens in collections, whereas the residue analyses were made on smaller samples of eggs taken for the purpose. Nevertheless, since it must be presumed that, during the period ($c.$ 1950–65) when DDT was widely used as a sheep-dip, Ravens must have been widely contaminated by this insecticide and its metabolite, DDE, the lack of significant eggshell-thinning in a fairly large population sample points to a degree of resistance in this species to the effects of these organochlorine residues. DDT and DDE have been proved to be potent agents of eggshell-thinning in birds, but evidently have more effect on some species than on others. It is difficult to avoid the deduction that the Raven is one of the more resistant species.

CLUTCH SIZE

Ellis *et al.* (1994) have reviewed the published figures for Raven clutch size, in both Britain and other countries. I have included data for different regions of Britain and Ireland in Table 15 in two groups: those from the

---

25. *A large and uniform Raven clutch of the spotted blue type; it was in the nest on the cottage roof shown on p. 152. Elan Valley, Radnorshire (photo: D. A. Ratcliffe)*

26. *A Raven clutch of the more heavily marked blue-green type. Lake District (photo: G. Horne)*

BTO's Nest Records Scheme (NRS) and separate figures available from regional studies. As there is some overlap between these two groups of data, I have not amalgamated them. The NRS series is limited to 384 records which could be regarded as certain full clutches. The other data sets in Table 15 are understood to be based on full-clutch records. The BTO records give an overall mean of 4.77 eggs. Separation into means for Northern Ireland, Wales, Scotland and England showed no significant differences in a range of 4.63 to 4.85. The other data sets gave, for Great Britain, a mean of 4.79 and a range of 4.57 to 5.11 ($n = 1027$).

Only the Orkney records give a clutch mean exceeding 5 eggs. For north Wales Allin (1968) published the figure of 5.17 eggs, but it is noticeable that his is the only data set completely lacking records of two or three eggs. I have examined Eric Allin's clutch records (which he submitted to the BTO) and find that there were 6 c/2 and 21 c/3; while some of these may have been incomplete clutches, it is likely that some were full sets and should have been included in his series. The 1968 north Wales mean is thus probably biased upwards and I have used a revised figure in Table 15.

Table 15 thus suggests that the available evidence is for a fair degree of geographical uniformity in Raven mean clutch size within Britain and Ireland. The combined data show that clutches of four, five and six eggs are the norm for the species here, amounting to 84% of all layings. Sets of seven appear to be most frequent in Wales – 7.5% of BTO records, compared with 2% elsewhere. Thomas (1993) recorded an exceptional clutch of eight eggs in Argyll. A nestful of ten eggs in Anglesey in 1934 (Allen & Naish, 1935) could well have been the layings of two females. In a small percentage of nests, from one to three eggs is undoubtedly the number being incubated; perhaps because the bird could lay no more, or because other eggs had been lost. A bird robbed of an incomplete clutch may sometimes lay the remainder and carry on incubation, instead of waiting to produce a full repeat laying. Some observers have assumed that successive clutches of two or three sparsely marked or unmarked eggs are the product of aged females, but there is no corroborating evidence for this view. Sparsely marked clutches of five or six eggs are by no means uncommon.

Clutch means for other parts of the world include 5.97 ($n = 45$; 33% were c/7) for south-eastern Oregon (Stiehl, 1985), 5.32 ($n = 31$) for south-western Idaho (Kochert, Bammann & Doremus, 1977), 5.3 ($n = 12$) for north Africa (Etchécopar & Hüe, 1967) and 4.7 ($n = 12$) for Finland (Haartman, 1969). There is no evidence that clutch size increases with latitude between the temperate and arctic regions, as it does in many widespread smaller passerines.

Marquiss, Newton and Ratcliffe (1978) noted that clutch size in Northumberland and the Southern Uplands was inversely correlated with laying date, i.e. the earliest clutches were the largest. Davis and Davis (1986) also found that clutch size, including repeats, declined through the season in central Wales: 25 clutches started in late February averaged 4.8 eggs, 48 started in early March 4.7, 16 started in late March 3.7, and eight started in early April 3.4 eggs. Clutch size in this sample also varied slightly with territory size: 86 clutches from nests under 2.5 km apart averaged 4.4 eggs, while

28 at 2.5 km or more apart averaged 4.9. In part this reflected an altitude effect, since nest spacings were wider at both higher and lower elevations, and closer at middle elevations. Below 250 m, 31 nests had mean clutches of 4.5 eggs, as did 36 nests above 350 m, while 47 nests between these altitudes averaged 4.3 eggs. Significance testing was not applied to these figures.

The central Wales clutch mean of 4.50 during 1975–80 is the lowest for any region of Britain, but Cross and Davis (unpubl.) give the figure of 4.79 eggs ($n = 29$) for the same district in 1986, and Cross (unpubl.) found 4.82 ($n = 39$) in 1992 and 5.06 ($n = 31$) in 1993. Incorporation of these additional data gives an overall mean of 4.69 eggs for central Wales during 1975–93, bringing it closer to the BTO figure for the whole of Wales.

The Raven consistently has a larger clutch size than its smaller relative the Carrion/Hooded Crow, in which three, four or five eggs are the norm and sixes infrequent in Britain (mean of 4.1, $n = 345$; and, in a series of 216 clutches, no c/6 reported, Cramp & Perrins, 1994).

Ravens usually lay replacement (repeat) clutches when the first laying is lost before completion or at an early stage in incubation. If incubation has proceeded beyond the half-way stage (10–12 days), the chances of a repeat laying diminish rapidly, but even pairs that have lost eggs incubated to full term or very small chicks have been known to lay again (A. V. Cross). Replacement clutches after initial failure have been the norm in southern Scotland, northern England and north Wales. In 82 instances of first-clutch failure in these districts that I was able to follow up, there were 63 repeat layings (77%). I excluded other failures where repeat clutches could have been laid in alternative haunts that were not visited or where shooting of the birds was evidently involved. Of these 82 failures, all but three were attributed to egg-collecting. A much lower incidence of repeats is reported in some other districts. Davis and Davis (1986) found that, of 269 pairs which laid eggs during 1975–80, 88 failed to rear young, but only 11 produced repeat layings (it was not stated how many failures resulted from clutch losses). Ewins, Dymond and Marquiss (1986) reported that, on Fetlar in the Shetlands, only two replacement clutches followed 23 failed breeding attempts. Occasionally, second repeats have been reported, i.e. third layings after failure of the first two (twice reported by J. Hutchinson in southern Scotland).

Many repeat clutches are the same size as the first layings; occasionally they are larger, but there is a tendency to slight reduction on average. In northern England and southern Scotland, two sets of comparisons showed repeat clutches to be 0.67 and 0.54 eggs less, on average, both differences being significant (Table 16). Although the sample was small, Davis and Davis (1986) also found a similar reduction in repeats (mean 3.9 eggs, $n = 9$) compared with first layings (4.5 eggs, $n = 114$). These nine repeats nevertheless averaged more than the mean of 3.4 eggs for 15 late first layings produced during the same period, late March to early April.

Davis and Davis (1986) support the views of Harlow (1922), Simson (1966) and Allin (1968) that individual female Ravens tend to constancy in clutch size in successive years. They found, in central Wales, that at 20 territories where the clutch size (excluding repeats) was known in three or four of the years 1975–78, at 14 it did not vary by more than a single egg, and in 17 by not more than two eggs. These data assume that survival and site fidelity tend to lead to territory occupations by the same females over a period of 4 years. I have examined my own records for instances where full first clutches of distinctive egg types (denoting particular females) were seen in the same nesting places in two or more years, over periods not exceeding 5 years. Out of 23 such paired records, 15 showed the same clutch size, six differed by one egg, and two differed by two eggs. Birds which lay seven eggs often do so every year.

It might be expected that clutch size in individuals would be responsive to annual variations in food supply and weather, and between different territories according to differences in their feeding value in any one year. In two areas of central Wales, Newton, Davis and Davis (1982) found that variations from year to year in the number of eggs laid per pair were significantly correlated with carrion supplies over the winter and spring during 1975–79. This supports an egg-collectors' old belief that clutches are larger in hard winters which produce especially high sheep mortality. Newton *et al.* also found a tendency to decreasing clutch size with increasing extent of afforestation in the vicinity of the nesting place. Clutch means decreased from 4.6 to 2.5 eggs as closed forest within 1 km of the nest increased from 0 to 61–80%; and from 5.1 to 3.0 eggs as closed forest within 3 km increased from 0 to 41–60%, but the data were not amenable to significance testing. In Northumberland and the Southern Uplands, Marquiss, Newton and Ratcliffe (1978) found no significant correlation between clutch size and extent of afforestation within 3 or 5 km of the nest; however, in those nest areas where over 25% of the former sheepwalk within 5 km was planted, egg-laying was later than in unplanted areas.

Allin (1968) reported no significant difference in mean clutch size of nests in north Wales below and above 305 m: 5.20 and 5.12 eggs respectively. This sample probably did not include enough high-level (>500 m) nests to be a rigorous examination of possible altitude effects on clutch size.

Dr Crick finds evidence of a long-term decline in clutch size in the BTO data set for the UK as a whole, from 5.0 in 1947–54 to 4.7 in 1986–93, but the apparent trend is not significant.

LAYING

The Raven is celebrated as one of the earliest of our nesting birds. A few pairs begin laying as early as the first week in February, and many have full clutches before the end of that month. The majority of breeding females are sitting by 24 March in nearly all parts of Britain and Ireland. Table 17 gives mean first-egg dates for various regions of Britain and Ireland, and shows that there is a tendency for this bench-mark in the breeding cycle to become later with distance north. The rather few records for Shetland indicate that laying is unusually late there, agreeing with the earlier comments by Saxby (1874). Yet, in the adjoining Orkneys, mean laying date is only a day later that in the Southern Uplands, although February clutches are rather infrequent (C. J. Booth).

Onset of laying is also retarded with increasing altitude. Allin (1968) found that, in 119 north Wales nests below 305 m, 42.0% of clutches were completed by 28 February and 76.5% by 10 March; whereas in 48 nests above 305 m only 10.4% and 37.5% of clutches were completed by these dates. Many of the lower-level nests were coastal, so that there may have been a maritime effect as well. Data from northern England and the Southern Uplands show a significant relationship between laying date and altitude, and Fig. 9 suggests that, on average, onset of laying is delayed by 1 day for every 50-m increase in altitude.

There is a wide spread of first-egg date in any district or at any altitude, so that mid-February layings can occur almost anywhere, and a few pairs will usually be found to delay until late March, regardless of location. For the population as a whole, laying is spread over a full month – a relatively long period. Occasional pairs appear not to lay until well into April – it is difficult to be sure that these are not repeat layings, but some of these late-nesters have given no sign of having laid earlier and failed. The data for northern England and southern Scotland show an unexpected and significant trend to earlier onset of laying between 1933 and 1991, by one day every 10 years, and almost a week over the whole of this period (Fig. 10). By contrast, the BTO data show no evidence of a significant change in laying

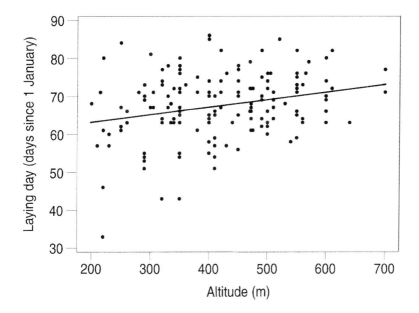

Figure 9. Delay in onset of Raven laying date with increasing altitude in northern England and southern Scotland. (Prepared by Alan Fielding.)
NOTES: Layday = 59.2 + 0.0202 altitude; R^2 = 6.1%. Data are for inland pairs, from the records of G. Horne, D. A. Ratcliffe and R. J. Birkett.

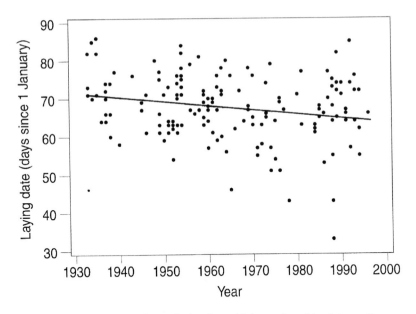

Figure 10. Advance in onset of Raven laying date with increasing altitude in northern England and southern Scotland. (Prepared by Alan Fielding.)
NOTES: Layday = 74.6−0.108 year; R^2 = 4.3%. Data are for inland pairs, from the records of G. Horne, D. A. Ratcliffe and R. J. Birkett.

date over the country as a whole from 1946 to 1993, while in Scotland there is a tendency (not significant) for laying to have become later since 1961.

Laying in regions farther north is clearly delayed by lower temperatures. With an average fledging date of 18–20 June, laying on the Varanger Fjord at 70°N in Arctic Norway begins around 10 April – a time when much of the land is still snow-covered. In Iceland, onset of laying varied between districts and years, but did not usually begin before 1 April and could be a week later (Skarphedinsson et al., 1990). In some temperate regions, laying is also later than in Britain and Ireland. Stiehl (1985) reported a spread of laying from 7 March to 16 May, with a median date of 4 April, for a Raven population at 1,350 m average elevation in south-eastern Oregon. Breeding is still further delayed in Spain, where Verner (1909) gave the average first-egg date as 20 April. In Portugal, Bobby Smith has seen pairs feeding young in the nest in late May. In north Africa, also, laying is said to be in April (Goodwin, 1986). Despite the warmer temperatures of these last two regions, the main food sources evidently become available later than in Britain and Ireland. Yet, in northern India, Ravens lay in January (Goodwin, 1986).

There are very occasional autumn nestings. A nest with eggs was reported in the Howgill Fells near Sedbergh, on 13 October 1945, but, when seen by J. H. Hyatt on 28 October, the eggs had been taken. On 1 January 1989, Richard and Barbara Mearns found an almost fledged young Raven in a nest on the Galloway coast. Laying must have been at the very end of

October, and the Mearnses point to the occasional finding of Rooks with young in November. Hurford and Lansdown (1995) mention a nest found with eggs in November at Mewslade, Glamorgan, in the 1980s.

INCUBATION

The eggs are normally laid at intervals of 1 day, though sometimes between 1 and 2 days. Goodwin (1986) says they are usually laid in the morning. Ryves (1948) stated that, in Cornwall, incubation begins with the last egg or occasionally the one before. Elsewhere it is usual to find that one or two eggs hatch a day or more after the rest, indicating that incubation begins before the clutch is complete, and Davis and Davis (1986) found that, in central Wales, it began most often with the penultimate egg. Davis and Davis found that six incubation periods, from laying of the last egg to hatching of the first, were 17, 18, 19–20, 20, 20–21 and 20–21 days. Other authors have given the usual period as 20–21 days (Witherby et al., 1938; Cramp & Perrins, 1994), so the shorter times suggest that some Ravens start incubating when two, or even three, eggs are still to be laid. From observation of hatching sequence in 12 clutches, Stiehl (1985) found that incubation began with the second egg in Oregon. Incomplete clutches may be covered by the bird, but can also be left untended, especially during warm weather, and Davis and Davis found that eggs were then sometimes concealed in the wool lining by the parent. Gwinner (1965a) noted that the female deliberately sinks the eggs into the nest lining during incubation. At least two birds have been found incubating lumps of rock which had fallen and broken their eggs.

Raven nests that have been robbed usually show a wool lining with a characteristically dishevelled appearance. This is evidently caused by the parent birds pulling at the nest lining, sometimes leaving part of it hanging over the edge of the stick-basket. Tony Cross watched a female repeatedly pulling at the lining of its nest on the day when its eggs were removed. Whether they do this in a forlorn search for the missing eggs, or out of a frustration akin to their redirected aggression when the young are examined (p. 178), is uncertain. Not all Ravens disturb their nest lining in this way and it is sometimes impossible to tell, from a single visit, whether a complete new nest is just ready for eggs or has already been emptied of them. The laying of repeat clutches usually begins about 2 weeks after the loss of first sets. In 17 instances, the range was from 10 to 16 days, with a mean of 13 days. This would give an average time to completion of second clutches of 18 days.

All observers agree that incubation is by the female alone, with the male playing the part of food-provider and sentinel. Ryves (1948) says that she usually leaves her eggs to receive the meal which he brings in his pouch, or sometimes in his bill. If she refuses to leave the nest, he feeds her on it. One sitting Raven nearly always flew to drink from a nearby pool after swallowing the food brought to her. Males have also been seen to bring water in their throat pouch. Ryves found that the incubating female sat very closely and for long periods, seldom leaving the eggs unless called off by her mate. When off them for a meal, she usually returned within 15–20 minutes at most.

Tony Cross filmed much of the incubation period at a nest in central Wales. The sitting female stayed alert and restlessly turned her head to and fro during the daytime. She now and then shuffled into a new position and occasionally turned the eggs. During the 11 days' observations, the female left the nest 11 times a day, on average (17 times was the most). These absences were usually for 5–10 minutes, but occasionally lasted for up to 24 minutes; they added up to over 11% of the daylight period. The fixed video-recorder did not show where the bird went or what she did while away, but there were several shorter disturbances when she appeared to chase away some intruding creature near the nest. Kites were constantly in the vicinity and some of the female's absences almost certainly involved interactions with this bird. The male fed his mate on the nest seven times a day on average, his visits often lasting only a few seconds, and seldom more than a minute. Sometimes the two birds flew away together. From the ninth day of filming, during times when the male was absent, an intruding male Raven appeared at the nest and copulated with the female, or attempted to do so. This happened on 4 separate days and twice on the last day. The eggs hatched, so that the female's frequent absences had no adverse effects. The incubation pattern was somewhat different from that described by Ryves, but this could have been due to individual variation.

The male may be away and out of sight for long periods during the day, but is otherwise about the nesting haunts, keeping a watchful eye for any intruders, be they humans, mammal and bird predators or other Ravens. When people approach to within a few hundred metres, the male may give a rather casual 'clonk' or two and take off to soar around at a distance. If their approach continues and is seen to be aiming at the nesting crags or trees, the bird's tone changes to a more insistently warning 'aark'. The sitting female may then decide to leave, launching unhurriedly from the nest and flying quietly away to join her mate. The two will often perch on some vantage point to await events, but they usually begin a demonstration in flight, soaring and flapping around and voicing protest with a barrage of angry croaks. The female may also choose – especially when incubation is advanced – to sit things out for longer, leaving only when the nest is under close approach. If the male is away, she is also frequently taken by surprise by humans at close quarters. The bird then leaves in a great flurry of loudly beating wings as it tries to distance itself as rapidly as possible from the intruders.

Ravens will seldom continue sitting once they see humans within a hundred metres, but their behaviour when disturbed at the nest shows great individual variation, in boldness or shyness, and in demonstrativeness. Occasional individuals become accustomed to the proximity of people and will continue to incubate fearlessly, even when they are close at hand. Behaviour is often a good reflection of how they are treated. Birds that have been shot at become extremely shy and refuse to come within several hundred metres while intruders are anywhere near the nest.

The eggs become even more glossy towards hatching, through absorbing oil from the bird's feathers. Ryves (1948) said that the sitting female begins to whiten the outside of the nest with faeces as incubation advances, but most of the nests I have seen stayed clean until the young began to foul them. The male usually has a favourite roosting place within 100 m of the nest and, on cliffs, this is often revealed by a conspicuous green streak similar to that below the nest, with fresh white droppings superimposed.

CHAPTER 10

Breeding: the young

Newly hatched Raven chicks are like those of other passerines – naked, blind, pot-bellied and helpless, apart from their ability to beg and gape for food, which they do in response to movement or sound. Their skin is flesh-coloured with a slightly orange tint while their throats are a bright carmine. They are treated with great tenderness by the female, which eats the eggshell, cleans the newly hatched chick and places its neck to rest on a remaining eggshell (Gwinner, 1965a). Gilbert and Brook (1924) noted that, when the chicks were newly hatched, the mother bird spent a good deal of time tugging at the nest lining for their benefit – 'making the bed' as the writers put it.

The small young are fed on half-digested matter regurgitated from the throat pouch of the male, often directly, but sometimes by the female who has received the food first. In a captive study, Gwinner (1965a) observed that, in the first 2 days, nestlings were not given meat, the parents preferring to feed them on invertebrates, with any hard parts removed. Mice and small chicks were pulled into very small pieces and often kept in the pouch before being fed to the young. The parents often drank water after filling the pouch, to soften food, while the young were 3–10 days old. Ryves (1948) found that the faeces of small young were generally swallowed by the old birds, though they were sometimes carried away. Other than during feeding time, small chicks are brooded fairly continuously, especially during cold weather, for they are tender creatures. In this task, the male sometimes relieves his mate, who flies away to exercise and feed on her own account, but is normally back within the hour (Ryves). Goodwin (1986) says that the male does not actually brood, but stands or crouches over the chicks.

NESTLING GROWTH

The skin colour darkens after a few days to a greyish cast, but with a purplish shade in places. A fairly dense but short, brownish down soon develops on the upper parts, upper wing and thigh. Pen feathers begin to sprout first on the wings and tail, but Thomas (1993) found that the chicks did not produce measurable primaries until 11 days old. The eyes open gradually but, since the chicks tend to keep their eyes closed, it is unclear when they can first see properly. By 3 weeks of age a thin covering of feathers has developed, the heavy bill has formed and the legs and feet are quite strong. The young are fed on small items or regurgitated food through the whole of their nestling period and there is none of the self-feeding that occurs in older young of the birds of prey. Ryves – to whom I am indebted for many of these observations during the nestling period – found that, at this stage, food was often first 'prepared' at a distant larder, where it was broken into small portions, pouched and then carried to the nest. A good deal of food was also cached by the parents at various stages of nestling growth, in holes among rocks or under stones and in crannies.

The male is the sole provider of food at first, and has to work hard in keeping his brood and mate supplied, but, as the chicks grow bigger, the female increasingly forages for them as well. Gilbert and Brook (1924), observing daily routine from a photographer's hide, found that the young were fed at frequent intervals from dawn until about 11.00 h GMT. There were longer intervals between feeds up to 15.00 h, with the slackest time during 12.00–14.00 h. Feeding intensified in the late afternoon and, towards dusk, food was brought every few minutes. Feeding intervals become longer with older chicks and food is given to them in more solid

27. *Raven feeding 2- to 3-week-old young in a rowan tree-nest. Wales (photo: Arthur Brook/National Museum of Wales)*

form. Heinrich (1990) noticed that small chicks did not beg until a parent gave a short grunting sound, but older young gaped as soon as an adult bird landed on the nest.

Observations at two nests in Idaho, with three and four young, showed that the adult Ravens visited the nests on an average 64 times a day, feeding the young on 70% and cleaning the nest on 34% of visits (Steenhof and Kochert, 1982). Parental attentiveness declined with age of chicks, from 102 visits a day when the chicks were 13 days old to 36 when they were 35 days old, but the number of feeding visits was relatively constant. Visits averaged 114 seconds, but nearly half were less than 15 seconds. The female made slightly more nest visits than the male, except at fledging time.

After the chicks are about 10 days old, they are brooded less and less during the day, except when the weather is wet or cold. Brooding is continued at night for some time and, on hot days while the sun is full on the face of a nesting cliff, the parents will take it in turn to shelter the young from direct exposure. Other observers have seen the adult birds bring water to their panting young on hot days (Gwinner, 1965a; Hauri, 1956). Females have also been seen to wet their underparts and cover young when they were suffering from heat, or to bore a hole through the bottom of the nest by way of ventilation (Goodwin, 1986). Brooding normally ceases after about 18 days and the parents then take it in turn to forage for food, one remaining on guard near the nest during the other's absence. Heinrich (1990) watched one female tenderly preening the plumage of its half-grown brood. Tony Cross found that, after the female at a nest was found dead below the tree, the male continued to feed the single 2- to 3-week-old chick. Instances of fostering of chicks have also been reported (see p. 200–201).

When night-brooding stops, both adults will roost close together on a ledge away from the nest. Nest sanitation ceases, for larger young eject their faeces over the outer edge, but only weakly, so that the whole outer structure becomes plastered with their liquid white excreta. The degree of 'white-washing' of the nest exterior is a good indication of the size of the brood and its state of development. On a hot day, the immediate vicinity becomes a noisome spot and there may indeed be a deterrent effect on potential predators. This would tend to counter the increasingly conspicuous appearance of an eyrie placed on dark rocks.

Young Ravens become well feathered at 4 weeks old, though their plumage continues to develop for another 2 weeks and they do not leave the nest until 45 days old on average (Davis & Davis, 1986; Thomas, 1993). A few do not leave until they are 50 days old. The time from egg

28. Brood of half-grown young in a tree-nest. Southern Uplands (photo: Brian Turner)

to flying for an individual thus averages about 70 days. During their last week or so in the nest, the young become more restless, hopping about and flapping their wings in exercise a good deal. Where a rock-nest has good ledges around it, the brood will hop and flap out onto these and sit there for spells, but then feel the need for the security of the nest and scuttle back to it. In trees they make their way out into the surrounding branches. They indulge in a good deal of squawking among themselves, in weak imitation of their parents' voices, and can call a lot when hungry. The parent birds often sit close to them on ledges or the cliff-tops, or in nearby branches or other trees, but they will sometimes depart together in search of food and leave the brood unguarded for a time. The large young beg for food with trembling wings and hoarse calls, and the parents insert their bills down the youngsters' throats in feeding them. If humans reach a nest with large young, they usually flatten themselves in the nest and stay silent and motionless, except for the flicker of the third eyelid.

The parental expressions of alarm towards intruders tend to increase with age of their offspring. If a nest is visited by people when the young are large, in areas where Ravens have not been shot at, the adult birds will often come and sit close to the disturber in order to give vent to their feelings. One of the pair is usually bolder than the other – it can be either sex – and may come within 10 m, barking angrily all the while and flying to a new perch with a loud 'woof-woof' of wings. Typically, one or both parents appear to work themselves into a frenzy of impotent rage, tearing up tufts of grass or perching in trees where they try to wrench off twigs. Now and then this redirected aggression includes pitching on the back of a sheep and proceeding to tear off wool, to the animal's discomfiture. Chris Rollie in Scotland was bombarded by chunks of turf and the odd small stone pulled up by a demonstrating Raven perched above him and, in a similar incident in Oregon, seven large pebbles were dropped onto an intruder, but these events were probably accidental rather than intentional. Although occasional bold individuals will brush past, there are no records of Ravens attacking humans in Britain or Ireland, and these are harmless outlets for frustration. Few people have seen distraction display in the Raven, but Macgillivray saw one bird with a nest pitch some distance away and 'tumble about as if mortally wounded'.

Where Ravens are persecuted, they tend to keep a respectful distance from intruders and remain well out of gunshot range, however much their young may appear to be threatened. Some will then voice their displeasure almost continuously from a safe distance, but others remain fairly quiet, even when their young are large. A few stay a long way off, giving only the occasional croak, or even disappear in silence. In the USA, Knight (1984) compared the behaviour of Ravens at nests on farmland, with houses and people nearby, and on uninhabited rangeland. He found that farmland birds flew at greater distances (mean 456 m vs 91 m), stayed farther away (315 m vs 74 m), and demonstrated less against intruders at the nest, all the differences being significant. Knight attributed the greater timidity of Ravens on farmland to persecution (11 of 26 farmland nests were

29. Raven feeding month-old young in a crag-nest. Wales (photo: Arthur Brook/National Museum of Wales)

destroyed, but none of 17 on rangeland), and believed the response had distinct survival value.

Potential predators, such as the fox, which approach the vicinity of a nest with young are often treated aggressively, with the parent Ravens swooping very low and perching close while calling a great deal, in an attempt to divert the unwelcome visitor. Such treatment does not deter foxes from raiding easily accessible nests with young.

The first flight of a young bird from the nest is often weak and unsure, with an ungainly landing. Ryves saw youngsters crash into the sea by flying too far out from the home cliff, or land on rocks at its base and be drowned by the rising tide. They are vulnerable to attack by foxes and other mammal predators at this stage, and Peregrines disturbed from their eyries by people have been seen to strike down young Ravens which rashly ventured on the wing nearby. The young birds also remain at risk to predation by Golden Eagles for some time, as the records of remains at eagle eyries testify (see p. 128). They are easy targets for gunners and any other unfriendly people, too. Of BTO ringing recoveries, 6% were of youngsters which died within a month of fledging, without having moved a measurable distance from their nest site (Table 5).

Because of sibling differences in development, a large brood tends to leave the nest over a spread of 3 or 4 days. The earlier broods take to the wing before the end of April, but the majority fly in May. After fledging, the family moves increasing distances away from the nest during the day, but often returns to it, or to the immediate vicinity, to roost at night. Gradually, they abandon the nest site altogether and move to pastures new in search of good feeding places. Orton (1948) noticed that the family groups tended to seek the remoter and higher parts of the mountains, and that two such parties would occasionally join up together for a time. Earlier territorialism is evidently in abeyance at this stage in the breeding cycle.

Although the adult birds continue to feed them for a time, the youngsters soon pick up invertebrates and other smaller food items. The parents increasingly leave food on the ground for them to pick up. The young are recognisable by their rather uncertain flight, high-pitched voices and sleek heads, for their throats do not develop the characteristic 'beard' feathers of the adult until the first moult. In close view, their plumage lacks the purplish iridescence of the adults. Lorenz (1970) saw newly flown young practising the seizure of prey by making two-footed grabs at medium-sized objects, holding their heads as far back as possible. Kilham (1989) described fledged juveniles playing together on the ground, facing each other, bills open and heads up, then leaping away with wings extended, bouncing about and turning this way and that. Sometimes they chased one another, or walked about, picking up sticks or other objects that took their attention. Young Ravens seem less inclined to play together on the wing, as recently fledged raptors often do, and they appear to develop their full powers of flight rather slowly. Family groups often drift around in a leisurely fashion or soar together gracefully in rising air currents. Young birds often show a rather naive curiosity towards humans, flying low and circling over reclining figures, and this may be their undoing. Among others,

30. *Raven with near-fledged young in a crag-nest. Wales (photo: Arthur Brook/National Museum of Wales)*

Saxby (1874) spoke of shooting Ravens by lying on his back and feigning dead, thereby enticing them within range.

The Raven family stays together through much of the summer, travelling about as a party and frequently reaching places far distant from the nest. The young birds gradually gain confidence and mastery of the air,

182 *Breeding: the young*

and are able to fend for themselves in feeding. Davis and Davis (1986) noticed that, as time advanced, they moved about as a small party of brood-mates, foraging and roosting quite independently of their parents. The family groups may still be together by late July or early August, or even later, depending on fledging date. But, after about 3 months, the youngsters become detached from their parents and move off to make their own way in the world, while the adult birds return as a pair to their nesting territory. The manner of this separation is still a subject for argument, as related in Chapter 5. Goodwin (1986) said that Raven families could stay together for up to 6 months after fledging, but this seems unusual. In Breconshire, Andrew Dixon has seen several small parties of juveniles without parents during August and September in areas where there were no nests.

BROOD SIZE

Table 18 gives figures for brood size in various parts of Britain and Ireland. On average, fledged brood size in the Raven is substantially less than clutch size. Even for the higher figures, of over three young, there is an average loss of about 1.5 eggs per clutch, taking an overall mean of 4.82 eggs. In regions with lower brood size, the average loss is of almost two eggs per clutch. Most Ravens do not rear chicks from all their eggs. Fledged broods of seven young are unknown in Britain and Ireland, sixes are quite rare (1%), and even fives are not common (10%). Even with the large brood size of 4.2 young reported from Malheur National Wildlife Refuge in Oregon (Stiehl, 1985), there was an average loss of almost two young from the mean clutch size of 6.0 eggs.

Some of the losses are from eggs which fail to hatch. Unhatched eggs remain in the nest with the chicks. Holyoak (1967) says that they are eaten by the parents within 12 days. I have not seen unhatched eggs in nests with chicks older than an estimated 12 days but, as the young become bigger, there is an increasing likelihood that they would crush any eggs remaining. The only reliable information from Britain or Ireland on the frequency of egg failure in the Raven is from the Dartmoor area of Devon, where G. Kaczanow found (by water testing) that in 100 clutches, 78% contained one fertile egg and 21% two fertile eggs. Stiehl (1985) found that in Oregon, one egg per clutch usually failed to hatch; his Ravens laid large clutches, averaging almost six eggs.

I have examined my own records for nests with young estimated at 4–12 days old; there were 60, of which nine contained unhatched eggs, two had crushed eggs and one a dead chick. Except for one nest with four unhatched eggs (and a live 12-day-old chick) no nest held more than a single bad egg, and failure of whole clutches is rarely reported. Allowing a mean clutch size of 4.8 eggs, these 60 nests had held an estimated 288 eggs, of which at least 14 (5%) failed to hatch. Many of these nests were examined only through binoculars, so that the indicated egg-failure rate may well be too low. Unhatched eggs can become buried in the nest lining or hidden under the

31. Raven with near-fledged young in a crag-nest. Southern Uplands (photo: Bobby Smith)

chicks, and so may not be visible except to close inspection. It is also possible that the adults may eat some failed eggs within a few days of the others hatching, so that they escape notice. Even so, I believe that egg failure in Britain and Ireland is usually less than the rates reported by Kaczanow and Stiehl.

While Allin (1968) surmised that egg failure and loss of younger chicks were about equally important in accounting for the disparity between clutch size and fledged brood size in north Wales Ravens, my own records suggest that the death of chicks at an early age accounts for the greater part of this difference. The chicks grow quite rapidly, but are prone to a high mortality in their first few days of life, and thereby subject to the process of brood reduction that is characteristic of the Crow family. From my own records for all districts examined, 19 broods up to 5 days old averaged 4.32 chicks – less than the mean clutch size (4.8), but appreciably more than the mean 2.67 young for 169 broods older than 5 days; the difference is significant. Alan Fielding notes that the best statistical model to describe the relationship between brood size and brood age is a non-linear exponential decay model (Fig. 11). This model suggests that brood size shows a reduction shortly after hatching, but that after this there is little change in brood size. Ewins, Dymond and Marquiss (1986) also found a marked drop in mean brood size, as between 'small chicks' (3.57, $n = 47$) and fledged young

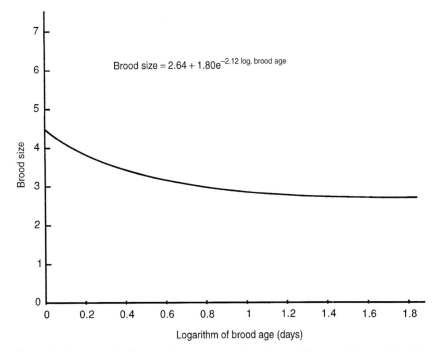

Figure 11. Relationship between brood size and brood age of Ravens. (Prepared by Alan Fielding.)

(3.19, $n = 21$). It is not unusual for only one or two nestling Ravens to survive out of clutches of five to seven eggs.

Ravens show a degree of the asynchronous hatching that was described by Lockie (1955) in Rooks and Jackdaws, and the sibling competition for food which stems from this. The process works as follows. When the eggs hatch serially, instead of more or less simultaneously, the graded ages of the chicks give them a differential vigour and attention-demand in begging for food. The oldest and largest commands the most attention and the parents will continue to feed it first, until it is satiated and subsides. The next most-demanding chick then takes its place and the others tend to be neglected until it, too, no longer craves attention. This continues, down to the smallest. When food is plentiful, all the older chicks become satisfied and the smallest receives its due share but, when feeding is too infrequent to satisfy the larger nestlings, their smaller brethren begin to go hungry. A chick weakened by hunger will presently cease to beg and, once this happens, it is finished, for the parents will feed only begging chicks. And so, depending on just how scarce food chances to be, a varying number of the brood will die, because of the competition from older members.

So runs the theory, anyway, with the further explanation that natural selection has determined that, in birds such as this, where the food supply is variable and the nestling period prolonged, it is more advantageous for a smaller number of fully nourished young to survive than a larger number of undernourished ones. While Ravens are certainly prone to appreciable brood reduction, the degree of asynchrony in hatching – as in onset of incubation – does not seem sufficient to account for the scale of reduction often found. Incubation often does not begin until the penultimate egg, and the whole clutch usually hatches within 2 or 3 days. Possibly females begin incubation earlier in seasons when food is short. Davis and Davis (1986) found that, in central Wales, the chick from the last egg often died and the brood size at the small-chick stage was commonly one less than the clutch. Most of the brood reduction occurs within the first week. Once the young are 7–10 days old, they mostly survive and mortality within the brood for older nestlings is quite low.

It is unusual for a species so seldom to realise the potential output of young that its clutch size could allow. Golden Eagles in Scotland nearly all lay sets of two eggs; in the west they seldom rear more than a single young, but in the east they often bring off two young. In the Raven, each female appears, in the vast majority of layings, to lay one or two extra eggs for no benefit. Davis and Davis (1986) found that, in central Wales, final brood size tended to increase with clutch size, up to five eggs; but in successful nests, the proportion of eggs that produced fledged young declined from 73% in clutches of three and four to 52% in clutches of six. Clutches of five averaged only 3.3 young, compared with 3.0 young for clutches of four and 3.1 young from clutches of six eggs. As the Davises remarked, there seemed to be no advantage in Ravens laying more than five eggs. Why has natural selection not trimmed Raven clutch size to the more normal expectations of chick survival, instead of allowing this uneconomic wastage of energy to continue? There are no obvious answers.

Fledged brood size varies somewhat between different districts of Britain (Table 18). The lowest figures are for Snowdonia, 2.27 ($n = 37$) during 1951–79 and 2.47 ($n = 32$) during 1978–81, and for the western Highlands and Islands, 2.62 ($n = 52$) during 1956–71. The highest are for north Wales (including many coastal Anglesey and low-level inland nests), 3.29 ($n = 154$) during 1946–67; the Southern Uplands, 3.31 ($n = 74$) during 1976–95; and Northern Ireland, 3.53 ($n = 237$) during 1984–96. There were no consistent differences between coastal and inland areas. The district with highest breeding density (central Wales) does not have the highest brood size (2.8). Interestingly, two inland districts have shown an increase in brood size during recent decades. In northern England the figure has risen from 2.80 during 1935–75 to 3.11 during 1976–95 and, in the Southern Uplands, from 2.95 during 1925–75 to 3.31 during 1976–95. The difference for northern England was significant, but that for the Southern Uplands only marginally so (Table 18).

Dr Humphrey Crick has analysed the large series of BTO brood-size records for the Raven ($n = 1,758$) spread over 1947–93, though these do not take age differences of young into account and thus may well give figures slightly higher than for fledged broods. The overall mean for the UK over the whole period was 3.10 young, with Northern Ireland showing a significantly higher figure of 3.64 compared with other regions (Table 18). Within the BTO records, there has been a general and significant decline in brood size, which is especially marked in Scotland (from 3.74 in 1953 to 2.85 in 1993) and Wales (from 3.45 in 1946 to 2.93 in 1993) (H. Crick, unpubl.). It is difficult to reconcile these findings with the evidence from Lakeland and the Southern Uplands of *increase* in brood size during recent years.

Marquiss, Newton and Ratcliffe (1978) found that, in the Southern Uplands and Cheviots, brood size was inversely correlated with the extent of afforestation in the vicinity of the nest in 1974 and 1976, but not in 1975. At those nest areas where over 25% of former sheepwalk within 5 km was planted, Ravens reared smaller broods than formerly. In central Wales, Newton, Davis and Davis (1982) detected a reduction in mean brood size from 3.0 to 2.6 as the proportion of closed conifer forest within 1 km of the nest increased up to 60%, but there was no effect when afforestation within 3 km was measured. Nor did brood size maintain the correlation with carrion supplies that clutch size had shown (see p. 168). Allin (1968), in north Wales, noted a suggestive but not statistically valid decrease in mean brood size, from 3.37 to 3.12, when nests below and above 305 m were compared. In central Wales, Davis and Davis (1986) found no variation in mean size of successful broods according to altitude.

Runs of records for particular nesting places in the Lake District showed no obvious tendency for any one haunt to have consistently higher or lower brood size than the general run of figures over periods of 6–12 years. No one place gave more than three consecutive years for large broods of four or more young, or for small broods of two or one. There was thus a lack of evidence for any appreciable and consistent difference in feeding value of breeding territories, or in parental performance, at least as regards brood size (but see p. 195).

Breeding: the young 187

Davis and Davis (1986) found that, in central Wales, mean brood size was significantly lower in 1978, at 2.2, compared with a more constant level of 2.8 to 3.1 in the other years from 1975 to 1980. Year-to-year variations from a minimum of 2.54 to a maximum of 4.0 young were found in Shetland, but the annual samples were too small to show significant differences (Ellis *et al.*, 1994). Years with harsh spring weather can affect fledging success and, in 1994, G. Horne found that 39% of Lakeland pairs lost first clutches or small young through a spell of very cold and wet weather at the end of March and beginning of April. The nesting of many Lakeland pairs again failed during cold, wet weather in April, 1996 (G. Horne).

32. *Brood of full-grown young in a crag-nest. Southern Uplands (photo: Brian Turner)*

In a small series of records from Lakeland and southern Scotland, brood size for known successful repeat layings was smaller, by an average of 0.54 young, than for first clutches for the same districts over the same period, the difference being significant (Table 19). Davis and Davis (1986) found, by contrast, that repeat layings were more successful than first layings by the population as a whole. Of 11 known repeats, nine produced flying young, with a mean brood size of 3.3, compared with 2.8 for first layings; however, the sample is too small to allow firm conclusions.

BREEDING PERFORMANCE

Breeding performance of a population can be measured in different ways. First is the output of progeny from those pairs which manage to rear young, with mean fledged brood size as the comparative index (as in the previous section). It is usually felt more appropriate to relate this output to the number of pairs which attempt to breed (i.e. those which lay eggs), and the proportion of successful pairs multiplied by mean brood size gives an index of breeding success (Table 20, column 9). But, since the breeding areas also usually contain some territory-holding pairs which do not lay, it is perhaps best to regard the proportion of all territorial pairs which rear young as the more crucial parameter. When this last figure is multiplied by mean brood size, it gives an index of productivity for the territorial population (Table 20, column 10).

CAUSES OF COMPLETE BREEDING FAILURE

I have dealt above with causes of partial clutch or brood reduction, in which infertility and food supply are major factors. It is unusual for whole clutches to fail to hatch, or whole broods to starve, and total failure of nests is usually a result of persecution. Egg-collecting has long been the major cause, but adults and/or young are sometimes killed by unfriendly shepherds, farmers and keepers. They may be poisoned, either deliberately or accidentally. Occasional nests fail through chance casual disturbance, as from campers, rock-climbers and hang-gliders, and fires running out of control. Failures through natural causes are rare, but occasionally happen through extreme weather and rock falls. Predation by foxes and possibly other ground mammals accounts for a few nests, and parents are occasionally killed by Peregrines and Golden Eagles. On the sea-cliffs, Fulmars are locally a fairly new hazard to nesting Ravens. Very little is known about the role of disease and parasites in the Raven.

EGG-COLLECTING AND ITS EFFECTS ON BREEDING PERFORMANCE

The proportion of egg-laying pairs that manage to rear young formerly varied considerably between districts. In the remote mountain regions of the Highlands and those rocky coasts with a prevalence of high sea-cliffs, probably a majority of pairs has always succeeded in bringing off broods safely. In some other districts, breeding success was consistently much lower; sometimes the nesting birds or their young were killed, but much more often the taking of eggs was responsible. Indeed, probably the commonest cause of breeding failure, once eggs have been laid, has long been egg-collecting. Certain areas suffered from an almost systematic 'farming' of nests annually, but in others egg-taking was more haphazard.

Despite the difficulty of reaching many eyries, the Lake District was one of the worst affected areas, even as far back as 1900. Abraham (1919) said that, 'A certain well-known collector who visits the Ullswater area every year has almost completed his hundred clutches of Raven's eggs'. During much of the period up to 1939, there were several collectors living in the district, who between them shared out the nests across the whole area. Many eyries were annually robbed of eggs, and often of repeat clutches as well; or eggs were taken so late in incubation that no repeats were laid. Some nests were pillaged relentlessly over a long period, though probably some in the remotest and most inaccessible crags always succeeded in fledging young. No precise figures are available, but probably the success of pairs which laid eggs was well below 50% in some years.

Similar treatment was meted out to the species in parts of Wales, southwest England, the Pennines, the Cheviots and the Southern Uplands. Anywhere that Ravens nested in small crags or low trees, where their eyries were easy to reach, they were especially subject to a frequent plundering by eggers, even if only the local boys. Egging continued with little abatement after World War 2, and may even have become intensified in some areas. In the Southern Uplands, out of 264 nestings that I recorded during 1946–69, at least 117 first clutches (44%) were lost, all but six to egg-collectors, though at least 97 clutches (37%) hatched. Other nests seen with eggs but not followed up almost certainly suffered further robberies, but some of the repeats (at least 37 laid, but at least 12 robbed) could also have hatched. Of the six other first-clutch failures, four were caused by different kinds of human interference and two by eviction by Peregrines.

By the 1970s, egg-collecting in Lakeland had declined appreciably, though it was still the commonest cause of nest failure. Geoff Horne has provided me with data from 1974–95, in which an average sample of 30 pairs was examined annually in this district. Mean breeding success for pairs known to lay eggs was 73%, including broods from repeat layings, and ranged from 54 to 88% in different years. Frequency of non-breeding territory-holders was not recorded, so that productivity has to be given for the breeders (Table 20). Most failures were attributed to egg-collecting, which still continues, albeit in a less systematic fashion than hitherto, and mainly by people coming from north-east England and Yorkshire. At least 15 nests were robbed in 1993. In the Southern Uplands, egging is nowadays also much reduced, so that breeding performance has improved markedly. In 1989 and 1994, two surveys collated by Chris Rollie showed 77% nesting success and a productivity for territorial pairs of 1.8 young (Table 20).

In Table 11 I have analysed the records of nesting success for the eight Lakeland and Southern Upland territories where there is a long run of observations since 1945. Despite the subjective impression that all these territories had suffered a good deal of robbery over the years, in only one of them was overall success of known outcomes less than 50%. Sometimes, repeat layings were successful; the figures are minimal because a number of repeat nestings were not followed through. These territories had relatively accessible nest sites, and the success rate of the more difficult nesting places in these districts is likely to have been still higher.

The most important question is what effect egg-taking has had upon Raven populations. Breeding success was undoubtedly depressed substantially during certain periods in the worst-affected districts, though much depended on the frequency of repeat clutches and their success rate. The remarkable stability of the breeding population, even in the most heavily robbed districts, combined with a frequent surplus of non-breeding birds in flocks, suggests that there was no discernible effect on the established

33. *Head of recently fledged young Raven, showing sleek throat plumage and nasal bristles. Southern Uplands (photo: Brian Turner)*

breeding levels. It may still be the case that Ravens were held back from expanding their range into adjoining districts, where they were few or absent, by a lack of recruits from the stronghold districts. More probably, though, such recolonisation of lost ground was prevented largely by serious problems there which greeted any new arrivals (see p. 215). Egg-collecting probably had no adverse effects on mortality of breeders, and may even have reduced the risks to them by lessening their vulnerability to destruction while at the nest.

The main result of egg-collecting thus seems to have been its nuisance effect – the spoiling of the pleasure that Raven-watchers enjoy by following the bird through its nesting cycle, and the frustrating of scientific observations on Raven biology, including ringing studies. Sadly, causing annoyance and disappointment to others seems with some eggers to be as impelling a reason for robbing nests as the actual possession of the eggs.

REGIONAL VARIATIONS IN PRODUCTIVITY

In Wales, Snowdonia was less popular as an eggers' stamping ground, though a few of the more accessible nests used to be pillaged. During

1951–79, I looked at 58 territories in late April and May: in five the birds were present but with no evidence of nesting, ten failed (eight robbed, two shot), and 43 (81% of laying pairs) had young. I was able to follow up the eventual outcome at only a few nests, but believe that most of those with young would succeed. Mean brood size for nests with large young was only 2.27, giving a productivity of 1.68. At a mixture of coastal and inland eyries in north Wales during 1946–67, Allin (1968) found 80% were successful, and productivity was high. For the later period 1978–85 in Snowdonia, Dare (1986a) found a probable success rate for pairs laying eggs of 73%, with productivity closer to my earlier figure (Table 20). Exactly the same figures were recorded for a smaller sample of pairs on the lower moorlands and enclosed hill farms in Denbighshire immediately to the east.

In central Wales, egg-collecting had declined by the period of the study (1975–80) reported by Davis and Davis (1986), but was by no means finished. An overall total of 67% of nestings were successful in rearing young (nine at the second attempt) (Table 20). Of the total of 97 failed nests, 18 were attributed to egging, while another 24 resulted from deliberate destruction by shepherds and others. Cause of failure was unknown in most of the rest, though some kind of human disturbance was usually suspected. In their resurvey of this population in 1986, Cross and Davis (1986) found that 65% of laying pairs reared young (three from repeats). Cross

34. *Young Raven overhead; the emargination of the primaries is less marked than in the adults. (photo: Tony Cross)*

(unpubl.), covering a still larger area of central Wales, recorded an overall success of 73% for nesting pairs in 1992–93. Productivity was similar to the north Wales levels.

In a study of Shetland Ravens in 1982–84, Ewins, Dymond and Marquiss (1986) found that, in a sample scattered over the islands, only 52% of breeding attempts were successful and productivity was the lowest of the figures in Table 20. In a later Shetland survey, Ellis *et al.* (1994) examined a non-random sample of 43 territories, representing 21% of the known total, in every year from 1987–93. For the laying pairs, successful breeding averaged 58% (range 47–69%); and productivity was 1.62 young per territory-holding pair. These Shetland results show the anomalous position that one of the densest British Raven populations is also one of the least successful, though large brood size raises productivity to a respectable level. Egg-collecting is ruled out as a significant factor for the species there. In the earlier study, Ewins *et al.* reported that, for the 37 of the 64 failures in which the cause was known, 21 resulted from human persecution, 14 from interference by Fulmars, and two from infertility. Ellis *et al.* found that persecution certainly occurred at five of the 43 territories and was suspected at several others but, of 127 nest failures, only six occurred after the eggs had hatched and only three of these through human interference. Much of the Shetland failure thus occurred at the egg stage but was unexplained and mysterious.

In some areas where the Raven suffers little or no interference in its nesting, it can achieve a high rate of nesting success. In looking at Peregrines and Golden Eagles in the western Highlands during 1952–71, I also recorded Raven nests that I came across. In 102 nests, two had eggs, 92 had young, and only eight had failed. It is possible that, as I was not particularly looking for Ravens, and as most of the observations were in April and May, I may have missed other nests that had failed at the egg stage. Nevertheless, it seemed likely that around 90% of clutches had hatched and that most of these broods would fly. Sixty-one broods were either on the point of fledging or had actually left the nest. Thomas (1993) recorded that, in mid-Argyll, 83% of 86 nesting attempts produced one or more fledged young. In Northern Ireland, high breeding success (85%) combines with large brood size to give unusually high productivity (Wells, 1996; see Table 20).

In other parts of the world where the Raven is little molested, it can achieve a high breeding success. The 20 or so pairs along the north shore of the Varanger Fjord in Arctic Norway appear to reach around 95% success annually. They are treated kindly there and persecution is unusual. In Iceland, even under an official programme of systematic Raven-culling, the heavy losses were mainly of non-breeding flock birds, and the breeders still achieved an astonishing overall mean of 83% success in three study areas (Skarphedinsson *et al.*, 1990).

It is only to be expected that Raven territories in any one district will vary in quality, especially as regards feeding value, according to a number of local factors, such as topography, climate, geology and soil, and management practices. Some will tend to be consistently better than others year by year, and these differences should be reflected in Raven breeding performance. There is so far rather little evidence that this theoretical picture

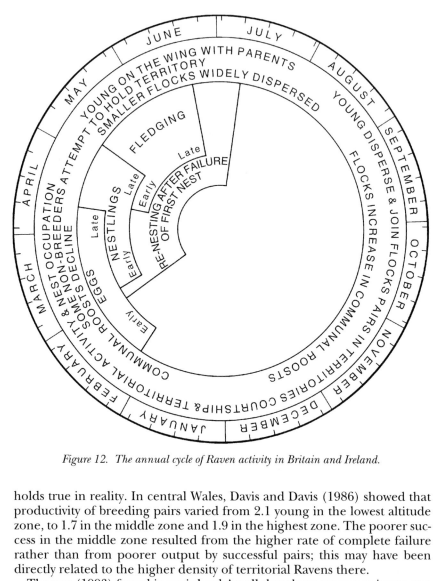

Figure 12. The annual cycle of Raven activity in Britain and Ireland.

holds true in reality. In central Wales, Davis and Davis (1986) showed that productivity of breeding pairs varied from 2.1 young in the lowest altitude zone, to 1.7 in the middle zone and 1.9 in the highest zone. The poorer success in the middle zone resulted from the higher rate of complete failure rather than from poorer output by successful pairs; this may have been directly related to the higher density of territorial Ravens there.

Thomas (1993) found in mainland Argyll that there were consistent correlations between habitat differences and both laying date and clutch size, which allowed predictions about these breeding parameters to be made. Habitat differences did not appear to influence whether breeding was attempted in occupied territories, nor the size of fledged broods, but there was an effect on success of nestings. Breeding attempts tended to be more successful near the coast and at lower altitudes, and as the extent of grassland within the home range increased, whereas success fell with increasing amounts of land under forest thicket or ling heather bog.

As the best evidence for this effect, King (1995) found that, for a group of 12 Raven pairs in the southern Lake District, territory quality was the main factor influencing nesting success within a single year. Quality was measured as territory size and nearest-neighbour distance (the larger the better), extent of wet heath, and food abundance (largely the availability of sheep carrion), while breeding success was measured as the rearing of young (or failure) and fledged brood size in individual territories. Until there is more information about the feeding ranges of neighbouring pairs of Ravens, it is, however, only an assumption that apparent differences in territory size within a single district are real and meaningful. Local differences in nearest-neighbour distance – which suggest territory-size differences – are more likely to reflect irregular distribution of suitable nesting places. Since neighbouring Ravens are believed to overlap in feeding range, their territories may not be exclusive and they may be able to equalise some of the local differences in food supply. While it is perhaps better to think of these areas as home ranges rather than territories, until more precise observations and measurements are available, the matter remains hypothetical.

Where egg-collecting is heavy, there can be marked differences in productivity between individual territories in the same district over time. These reflect variations in quality of territories in regard to nest-site security, itself largely a matter of the height and steepness of nesting cliffs, and their remoteness, or the accessibility of tree-nests. Thomas (1993) found that this connection did not hold in Argyll, where Ravens were little disturbed at the nest by humans. Determined persecution by keepers and hostile farmers, especially by use of poison, can equally make nest security irrelevant.

Annual differences in productivity for the same area have not yet proved to be significant. Davis and Davis (1986) found a range of 1.5–2.5 (mean 1.9) young for all breeding pairs in central Wales during 1975–80, while Ellis *et al.* (1994) found a range of 1.2–2.1 (mean 1.6) in Shetland during 1984–93.

The importance of productivity as a factor in population dynamics is discussed in the following Chapter.

It is possible that Ravens bred in captivity in earlier times, such as during the Roman era, but there appear to be no records. On Lord Lilford's estate in Northamptonshire in 1894, a tame pair were allowed their freedom and reared four young in a beech tree-near the house. The species seems to be somewhat temperamental and, although breeding is often initiated by paired captive birds, it often ends in failure.

CHAPTER 11

Territorialism and population regulation

After I had been looking at nesting Ravens for some time, I was struck by their constancy in numbers, year by year. My first observations were in Lakeland and the Southern Uplands but, when I went to Snowdonia, the same appeared to hold good there too. Other observers had noted this stability of breeding population as well and sometimes their records went back for many years. Every year the bird was famously faithful to its breeding haunts, so that numbers could be checked readily and showed a near-constancy that became taken for granted. It was the rule for nesting places to be regularly occupied and eggs laid, and only a small number showed breaks in tenancy or breeding for more than the odd year, or were occupied only once in a while. Populations in different districts seemed to be held at some ceiling, which they neither rose above nor fell below, except by only minor degrees. This stability in numbers was also true of the Peregrines which shared so many of the Ravens' nesting crags. Macgillivray (1837) had long before made the perceptive comment (see p. 98) that Raven breeding numbers did not change, despite good output of young over a long period, and adjoining pairs remained the same distance apart. So, although Raven populations across the country had clearly been much reduced during the nineteenth century, in some of the wilder districts where they had survived, their breeding numbers had become relatively stable. This status was intriguing in the light of other noticeable features, namely, that:

1. Some of these districts had winter flocks of Ravens which were additional to the established territory-holders, and smaller flocks or scattered pairs of non-breeding Ravens were often around during the spring when the settled pairs were nesting. There was also evidence that Ravens which lost their mates at nesting time often remated quickly. Clearly, it was not a shortage of birds which limited the breeding population. It also appeared that differences in breeding performance between districts had no effect on this stability in breeding population. Whether the output of young was low, as in parts of the Southern Uplands, or high, as in Snowdonia, had no effect on the number of nesting pairs in each succeeding year.

2. In the more rugged uplands, such as Snowdonia and Lakeland, there were plenty of high and apparently suitable cliffs that were not occupied by nesting Ravens. Moreover, at different times, many pairs used two or more quite separate cliffs as alternative nesting places. In any one year, a good many known nesting cliffs were thus untenanted. The evident ceiling on numbers could not therefore result from limitation by lack of suitable nesting places in these favourable districts, though there was a clear limit to numbers of cliff-breeders in other areas where suitable rocks were scarce, such as the Pennines, Cheviots and some parts of the Southern Uplands and eastern Highlands. Considering 1 and 2 together, although there was evidently scope for breeding numbers to increase in some districts, they did not do so, and a surplus of potential nesters was held within a non-breeding sector.

3. The other striking feature was the regularity of spacing of nesting pairs in the more rugged districts, where there was a fairly free choice of suitable nesting cliffs. Contiguous breeding pairs spaced themselves out at roughly the same distances from their neighbours, giving a relatively even distribution pattern overall (see Fig. 13). Any marked departure from randomness in distribution has to have a reason. An even dispersion of active nests is a characteristic of strongly territorial species that repel others of their kind who come within a certain minimum distance. Each pair thereby establishes dominion over an exclusive area, sometimes with observable defended boundaries approximately half-way towards its various neighbours, and the whole district occupied by the breeding population becomes parcelled out into such territories. This is true of predators such as the Common Buzzard (Picozzi & Weir, 1976) and Short-eared Owl (Lockie, 1955), and of many solitary nesting birds in general (Howard, 1920).

198 *Territorialism and population regulation*

4. Although the occasional nesting bird or pair was killed by sheep-farmers or gamekeepers, in the districts with good Raven numbers there was no evidence of widespread persecution, which could produce heavy mortality among the established breeders and thus a heavy demand for replacements of breeding age in the population. From the evidence of distinctive eggs, it seemed that many females, at least, survived for several years.

Figure 13. Breeding dispersion of Ravens in the Lake District.
NOTES: Alternative nesting places within a territory may be joined.

SURPLUS NON-BREEDING POPULATIONS

The evidence for floating groups of non-breeding Ravens and their distribution is given in Chapter 6, though they are most obvious outside the nesting season, when aggregation into flocks is most marked. Non-breeding populations are present in most inland districts, albeit in varying size, and are least obvious in Snowdonia, southern Scotland and Argyll. In Shetland, Ewins, Dymond and Marquiss (1986) found nearly 400 birds occupying territories, and probably an equal number of flock birds, in the early 1980s. The tendency for many flock birds to be paired suggests even more that they represent a surplus of potential nesters which could enlarge the breeding population if circumstances allowed. If Ravens do not breed until the age of 3 years (see p. 215), it is possible that these flocks are composed largely, or even entirely, of birds below this age, so that they are not immediately capable of nesting. In that case, the appearance of a potential surplus might, to some extent, be illusory. Such a situation nevertheless seems improbable. While yearling birds might be expected to be the most numerous age-class among the non-breeders, it is highly likely that birds up

to at least several years old will be represented as well. There is some evidence for this (see p. 103).

In some districts, the non-breeders become well dispersed during the breeding season. Some roam around in small parties, while others detach themselves from the winter groups as pairs or single birds. Davis and Davis (1986) found that their study area in central Wales held a large population of non-breeders in addition to the territorial pairs, and perhaps equal to them in numbers. These additional Ravens tended to frequent places lacking nesting pairs. Sometimes, such extra individuals or pairs stay loosely attached to certain areas, or to cliffs or trees, and they may attempt to set up territories of their own or even build nests. Perhaps some of them are biding their time and awaiting their chance of mating with a widowed bird among the settled breeders.

Some of these spare potential breeders probably do not hang around awaiting the chance of filling gaps in an established population, but make their way into apparently suitable areas where there are no settled Ravens to keep them out and there attempt to set up a new nesting territory of their own. When these attempts succeed, the breeding population expands, but the lack of expansion into many such areas suggests that they must often fail. Some of the eastern grouse-moors appear in fact to be veritable traps, draining away much of the surplus population through the high mortality afflicting birds which reach these areas. Again, as with Peregrines, the attractions of certain cliffs for prospecting nesters can prove to exert a 'sump effect' that causes a high wastage of new birds.

RAPID REMATING

The belief that floating populations of Ravens in the nesting season consist of, or contain, a surplus of potential breeders, is supported by the evidence for the rapid remating of the survivor if one of a breeding pair dies. This trait is well known in the Peregrine (Ratcliffe, 1993), but appears also to be well developed in the Raven. Macgillivray (1837) long ago remarked upon it from his observations in the Outer Hebrides: 'When a Raven has lost his mate during this [breeding] season, even after the young ones are far advanced, he is observed to procure a stepmother for them with great celerity'. Macpherson (1892) quoted a story by Dr A. C. Gibson about the Ravens of Kernal Crag in the Coniston Fells of Lakeland:

> On this crag, probably for ages, a pair of Ravens have annually had their nest, and though their young have again and again been destroyed by the shepherds, they always return to this favourite spot; and frequently when one of the parents has been shot in the brooding season, the survivor has immediately been provided with another helpmate . . . It happened, a year or two since, that both the parent birds were shot while the nest was full of unfledged young, and their duties were immediately undertaken by a couple of strange Ravens, who attended assiduously to the wants of the orphan brood, until they were fit to forage for themselves.

Morris (1895) described how the widow of a male Raven, shot in the morning, had, by 15.00 h, found a new mate who tended the young as though they were his own; but in another instance, where the female was destroyed, a new 'step-mother' attacked the young and tipped them out of the nest. Newton (1864–1902) found that, when the male of a pair of Ravens in the Breckland was killed in 1854, the sitting female deserted her nest, but remated, built another nest and reared young during the same season. In Cornwall, Ryves (1948) recorded that, when the male of a pair of Ravens with young was found poisoned, the female continued feeding the young alone but, by the third day, a new partner was present and helping to feed the nestlings. The brood safely fledged and the new mate remained with the family just as a genuine male parent would do. On 6 April 1938, P. Meeson was taken to an Argyll nest by a stalker who had shot the incubating female shortly before, in March, and a new female flew from the same nest, which contained erythristic eggs as the second clutch. On 10 April 1958, I found a female Raven dead, evidently poisoned, on its nest of eggs on a Lakeland crag. It was difficult to tell how long it had been there – possibly up to 2 weeks – but a pair of Ravens was croaking overhead, the male evidently having remated, though they appeared to have forgone any further attempt at nesting that year.

Rapid remating in the Raven is also reported from North America. Stiehl (1985) found that, in Oregon, after a female was found shot near the nest and the 16-day-old brood destroyed, the male remated and, using the same nest, the new pair successfully reared a brood.

I cannot do better than round off this account by giving two quotations. The first is from Gilbert and Brook (1924):

> They pair for life, and are exemplary spouses and very good parents indeed. If one of the old birds is shot or dies there is no nonsense about going into mourning. The survivor gets a mate, and it is remarkable how soon a new lover is found; but once found they seem to be good lovers, constant and faithful unto death.

The second reference is from Bolam (1912):

> The well-known ability of such rare birds as the Raven to find another sharer of their joys and woes, at very short notice, when one of a nesting pair happens to meet with an untimely end, must always remain a matter for marvel to those able to appreciate the fact, and has more than once been touched upon in verse. With one such quotation this chapter may be closed:

> The Corby, up in his lonely cliff, is crouping, in sorrow, his lane;
> For his matie's lying stark and stiff, doon-by, at the wetter-gate stane;
> But brawly he kens he manna greet, fu' lang o'er her sorrowfu' fate,
> But maun get a new wife to tak' her seat on the eggs, or it's ower late:
> An' the morrow's morn, ere peep o'licht, to the distant hills he'll ha' flown,
> An' be back hame afore the nicht, wi' anither to ca' his own.

Rapid remating of uncommon birds may be a marvel to the human mind, but is perhaps fairly straightforward to the birds themselves. Ravens are no doubt well aware of where others of their kind are to be found, even if they are not in the immediate vicinity. Non-breeders, whether in parties, pairs or alone, are often not far away from the nesting pairs and probably on the look-out for opportunities of this kind. How the whirlwind courtship takes place, with what differences between the sexes, and whether birds are stolen by splitting up non-breeding pairs is a matter for conjecture. There must be obvious advantage in non-breeders readily making themselves available, even if this first involves assuming the domestic duties of their predecessors, since their place in a breeding territory will be assured for the next year and they are potentially long-lived birds. Reproductively immature birds might also gain entry to the territorial population in this way.

The significance of rapid remating as a population feature is that it strongly suggests that the non-breeders must await their turn to breed until there are gaps for them to fill among the established nesters. This implies in turn that the size of the breeding population is not limited by lack of recruits, since there is an evident surplus in waiting, but that numbers are held to a ceiling by the behaviour of the established breeders themselves.

THE CEILING ON NUMBERS

Relative stability of Raven breeding populations is reported for many districts, though it is seldom that all known territories are occupied in any one year. In central Wales, Davis and Davis (1986) recorded from 88 to 100% occupation of 70 territories during 1975–80. Another seven previously known territories in their study area remained vacant during this time. A resurvey in 1986 showed 61 (85%) of these territories to be occupied (Cross & Davis, 1986). In 1992–94, survey of a wider area gave 329 out of 345 territories (95%) occupied (A. V. Cross). In Lakeland, during 1945–62, 248 visits to 65 different territories showed 225 occupations (91% overall); the largest number of desertions was four out of 33 in 1957 and four out of 37 in 1959. Earlier records (1900–1959) for this district showed that, out of a total of 72 territories, 60 were regularly used, suggesting that the maximum scope for fluctuation was only 9% around a mean of 66. Since not all the irregularly used territories would be likely to be deserted in any one year, actual population fluctuation was probably less than this. In the Southern Uplands, 265 visits to 40 different territories during 1946–62 gave 235 occupations (89% overall). Nine of the 40 territories were deserted in at least one year, but five out of 23 (78% occupation) was the most recorded in one year (1961), pointing to a maximum fluctuation of 11% below the mean. Ellis *et al.* (1994) found that, in Shetland, an average annual sample of 41 territories (out of a known total of 206) showed 67–93% occupancy during 1984–93, a fluctuation of 16% around the mean of 80 – more than in other districts. (See also Table 20.)

In any large Raven population, and in any one year, there is usually a small proportion of territory-holding pairs which does not lay eggs. Some of

35. *Demonstrating Raven about to land. Wales (photo: Tony Cross)*

these may build nests in known breeding places but often leave them uncompleted. Most of such pairs stay around throughout the spring. Ellis *et al.* (1994) found that, on Shetland, an average of 92% of territorial pairs laid eggs during 1987–93, there being no significant differences between years. In central Wales, Davis and Davis (1986) recorded that an average of 88% of territorial pairs laid eggs during 1975–80, with variations from 83 to 92% between years. They thought it likely that most of the non-layers were mature birds, unable to lay in some years through food shortage, social pressures or other problems. Nest-building by them was frequent; some pairs repaired two or three nests, and one as many as five. There were eight non-breeding instances in 1978, the least successful year, when food was scarce. In the same study area, Cross and Davis (1986) found that at least 52 out of 61 (85%) territorial pairs laid in 1986; over a wider area of central Wales, A. V. Cross reported 319 out of 329 (97%) territorial pairs laying during 1992–94. Non-laying through old age of the female, noted by Harlow (1922), seemed to be only an occasional possible factor in central Wales.

Even territories that have had regularly nesting pairs for many years will occasionally have a blank season when the resident pair appears not to lay. It is always difficult in such cases to rule out failure of an unseen nest, or breeding in an undiscovered alternative locality, but some are undoubtedly genuine instances of non-breeding years. Breeding is usually resumed again, as normal, in the following year.

Now and then, a new pair of Ravens will occupy a nesting place between two regular pairs, breed for a year or two, or even several, and then drop out again but perhaps stay around as non-breeders. Sometimes, an extra pair, or a single bird, will appear about an alternative cliff when the territory-holders have an active nest elsewhere, or they may frequent crags where breeding is previously unknown. These apparent interlopers are loosely attached to such places and seldom make attempts at nest-building. Such sporadic or incomplete nestings and occupations seem to represent an intermediate stage between the itinerant non-breeding stock and the resident breeding population – a kind of tension zone in which subordinate would-be incomers are held at bay or squeezed out again by the more dominant well-established nesters. The situation points to an active exclusion process that holds breeding numbers to a fairly well-defined ceiling.

TERRITORIALISM AND VARIATIONS IN BREEDING DENSITY

The evidence that the Raven is a highly territorial bird in its nesting behaviour is discussed in Chapter 5. While it is less clear whether Ravens actively defend the whole of the area around the nest, to the point of establishing boundaries with neighbours, the regularity of nest-spacing, when suitable breeding places are freely available, suggests that these areas are justifiably regarded as nesting territories. Territoriality in nesting dispersion thus appears to be the immediate process limiting breeding density and, hence, overall numbers in these districts (Ratcliffe, 1962). To put it another way, territorial behaviour establishes the ceiling in the breeding population within an area and prevents further increase above this. Gothe (1961), studying a re-establishing Raven population in beechwood habitat in the Mecklenburg area of northern Germany, came to much the same conclusion about the role of territorial spacing in limiting breeding numbers.

Territorialism has evolved because it serves some further purpose. There has to be some advantage to individual birds in the regular spacing of nesting pairs and in the particular ceiling on numbers which this achieves. This advantage may be deduced from the related circumstances of the species' lives. Eliot Howard (1920), who first propounded the theory of territory comprehensively, connected it with food and suggested that an exclusive area for each pair of birds helped to safeguard an adequate food supply. In the Raven, there is no evidence that territories centred on the nest are maintained as exclusive feeding areas, and the indications are rather that nesting Ravens overlap a good deal in their foraging ranges (see p. 93). There is, nevertheless, good reason to connect the average size of Raven territories within a largish district with the scale of local food supply. The main pointer is that breeding density and, hence, spacing distance and territory size, vary in close parallel with the availability of food.

This simple correlation is clear from a comparison between the deer forests of the western Highlands and the sheepwalks of other hill districts. Within Scotland, breeding density decreases from south to north in inland districts. In parts of the Southern Uplands, density was moderate up to 1970, but even towards the southern fringes of the Highlands it has remained quite low (Table 22). In Wester Ross and Sutherland, despite the frequency of suitable cliffs, mean nearest neighbour distance and breeding-territory size are 6.1 km and 61.9 km^2, and 7.2 km and 87.3 km^2, respectively. Both districts are barren deer-forest country, though usually with some sheep as well. Although carrion is often plentiful, the Raven has a serious competitor here in the Golden Eagle, which is widespread and as numerous as the Raven (Fig. 8). Contrasting with this situation are the northern island groups of Orkney and Shetland, where there are no Golden Eagles and most Ravens are in coastal habitats, though able to forage inland as well. Shetland has one of the highest Raven breeding densities in Britain, at least 150 pairs averaging about 10 km^2 of land each (Ellis et al., 1994), and levels on Orkney are almost as high (Table 22).

An example of the connection between food supply and breeding density is found within the Shetland Isles, where Ewins, Dymond and Marquiss (1986) state that 12 nests within 5 km of a large garbage dump had an average nearest-neighbour distance of 1.8 km, whereas 124 nests scattered over the rather less productive generality of Shetland had the significantly lower value of 2.5 km. The high densities over the Shetland and Orkney groups are explained rather differently. Shetland has extensive and mostly infertile moorlands which carry large sheep stocks, with a fairly high mortality, and so there is a plentiful supply of carrion. On Orkney, cattle-rearing on enclosed grassland predominates and sheep are mainly on improved pasture. The moorland has comparatively few of these animals, and so large mammal carrion is scarce, but there is a great abundance of rabbits and hares. Both island groups have huge sea-bird colonies that must increase the bounty of the coast.

Over much of the West Country, the Welsh hills, Lakeland and parts of the Southern Uplands, breeding density is very high or high (Table 22). These are districts managed largely for sheep, where carrion mutton has long been plentiful, especially at lambing time. Central Wales has the densest Raven population of any inland part of these islands and this is the more remarkable since it also has one of the highest densities of nesting Buzzards (Newton, Davis & Davis, 1982; Davis & Davis, 1986). The explanation given is that the rather mild upland climate and relatively good productivity of the pasture allows large numbers of sheep to be overwintered on the open hill, but this also results in a fair mortality and carrion supplies well scattered over the range. Newton et al. estimated an average of 190 kg/km^2 of sheep and lamb carrion per annum, mostly in late winter and early spring, in their study area. Ravens had to compete with foxes, dogs, Buzzards and Kites for this food source, so that only a part of it was available to them. Some breeding pairs may also have benefited from the proximity of rubbish dumps and abattoirs. The relatively large extent of improved hill ground also gives high numbers of moles, earthworms and insects. The warmer climate, compared with regions farther north, may also benefit the birds more directly.

Perhaps the most telling piece of evidence to connect Raven breeding density with food supply is that, eventually, some populations lost their stability, but in parallel with change in sheep numbers. As afforestation spread extensively across the Borders and Southern Uplands, and the sheep were removed from the planted areas, so Raven nesting pairs steadily dropped out (Marquiss, Newton & Ratcliffe, 1978). By 1981, across the whole region, at least 39 out of an original 120–125 pairs (during 1946–60) were believed to be lost to afforestation (Mearns, 1983). At least eight former regularly nesting pairs also stayed around as non-breeders. A few displaced pairs have since returned, but there has been a permanent reduction in breeding density in Galloway, where numbers were once the highest in the entire region. Here, it seems that the squeezing-out process described on p. 204 has eliminated pairs, at least as breeders, as food supply deteriorated. In the heavily afforested Cheviots, the Raven hovers on extinction. In central Wales, Davis and Davis (1986) thought that persistent non-breeding by a few pairs could be associated with high levels of afforestation in the surrounding area, though there was no evidence of population decline. The details of these afforestation effects are discussed in Chapter 12.

Conversely, Dare (1986a) found that Raven breeding numbers increased in Snowdonia by at least 40% – perhaps considerably more locally – between 1951–61 and 1978–85, evidently in response to increased stocking of sheep on the hills there. While there have been large increases in Raven population in many parts of Wales during the present century, these have been mainly a response to decrease in persecution. The situation in Snowdonia has been different, for here numbers were quite high, but evidently stable and appearing to be at saturation, during the 1950s. The increase has involved an increase in density, from 5.5 to 9.5 pairs per 100 km^2, and contraction of nearest-neighbour distance from 3.15 to 2.00 km (Table 22). Snowdonia nowadays has a Raven breeding density almost as high as central Wales. This is a more rugged district, with mostly less productive pastures, and it appears that a high sheep-stocking rate produces high mortality and abundant carrion (see Chapter 12).

Marquiss, Newton and Ratcliffe (1978) found that Raven breeding density was correlated with altitude in the Southern Uplands and Cheviots, but this may have been because cliffs are most numerous in the higher hills and availability of suitable nesting places ceases to be a limiting factor there. Moreover, the greater the number of cliffs, the larger the number of accidental sheep deaths and, hence, the carrion supply. Some of the lofty cliffs of the more rugged mountains are veritable death traps for sheep – I once counted nine recently fallen beneath one Snowdonian precipice – and this creates an extra element of mortality in comparison with the gently contoured uplands.

The Lake District is very similar to Snowdonia ecologically, and once had much the same Raven density, but has shown a much lesser increase during the last 25 years – by about 15 pairs during 1980–93 over a previous average level of 69 pairs (22%). This suggests that either sheep stocks in Lakeland have not increased to the same extent as in Snowdonia or improved management has produced a greater reduction in mortality – or both.

In both Snowdonia and Lakeland, the increases have taken place in two different ways. Some new pairs have appeared around the edges of both districts, or in areas where there had seemed previously to have been gaps in occurrence. Sometimes these involved occupation of quarries or small crags that had earlier been ignored or used only occasionally. Such increase is an expansion of breeding *distribution*. The other manner of increase has been by the successful intrusion of second pairs into territories once held by single pairs, involving a splitting into two smaller territories. This has happened at least 12 times in Snowdonia and six times in Lakeland, and is the most convincing evidence of an increase in breeding *density*.

The high breeding density in parts of the West Country is intriguing. That on the rugged Atlantic coast of Devon and Cornwall, reported in 1932, has been maintained, and occurs also on parts of the south Devon coast; it evidently reflects a permanently productive coastal habitat. In some inland parts of Devon, nesting Ravens have since increased to high density (*c.* 10 pairs/100 km^2, G. Kaczanow), mostly nesting in trees or in valleys draining from the uplands of Dartmoor and Exmoor, but also in rather ordinary rolling farmland country devoted largely to stock-rearing on enclosed pastures (P. J. Dare). The feeding habits of these birds have not been studied closely, but Dare notes that they forage little in the valley bottoms and resort mainly to higher, more open farmland along the numerous hill crests and on uncultivated ground where sheep carcasses are often available.

Table 22 gives figures for breeding density in various districts of Britain and Ireland, together with a few for regions overseas.

Several districts have shown examples of unusually close nesting, with simultaneously occupied nests only 350–900 m apart (Table 23), but these seldom persist for long or involve more than two pairs at a time. Such relaxation of normal spacing distances seldom appears to be sustainable. Yet there are some remarkable instances of continued high-density breeding on certain small offshore islands. On Lundy Island in the Bristol Channel, two to four pairs of Ravens usually breed, but in 1966 and 1967 there were an incredible nine pairs. The last figure averages a land area of 0.47 km^2 per pair or, if scaled up, is equivalent to 213 pairs per 100 km^2! Even the more usual density is quite high. The Pembrokeshire islands support 12 breeding pairs of Ravens at 0.58 km^2 per pair, another extremely high density. These and a few other examples are detailed in Table 23. Such very high densities occur over only quite limited areas. When two nests are very close together, they are usually out of sight from each other.

Some of these islands are close enough to adjoining mainland or bigger islands for Ravens to fly there to forage, but Lundy, Fair Isle and St Kilda are respectively 18, 38 and 64 km from land. Their common feature is that they are important sea-bird breeding haunts, except perhaps for Lundy, Colonsay and Clear Island. Most of them also lack foxes as competitors. It thus appears that, under conditions of unusually copious food supply, territorialism almost breaks down and Ravens become extremely tolerant to the close proximity of their neighbours. Land area also becomes somewhat irrelevant under these conditions, with a regular breeding pair on the tiny island of Grassholm, which covers barely 10 ha but holds a huge colony of Gannets. Gannetries usually have numbers of dead Gannets lying below their nesting ledges, offering an abundance of food (photograph, p. 38).

In other countries there is evidently a general correlation between Raven breeding density and food supply. The thin scatter of pairs through much of the boreal forest regions of Fennoscandia matches a rather unproductive habitat, whereas the higher numbers in some coastal districts connects with better feeding opportunities. The highest density recorded on mainland Europe is 18.7 pairs/100 km^2 for Wolgast, in the Baltic area of northern Germany (Sellin, 1987). The extremely high density of 35.6 pairs/100 km^2 reported by Nogales (1994) on El Hierro, one of the smaller islands of the Canaries, is an average for the whole area – 99 pairs in 278 km^2 – in 1987. Nogales commented that territories showed a clumped distribution, with a density of 128 pairs/100 km^2 in one area. Significantly, there were no other Corvidae or carrion-feeders on the island. Raven density on the largest island of the Canaries, Tenerife, was only 3.4–3.9 pairs/100 km^2. Nogales reported that M. N. Kochert had informed him of a still higher average density in the Snake River Canyon, Idaho, USA, with 98 pairs in 135 km^2 (72.6 pairs/100 km^2). This density was, however, calculated by encompassing a narrow band that circumscribes the canyon, to give an area measurement (Michael Kochert). Within the Snake River Canyon, the distribution of nesting Ravens (and various raptor species) is almost linear and semi-colonial, but the birds radiate out to a considerable distance beyond the canyon rim in order to feed.

Correspondence between levels of breeding density and food-value of habitat has been shown for other bird types as diverse as Red Grouse (Moss, 1969), Ptarmigan (Watson, 1965), Dunlin (Holmes, 1970) and Sparrowhawk (Newton, 1986). The match is one that might be expected. It is, however, easy to develop circular reasoning about the relationship between breeding density and food supply, and to overlook the possibility that other factors may, in some areas, play an important part in determining the size of Raven nesting population. Table 22 might suggest, for instance, that food supply in central Wales is almost ten times as high as in Sutherland, but in the latter there is the complication of Golden Eagle competition. Shortages at certain critical times of year may also be important in some districts. Food supply in a bird such as this is extremely difficult to measure accurately and this leads to some fairly large assumptions based on rather superficial evidence.

MECHANISM OF TERRITORIAL SPACING

In trying to explain stability of the breeding population in both Raven and Peregrine (Ratcliffe, 1962), I proposed a modification of Howard's view of the food-value of territory. My suggestion was that territorialism provides a balancing mechanism between breeding density and carrying capacity over a much larger area than a single territory. By limiting the number of breeders, the birds' spacing behaviour imposes a ceiling on the capacity of the population for increase and, hence, also limits the demands that it could make on food supply over the whole area occupied. Within this broad adjustment, other factors provide a finer tuning of breeding performance, according to smaller-scale variations in food supply, either between one territory and the next, or between different years (see Chapter 9).

The logic of this number-limiting mechanism is inescapable, but the still unresolved mystery is how the strength of spacing behaviour has become so closely related to differences in carrying capacity as to produce variations in ceiling of breeding density between districts. I posed the problem over 30 years ago, but understanding is hardly any closer. Does the birds' perception of differences in food supply affect their behaviour – e.g. through change in aggressiveness? Is it a simple reflection of the variable distances they need to range in satisfying their feeding requirements, or are there fixed variations in innate responses that are subject to selection? The evidence given above is that spacing behaviour can change with time, as food supply enlarges or decreases. While it is not certain that the identical birds are always involved, they sometimes appear to be. This suggests that such behaviour is sufficiently flexible to enable individuals to adapt to any alteration in their food supply. Splitting of territories by intrusion of a second pair implies that either the original owners have become more tolerant or newcomers have become more aggressive, or both. Conversely, squeezing out of established pairs from a previously 'saturated' population suggests an increase in territoriality on the part of other pairs, or a greater willingness by some to drop out of the breeding population.

Genetic selection seems unlikely and Newton (1986) has found that, in the Sparrowhawk, where the same question arises, there is free interchange of individuals between adjacent high- and low-density populations. For spacing to be a simple measure of feeding range and frequency of contacts between neighbours would seem to require a degree of exclusiveness in Raven feeding behaviour that is not borne out by observation. I feel intuitively that the individual birds (or pairs) do adjust their spacing behaviour according to their perception of food availability but, until someone either proves or disproves this notion by critical research, it must remain an open question. The adjustment response could be innate and merely the result of natural selection giving the maximum advantage to individuals behaving thus. There is also the possibility that the best territories may be taken by the more dominant or better-quality birds. Heinrich (1990) suggests that the behaviour of non-breeding flock Ravens in alerting their fellows to food sources may be related to mate selection of the most able providers. Until there is more definite evidence on this topic, it is best to say nothing further. This is another field in which circular thinking readily develops.

LIMITATIONS OF UNSUITABLE NESTING HABITAT

Given that there are many districts where the Raven is mainly or exclusively a rock-nester and ignores the numerous opportunities for tree-nesting, its presence as a breeder is severely constrained in areas where suitable nesting crags are scarce. This is the case in parts of the Pennines and in the Cheviots, much of the Southern Uplands, some of the eastern Highlands, and the less rocky ranges of the Irish mountains. Numbers, distribution and breeding density are here all limited at levels far below that which available food supply would seem to allow. In some places, quarrying has resulted in suitable cliffs, which were previously lacking, and so allowed breeding numbers to rise accordingly.

Looked at another way, it is only possible to make the connection between breeding density and food supply in areas where this other aspect of habitat limitation does not operate. Yet identifying the point at which shortage of suitable cliffs begins to reduce the potential density is quite difficult. There are, for instance, parts of the Lake District where nesting Ravens are not quite as thick on the ground as in other parts – but might this not be an effect of locally varying food supply rather than frequency of cliffs?

In inland areas the distribution of suitable cliffs is seldom such that Ravens can spread themselves out in a completely regular pattern. There are nearly always local scarcities that produce wider than average spacing of pairs in some parts of a district, and this may be compensated by clumping of pairs in other parts (see Fig. 13). Dare (1986a) examined this tendency statistically and found that dispersion of Raven pairs was slightly more regular in the Migneint-Hiraethog area of north Wales than in Snowdonia. Where good cliffs are concentrated in certain areas, this can produce a distinct clustering of Raven pairs, which appear to relax their normal territorial spacing locally yet without increasing breeding density overall. Sea-cliff-nesting often leads to a linear distribution of Ravens, and here the average spacing between one pair and the next (nearest-neighbour distance) may be either similar or quite different from that between pairs in adjoining inland areas (Table 22).

When the possibility of nesting in trees is allowed, many other parts of Britain and Ireland could support nesting Ravens. The species has, indeed, shown its capacity for spread in south-west England, Wales, the Southern Uplands and south-east Ireland by adapting to tree-nest sites in many areas. In some places, its use of man-made constructions and quarries gives further scope for establishing new territories. Tree-nesting appears to be an adaptation to reduced persecution that at once allows the potential for an enormous expansion of breeding range, but would favour increase in density only when suitable rock sites were limiting.

Kochert et al. (1984) and Steenhof et al. (1993) report an interesting situation in the western USA, where the towers of a new 500-kV transmission line built across Idaho and Oregon in 1980–81 were rapidly colonised by nesting Ravens. Beginning with one nest in 1981, there were nine in 1982, 39 in 1983 and 55 in 1984. Numbers of nesting pairs appeared to stabilise, at 80–81 pairs, by 1987–1989. Nests varied somewhat in spacing, the closest two being 300 m apart. This chance provision of new nesting habitat enlarged the regional breeding population, by adding many extra pairs which could not otherwise have bred but not at the expense of the established rock-nesters.

MORTALITY, POPULATION TURNOVER AND CAPACITY FOR SPREAD

While the BTO's ringing recoveries record a variety of causes of death, they throw uncertain light on the frequency of these. Most of the rings returned are from Ravens found dead, injured or sick by non-ornithologists, and a good many are from birds that are somewhat decomposed. Some rings are sent in by people who have themselves destroyed the birds and who, in Britain (where the species has been legally protected since 1954, except in Argyll and Skye, where it was not protected until 1981), will tend not to reveal the real cause of death. Moreover, the rings from some illegally killed Ravens are not reported at all. Mortality from shooting, poisoning and trapping is thus likely to be substantially under-reported.

Thirty-three Ravens were found sick or injured and many of these died later or were in unknown condition. A further 20 birds were cared for and later released to the wild, since their carers evidently thought they were well enough to fend for themselves again. Remarkably, 16 of these 20 released Ravens were in Ireland. Their further chances of survival were doubtful and probably it is best to regard all recovered birds, except the seven marked individuals (identified at large but not handled), as lost to the population. One record for an individual reported as recovered twice was also excluded. This gives a total of 525 birds on which to consider causes of mortality. In 325 of the 525 recoveries (62%), 'found dead' is all that is known about the circumstances. Another 94 (18%) were described as shot or possibly shot, and this was the commonest known cause of death. Other named causes were: poisoned or possibly poisoned (25); trapped (5); road casualties (9) (1 railway casualty); taken by animals, including other birds (9); hit wires or cables (6). The Ravens taken by animals may well have been sick or injured already, especially in the case of ground-predators, such as the fox, and isolated instances of killing by Carrion Crows and Herring Gulls. Two birds found close to Peregrine eyries may, however, have been struck down in flight. The role of disease in Raven biology appears to be little known.

When the Raven recoveries are grouped according to month (Table 7, Fig. 14), there is a clear concentration of records during the three main breeding months of March–May – 43% of the total for the year. Separation of the records into two age-classes, up to 365 days (i.e. largely birds in their first year) and above 365 days (mainly older birds), shows a marked divergence. Only 25% of first-year birds were recovered during March–May, whereas, for older birds, the figure was 59%. The differences are significant as regards both age and the seasonal bias for older birds. The high recovery rate in May of the first year reflects the vulnerability of nestlings and newly fledged young; while the figures for April of the first and second years should be shared, since many young are still less than 1 year out of the nest by their second April. The hard weather of winter does not appear to cause increased mortality. While behavioural stress may be increased during the breeding season, e.g. by competition for territories, the consistent marked rise in death rate of older Ravens during March–May looks suspiciously like increased vulnerability to persecution as birds attempt to take up nesting territories and breed. Although it is claimed that Ravens do not breed until they are 3 years old (see below), second-year birds are quite likely to try staking claims to a prospective nesting place before reaching reproductive maturity.

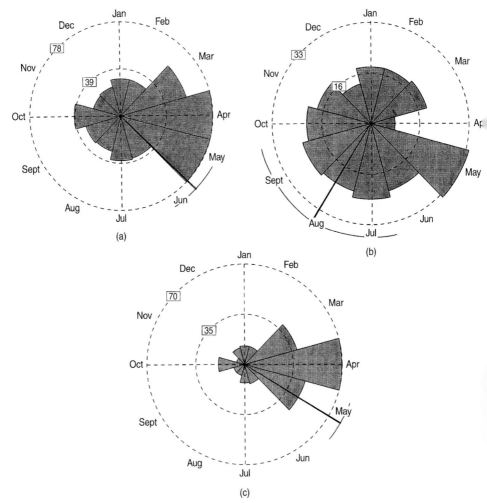

Figure 14. Recovery months of ringed Ravens: (a) all ages; (b) year-1 birds; (c) year 2+ birds. (Prepared by Alan Fielding)
NOTES: Based on recoveries recorded in the Ringing Scheme of the British Trust for Ornithology. Narrow lines to outer arcs show mean time of recovery. Scale varies between diagrams as indicated by the numbers.

There is little information on turnover rates in British and Irish Raven populations. The only estimates of mortality come from the BTO ringing recoveries, and a life table based on these is given in Table 24. Its validity depends on the assumption that recovered individuals are a random sample of the ringed birds – a condition by no means certain, but this is the best information available. I am grateful to Alan Fielding for his analysis of the data, which is limited to the years 1923–81, since the sample of birds ringed in the succeeding 12 years (the age of the oldest birds) up to 1993 may yield further recoveries in future. The data show that Ravens have a typically high first-year mortality (47%), with a rate only slightly lower over the succeeding 5 years. In simple terms, half the birds are dead by the age of 1.13 years (median survival time), and few live beyond 5 years (Fig. 15). When the data for 1923–81 were considered over four successive periods, there was no evidence of change in mortality with time.

These are overall figures for those regions where ringing of the bird has taken place, but they probably reflect broadly the death-rate picture for the Britain and Ireland population as a whole, beginning with the output of young. While there is a high wastage of new recruits, this needs to be put in perspective. If – to take only the lower figure for productivity in Table 20 – each of 4,300 territorial pairs rears an average of 1.6 young per annum, there will be an increment of around 1,000 birds to the British–Irish population each spring. After another year, over 3,500 of these will still be alive, so that, if the population remains stable, nearly another 3,500 older birds will have died also. The absolute number of birds aged 6 years or more will number several hundreds at any one time.

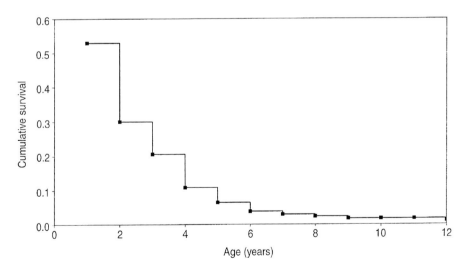

Figure 15. Survival curve for all ringed Ravens up to 1981. (Prepared by Alan Fielding.)
NOTES: Based on recoveries recorded in the Ringing Scheme of the British Trust for Ornithology.

No detailed marking studies have been made to examine the likelihood that mortality is much lower within established adult breeding groups than among the non-breeding population, which includes many juveniles, some of them itinerant. In the Lake District and Southern Uplands, producers of distinctive eggs have often lasted at least 5–7 years, but in some localities there was a more frequent change of egg type, and sometimes there was clearly a new female by the following year.

For females, this evidence of distinctive egg types also suggests that the same individuals continue to occupy the same territories for as long as they survive. The repeated occurrence of erythristic eggs in certain localities was perhaps the most convincing proof of this site fidelity, though in different regions two birds which laid these unusual eggs also moved to adjoining territories (R. J. Birkett, R. Roxburgh), showing that shifts can happen. On the whole, females appear to be faithful to their chosen haunts every year. And, since the Raven is generally believed to pair for life, the males must be presumed to be equally constant to their breeding places.

Many clutches are not particularly distinctive and reliance on egg type to identify individuals is no substitute for ringing or marking. In any case, it says nothing about males. Some individual Ravens have behavioural quirks that allow recognition year by year, but this, too, is an erratic and unreliable means of identification.

The Raven has the reputation of being one of the longest-lived of all birds. Finding acceptable evidence to support this belief is quite difficult. Some of the old tales about the the extreme age of Ravens are clearly absurd and belong to the bird's mythology. Smith (1905) suggested that some of them may have arisen through observations of the bird's strong fidelity to its nesting haunts, carried back through several generations of recollection and with the assumption that the same individuals had maintained this tradition. Smith said that the eminent ornithologist J. H. Gurney told him of a Raven that was given away at the age 60 years and lived for another 20 years; this bird was said to begin laying eggs near the end of its life! A bird shot near Stockholm in 1839 had in its beak a plate with the date 1770 engraved on it, suggesting that it was at least 69 years old (Gordon, 1915). E. C. Phillips claimed to know a captive bird that was 50 years old and still living.

Smith said that two of his tame captive Ravens died 'a perfectly natural death' at 17 and 22 years. Birds at the Tower of London mostly lived for 20–25 years, exceptionally 44 years (Heinrich, 1990). In the wild, expectation of life is considerably less and the oldest ringing recoveries so far are of four birds aged 12+ years (Table 24). In Lakeland, a bird at the nest in 1995 was wearing a clip 4 ring last used in 1974 (G. Horne). In North America, Kennedy and Walker (1988) state that the oldest recorded captive Raven lived for 29 years, and that banding records give 13+ years as the longest life-span for a wild bird. Until there is more reliable evidence to the contrary, I prefer to regard these more modest figures as an indication of the Raven's potential maximum age, i.e. 25–30 years. This rather academic figure is, in any case, hardly relevant to natural mortality in the wild.

Age of reproductive maturity is important to these questions of population turnover. Davis and Davis (1986) think it likely that most Ravens do not breed until they are at least 3 years old. They mention that a captive pair held by Neale (1901) part-built a nest at 3 years old and laid for the first time at 4 years old. Heinrich (1990) also states that the species does not breed until 3 years old. Coombs (1978) says that Ravens reach sexual maturity and begin to breed in the spring of their third calendar year. It would be good to know what these last two statements are based upon. Tony Cross, in 1996, found two marked 2-year-old males breeding with females of unknown age in Shropshire, but only one of the pairs was successful, rearing a single youngster. In Iceland, Skarphedinsson *et al.* (1990) recorded that a female returned to its natal territory at 1 year old and then bred successfully at 2 years old, while Kochert, Bammann and Doremus (1977) found that, in Idaho, a 2-year-old female bred but failed. By contrast, a wing-tagged male Raven on Mainland Orkney did not begin breeding until it was nearly 6 years old; it had been seen in each year, often with another bird, but in various places, and in the previous spring was in a roost of non-breeders (Booth, 1986).

Clearly, some Ravens can breed at 2 years old, but it is the average age of first breeding which matters most in population dynamics. This is simply not known for the Raven. Much may depend on the opportunities for occupying a breeding territory or mating with a widowed bird already holding one, and this will be influenced by the mortality rate in an established nesting population or the prospects for spread into new areas.

Judging by both the general stability in numbers of breeders and the frequent occurrence of non-breeders, an annual productivity commonly of at least 1.6 young per territorial pair is quite adequate to provide full replacement of normal losses and maintain population. Indeed, it is probably not productivity that limits the increase and spread of the Raven beyond its existing strongholds so much as the increased mortality that the birds suffer when they try to push beyond these well-established limits. The marked increase in recoveries of ringed Ravens during the breeding period March–May (Fig. 14, Table 7) suggests that mortality increases as birds attempt to occupy new breeding places. The tendency for numbers in communal roosts to drop appreciably during the spring and summer may partly reflect a dispersal of pairs seeking territories.

Skarphedinsson *et al.* (1990) have shown that even a massive government-sponsored cull of Ravens in Iceland had no discernible effect on stability of breeding population. During 1981–85 an average of 4,116 Ravens was killed each year, mainly by year-round use of narcotic baits near garbage dumps and fish factories frequented by non-breeding flocks. Smaller numbers were also shot, including at nesting places, but the performance of the breeding population remained good, with 83% of pairs successful in three study areas. Recovery rate of 587 ringed birds in these areas during 1981–85 was 42%, and it was estimated that 87% of the annual reproductive output of Icelandic Ravens was lost to human persecution during this period. A mean number of 1,738 Raven pairs was needed annually to compensate for this loss, through their breeding output, out of an estimated Iceland breeding population of 2,000 pairs. Yet occupancy of nesting territories remained stable throughout the study and even non-breeding flocks showed little evidence of decline except in one area.

This confirms that Ravens can withstand an enormous drain on their non-breeding populations with little or no effect on the established breeding strength, but it does nothing to alter the evidence that systematic persecution can prevent occupation of breeding areas and thereby restrict breeding distribution below its potential range. There must be a good many areas where food supply is quite sufficient to sustain breeding Ravens, but persecution prevents their establishment. Yet there are also likely to be many parts of Britain and Ireland, especially in the agricultural lowlands, where food supply is now doubtfully adequate to support the bird. If coastal Ravens could live largely off the sea-shore, why have they not recolonised the sea-cliffs of Kent and Sussex, Yorkshire, Angus and Aberdeen?

The Raven's rather limited movements suggest that, even if the inhibitory effects of persecution were lifted generally, it would not rapidly reoccupy suitable areas distant from its present strongholds, but would move out more slowly from these, a little at a time. The non-breeding flocks would seem to be a source of recruits to prospective new breeding areas, but the presence of already settled breeding pairs may be a stimulus inducing others to stay and make an attempt at nesting – so that, when established pairs are absent and this encouragement is lacking, colonisation takes longer. The Idaho power-line study shows a rapid increase under optimal conditions.

In his study of Ravens in mid-Argyll during 1989–91, Thomas (1993) was puzzled to find that certain areas of apparently quite suitable upland habitat were unoccupied, producing gaps in the regularity of distribution of territorial pairs. He estimated that breeding population was only 50% of the maximum possible. Applying a geographical information system to measurements of habitat diversity, in order to predict suitability of any area for Raven occupation and breeding, failed to identify reasons for these local absences. Thomas noted a scarcity of 'floating' birds throughout the district and felt that the most reasonable explanation of the patchy breeding distribution was the lack of recruitment to the population and he believes that poisoning may have been a contributing factor. By contrast, on the adjoining island of Mull, there were many surplus, non-territory-holding Ravens (Fielding & Haworth, 1995).

Although Raven assemblages are somewhat irregularly distributed, the total number of birds in them, and in the dispersed pairs of non-breeders attempting to establish territories, represents a large reservoir of population with the potential for a great expansion of range, if conditions allowed. The total number of non-breeders in Britain and Ireland seems likely to be at least as great as the number of breeders. The overall Raven population of these islands can thus be put at a minimum of 18,000 birds and appears to be relatively stable at present.

CHAPTER 12

Ravens in the modern scene

This chapter looks at recent changes in distribution and their causes. Compared with the beginning of the century, the Raven in Britain and Ireland is now, overall, in a far healthier position. The large increases in breeding populations in Wales, the West Country, and south-east and north-east Ireland, have restored the fortunes of the bird over large areas. Between the two *Atlases of Breeding Birds* surveys it nevertheless appears that declines in Scotland and north-east England have almost cancelled out further increases. There are now signs that the widespread decrease in Scotland may be undergoing reversal, at least in the south, so that the overall situation is delicately balanced. It is important to understand the underlying reasons for these changes, and the present status of

Ravens in different regions, with their implications for conservation of the species.

GAMEKEEPERING

The contrast between the abundance of the Raven in parts of Wales and the West Country and its scarcity in somewhat similar country from the Pennines to eastern Scotland is so striking that one might almost suppose two different forms of the bird to be involved. The real explanation is, however, much simpler.

There are good reasons for believing that game-preserving, which had such a profound influence in cutting back nineteenth-century distribution and numbers, continues to exert appreciable adverse effects. Nearly all the districts where good breeding populations of Ravens occur are those with little or no interest in the preservation and shooting of game. In 1957, Norman Moore published two almost exactly complementary maps of the distribution of gamekeepers and Buzzards in Britain, and interpreted this matching presence and absence as cause and effect. Precisely the same correlation is still found in the case of the Raven.

The Raven's failure to establish itself in the marginal lands and foothills of the Pennines, Cheviots, Southern Uplands and eastern Highlands – mostly appearing essentially similar in ecological character to those of Wales and south-west England where the Raven is so numerous – can be readily explained only by assuming an inhibiting degree of persecution. This may be partly because farmers in the more northerly areas are less able to accept birds like Ravens on their land, but it also matches a greater concern for the preservation of game in general. Probably many of the lowland areas where game interest is mainly in Pheasants and Partridges are no longer suitable habitat for the Raven in terms of food supply. While they would doubtless be unwelcome and greatly at risk there, it is in upland districts where Red Grouse are preserved that the evidence points especially to a deterrent effect of gamekeeping. Why, it may be asked, are the grouse-moors singled out in this conclusion? The answer is that Ravens have remained widespread throughout almost all of the upland sheepwalks and deer forests wherever there is a lack of interest in Grouse. Conversely, on all the main upland grouse-moors, the bird remains sparse or absent, or has even declined further in recent decades. Fig. 16 shows the current distribution of grouse-moors in Britain; it should be compared with the map of Raven distribution in Fig. 2.

Gibbons *et al.* (1995) also diagnosed the close negative correlation between Raven (and Buzzard) distribution and occurrence of grouse-moors, but were reluctant to claim that this was a result of persecution, since they felt that moor management and availability of food and nest sites could also have been involved. My own view is that the last three factors can largely be dismissed as an explanation for the absence of Ravens from grouse-moors. Many – probably most – grouse-moors carry at least some sheep nowadays, and appear to be quite suitable Raven habitat as regards

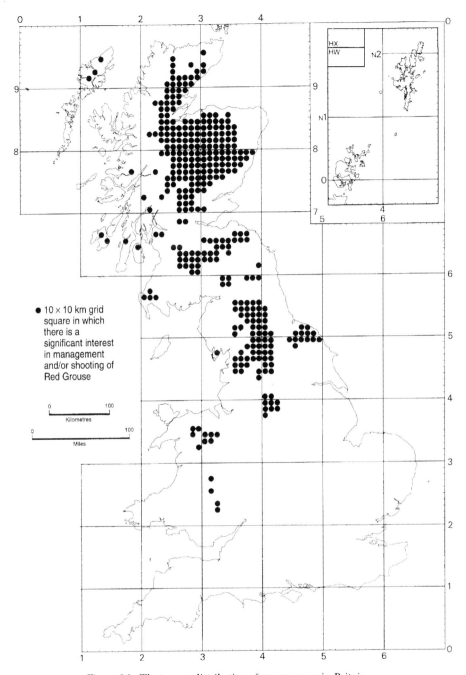

Figure 16. The present distribution of grouse moors in Britain.
NOTES: The map is based on personal observations, further records supplied by D. B. A. Thompson and D. Horsfield, and data from maps published by Hudson (1992). Records for Scotland may be incomplete, but the general pattern of distribution is clear enough. Ireland is omitted as no data are available. The map shows 10 × 10-km grid squares in which there is currently an interest in the management of Red Grouse.

food supply. The Bowland Fells, the Crossfell range from Stainmore to Tindale, and the Moorfoot Hills are good examples. It is true that most of these moors are gently contoured terrain with a general scarcity of good crags to provide suitable rock-nesting sites. Most of them nevertheless have suitable tree sites, especially around their lower edges and in shelter belts and clumps – situations which are widely used by Ravens for nesting on some of the less rugged sheepwalks in Wales and south-west England, where there are few, if any, gamekeepers (see photograph, p. 227)

The status of the Raven in the Pennines is particularly revealing. The bird was eradicated from the Peak District – a famous grouse-moor area – well before 1900. After a long absence, a single pair returned to nest in 1968, evidently successfully, but breeding did not persist and the *New Atlas of Breeding Birds* (1988–91) (Gibbons, Reid & Chapman, 1993) shows complete absence, although two or three pairs are reported back in 1996. The Bowland Fells on the western flank of the Pennines and the North York Moors to the east are notable grouse-moors that have remained almost devoid of nesting Ravens during the whole of the twentieth century. The same is true of the Pennines in Northumberland and Durham, and in the eastern parts of the Yorkshire Dales – all major grouse-moor areas to this day. The species has, in fact, even since 1950, shown an almost complete fade-out from the Pennines that is worth looking at in detail. Here, matters of land management, food supply and nest-site availability certainly *were* once suitable for the Raven; since they have not obviously changed, some other factor must have done so.

Not that the Raven was anything but a sparse Pennine breeder since 1900, with only 20 known breeding territories along the whole section between the Tyne and Aire Gaps – a distance of 135 km and an area of around 5,000 km^2. The small population was evidently at its peak in the early 1960s, and even then numbered only nine to ten pairs, with half the nesting places occupied only sporadically. This contrasts with 70–90 pairs in the *c.* 2,000 km^2 of the Lake District immediately to the west. The northern Pennines have even held a substantial non-breeding population in a regular communal roost on sheepwalk ground of the Eden Valley scarp, so that there has been no lack of potential colonists. The high recovery rate in the Pennines of Ravens ringed in Lakeland (Fig. 7) is another suggestive pointer to low survival prospects on the grouse-moors here. Even some good crags tended to be irregularly used and tree-nesting is exceptional. At most, only two, or possibly three, pairs were known to nest in 1995, but in 1996 four pairs bred. The breeding territories are as follows:

> Territory 1 is amid grouse-moor in the Geltsdale–Tindale area. It was subject to keeper persecution during the 1920s and 1930s, and nesting was irregular. More regular nesting occurred from 1946 to 1974, but the eggs were taken with equal regularity and only a single brood was known to be reared during this time, in a year when the birds moved to a new and inconspicuous site. The territory was deserted during 1974–83 and its subsequent reoccupation has been sporadic and with

very little success, even though the RSPB has established a reserve over the area (E. Blezard, R. Graham, G. Horne, J. Miles).

Territory 2 is also within grouse-moors north of Hartside. The small crags were used by Ravens around 1910 but were deserted during 1923–33. A pair nested during 1934–37, but was unsuccessful each year. A keeper was found to have trapped a Buzzard on its nest in the same locality in 1928, and the failure and subsequent disappearance of the Ravens was attributed to his activities (E. Blezard). Half a century then passed before a pair appeared again, but one reoccupied the place in 1996, and reared three young.

Territory 3 is on the Crossfell range, actually on sheepwalk, but within 2 km of extensive grouse-moors. It was sporadically occupied during 1920–50 (E. Blezard), and in 1946 large young were found dead below the nest. There was a brief spell of occupation from 1958 to 1964, and again in 1968, but it has been deserted since then and Ravens are only irregularly seen in the area (D. A. Ratcliffe, P. Burnham, T. J. Wells, G. Horne).

Territory 4 is on grouse-moors in the upper south Tyne catchment above Alston. Although there were occasional earlier reports of occupation, the only recorded nest in recent years was one with eggs seen by G. Horne on 13 April 1973. The locality has been visited frequently since then but no other nesting is reported.

Territory 5 is in Upper Teesdale, and was recorded by Nelson (1907) and Temperley (1951) as a traditional breeding haunt. Attempted breeding was fairly regular, though mainly unsuccessful, during the first half of this century. I saw a nest with eggs in 1955, but it failed, and in 1959 and 1965 only old nests were found. From 1967 onwards the territory was monitored by Ian Findlay, Warden of the National Nature Reserve that included all the nesting places, but remained under management as grouse-moor by the Raby and Strathmore estates, which retained ownership. Raven pairs were seen in every year from 1967 to 1984, but managed to lay eggs in only 2 years; both nests were destroyed. In 1985 a pair nested in sight of the Warden's house but were shot when the eggs were about to hatch. The next nesting was in 1988, when the adults were poisoned after their eggs had hatched. It was not until 1991 that a pair finally managed to rear two young – the first success for 40–50 years according to local knowledge – though, after the first day on the wing, none of the birds was seen again. A pair was about during 1992–96, but eggs were laid in only 2 years and the nest was robbed both times.

Territories 6 and 7 are in the Westmorland Pennines above Appleby; they are on sheepwalks but close to extensive grouse-moors. In 1925 they were used as alternative nesting places by one pair. In 1950 both territories were at first unoccupied, but the pair which nested in Territory 5 lost their eggs and moved to Territory 7 for a successful repeat nest. From 1962 to 1973 R. W. Robson found separate pairs nesting in each territory, both usually successfully. At some later time one pair dropped out, and Territory 7 became the most constantly occupied. After complete desertion during 1992–95, a pair bred successfully again in 1996 (T. J. Wells, G. Horne).

Territory 8 lies on sheep ground in the hills immediately behind Kirkby Stephen. It was sporadically used during 1920–30 but, after a long desertion, had a continuous spell of occupation from 1961 to 1973. During this time there were intrusive attentions by grouse-keepers coming from Swaledale, resulting in nest failure, but after protests to the police these ceased (W. Thompson). During their last few years, the Ravens took to tree-nesting lower down the valley, but then suddenly disappeared and none has returned.

Territory 9 is on the scarps above the Mallerstang valley, immediately abutting the extensive grouse-moors of Swaledale. A pair attempted to breed in most years during the inter-War period, despite a good deal of persecution by a man with an interest in game who lived in the valley (W. Thompson). They occasionally reared young during the 1960s and were still present in 1977, but then disappeared and there was no breeding during the 1980s. A pair returned and bred successfully from 1990 to 1993, but had gone again in 1994–95 (C. Armitstead).

Territory 10 is on the grouse-moors at the head of Swaledale. It was used occasionally up to at least 1975 but has been abandoned since 1980 (D. A. Ratcliffe, C. Armitstead).

Territory 11 is also on grouse-moors near Swaledale head. There are only two records of its use during the present century (*The Naturalist*, J. Tullett).

Territory 12 is on sheepwalks on the northern part of Ingleborough. I saw a nest with eggs in 1955 but have no other information and it appears now to have been deserted for many years.

Territory 13 is on sheepwalk on the south side of Ingleborough. It was occupied up to 1984 and from 1993–96 (D. Simpson).

Territory 14 is on Penyghent, on sheepwalk with some grouse interest, and was regularly occupied up to 1973 but has since lapsed (D. A. Ratcliffe). It was much disturbed by walkers and this may have been an important factor in its desertion.

Territory 15 is at the head of a branch of Wensleydale, on the edge of grouse-moors, and has been tenanted only sporadically since 1955. The last known nesting was in 1977 (C. Armitstead).

Territory 16 is at the head of Wensleydale, on mixed sheep/grouse-ground. When first reported, in 1975, it appeared to be a new nesting place. After 1976 there were no further records (C. Armitstead).

Territory 17 was evidently a 'once-off' nesting place, reported only in 1971 (C. Armitstead). It is mainly sheepwalk east of Sedburgh.

Territory 18 is in the limestone area near Malham, but within reach of grouse-ground. Young were reared in 1955, but the territory appears to have been deserted for some time.

Territory 19 is a 'once-off' site in the Lancashire portion of the Bowland Fells, where young were evidently reared in 1987 (M. Demain). Only one other recent nest (which failed) is known on these extensive grouse-moors (P. Stott).

Territory 20 is a small plantation on the Durham grouse-moors near Barnard Castle, where a tree-nest was occupied for the first time in 1995, and produced a single fledged young. In 1996, three young were reared in a crag not far away, evidently by the same pair (I. Findlay).

There is similar circumstantial evidence about the adverse effects of grouse-preserving from other regions.

The remarkable increase in Ravens in Wales includes eastern upland districts where there has been a substantial decline in interest in grouse-preserving in recent decades. The moorlands of Radnor, Montgomery, east Merioneth and Denbigh all include once-important grouse areas where birds such as this received short shrift (see Bolam, 1913). Poison is widely used by farmers against foxes and Crows, and contributes appreciably to Raven mortality, but does not appear to limit breeding distribution on the sheepwalks.

Across the Southern Uplands, Ravens have a mostly sporadic history of recent occupation in those areas where Grouse have continued to be preserved: in Wigtownshire, Ayrshire, the Leadhills area, the Pentlands, Moorfoot Hills and Lammermuirs. Most of the breeding population is on unkeepered ground. In the Langholm hills of Dumfriesshire, a pair of Ravens bred every year from 1924 to 1979. In the late 1970s there was a renewed interest in grouse management in the area and poisoned egg-baits began to appear on the moors. The Ravens ceased nesting in 1980–81, and evidently disappeared from 1983 onwards. A pair has returned to nest within the last few years, coinciding with an embargo on killing predatory birds in the area. A rapid increase has, indeed, followed this amnesty, with another three pairs nesting in the area and two more holding territory in 1996 (S. Redpath, R. Corner).

In the Highlands, Douglas Weir (1978) regarded a decline in Ravens on Speyside as the result of keepering activities. He found that 22 breeding territories, which during 1964–68 held an average of 16–17 pairs (with 10–11 usually successful), declined by 1977 to five pairs (only one or two successful). The widespread use of modern poisons on meat-baits began in this area in 1969, after which dead Ravens, moved or concealed by man or lying near whole or part carcasses of rabbits and hares, were found in eight of the 22 territories. Picozzi and Weir (1976) also quantified the effects of use of these poisons on a Speyside Buzzard breeding population. This included a total of 223 Buzzards killed on just four estates in 1968. When examining Peregrine nesting cliffs in the grouse-moor country of the eastern Highlands from 1961 to 1971, I found that 12 separate localities held old Raven nests, though no Ravens were present at the time in any of them. The birds themselves did not last long, but their nests, being durable structures, remained to tell the story of these failed attempts at occupation.

Probably poison is the biggest hazard on these keepered moors and, where put down regularly, no doubt mops up prospecting Ravens almost as fast as they appear. Ravens face other serious risks on the grouse-moors, nevertheless. Some are also quite deliberately shot, usually at the nest, when the sitter can often be surprised at close range and offers an easy target. Just after World War 2, Ravens made several attempts to establish tree-nesting sites in the Langholm hills, Dumfriesshire, in places where the bird was previously absent, but all were unsuccessful. The tree-nest on p. 149 represented one such attempt, in 1946. The shepherd on the ground happened to mention this nest to a local gamekeeper, who came and hid up in the nesting gully until the incubating bird returned and then shot it. There was never another tree-nesting attempt in this locality; three others that I knew of in the district during 1947–58 were all once-off tries. Illegal gin-trapping is evidently less often employed to kill Ravens and other bird predators these days, for it is more readily detected than other methods. Cage-traps to catch Crows alive are legally used, but Mick Marquiss notes that Ravens are very rarely caught in these.

As with the Peregrine, good Raven nesting cliffs on grouse-moors can create a 'sink', drawing in an endless succession of new birds in futile attempts at occupation as their predecessors are removed. Potential tree sites may also be attractive, though this is less obvious. On many grouse-moors Ravens are seldom even seen by casual observers. Just how many turn up and are quickly and quietly put down is something only the keepers know, and they mostly do not tell. The number of Pennine ringing recoveries shows the attractions of grouse-moors for Ravens, and the absence of breeding birds from the vicinity of most recovery sites is some indication of this wastage of potential colonists. Where Raven breeding areas abut lowland country in which there is a strong interest in the rearing of pheasants, as in the Welsh Borders, there must also be an adverse pressure against their eastwards spread. Together with the Buzzard, they are at high risk of a hostile reception whenever they seek to expand beyond the relative sanctuary of the sheepwalks.

There have been *some* improvements over this unacceptable warfare against some of our finest birds. The recent increase of breeding Ravens in the low hill ranges between Dumbarton and Stirling is attributed by John Mitchell to a great reduction in illegal use of poison over the last few years – as evidenced by the lack of poisoning incidents reported to the RSPB in the north Clyde area in 1994–95. Similarly, Mick Marquiss believes that decreased use of poison accounts for the increase in non-breeding Ravens in areas of the north-east Highlands, where nesting has almost ceased. It appears that the widespread publicity given to illegal poisoning, the disgrace of prosecution, and the heavy fines handed out by some courts, are having some beneficial effect.

There is a long way to go, nevertheless. The RSPB magazine *Birds* for August–October 1995 mentions a report which provides a detailed analysis of 170 known incidents of illegal persecution of birds of prey in Scotland in 1994. Poison and poisoned baits were the most frequent method of destruction, and most reported incidents were from areas of grouse-moor. A similar report for England, Wales and Northern Ireland showed 59 incidents of illegal destruction of birds of prey in 1994. Despite all the bland assurances by the land-owning establishment that the problem is being successfully addressed (or is greatly exaggerated), it appears that progress is local and limited. One Galloway land-owner, recently prosecuted for putting down illegal poison, declared – as if in mitigation – that nearly all the farmers in his area were doing the same. Added to this is the present clamour from the 'sporting' fraternity over the alleged depredations of several bird-of-prey species among game-birds and homing pigeons. The game-preservers appear currently to be orchestrating a campaign for relaxation of the legal protection for at least some birds of prey, so that they may be legitimately 'controlled', i.e. killed. This could be a slippery slope on which there might well be no stopping the descent. The Magpie is added to the list of the accused, as an alleged destroyer of songbirds. Far from enlightenment in these matters spreading, it seems that, if some people had their way, things could head backwards into the Victorian age.

I therefore remain convinced that the Raven is held back from realising its potential distribution in Britain – and probably Ireland, too – by the hostilities of those concerned with the management and shooting of game. Unless, or until, there is an appreciable reduction in the persecution of the bird by these people, this will continue to be the case. The bird would steadily re-establish as a breeder in many areas where it now has only a tenuous foot-hold or occurs only as a visiting non-breeder, if only it were given a chance. Whether it will ever do so will require the kindlier views that evidently prevail in Wales and the West Country. Much is likely to depend on the attitudes of future governments to field sports in general and on their willingness to enforce the law.

AFFORESTATION

Ravens are forest-dwelling birds in many parts of their world range. They are, for instance, widely distributed through the vast boreal forest regions of both Eurasia and North America. In Britain and Ireland, many of their earlier lowland haunts were in at least partly wooded country, where they nested mainly in trees. Even today, a good many pairs breed in woodlands of one kind or another. How is it, therefore, that the recent readvance of forest over many parts of Britain has mostly not been to the benefit of the Raven?

The answer lies in the kind of woodland created by modern silvicultural practice and its effects on the Raven's food supply. The natural boreal forest – or even its more managed parts – mostly consists of stands where the stocking density of the trees is sufficiently open to allow good light penetration and the presence of a continuous layer of vegetation – low shrubs, herbs, mosses and lichens – on the forest floor. This provides feeding habitat for a variety of mammals, from small rodents up to large deer, and many different kinds of birds. The trees are often spaced widely enough for larger predators, such as Buzzards, Ravens, and even Golden Eagles, to hunt and scavenge beneath the forest canopy. Moreover, in many regions, there are frequent natural breaks in the forest cover, especially where impeded drainage produces bogs and marshes, often of great extent, on which tree growth is sparse or absent, and here birds needing open ground can forage freely. In dry regions the forest may be naturally somewhat sparse, forming a 'park woodland' which affords good hunting opportunities to predatory birds.

Most of the forests re-established in Britain under the afforestation programmes of the Forestry Commission and, later, the private forestry operators, have not been of this kind. Since 1920, some 1.2Mha (12,000 km^2) of new forest have been planted in Britain, mostly in upland sheepwalk districts inhabited by Ravens. The bulk of these plantations are of conifers – Sitka spruce, lodgepole pine and larch – grown in artificially close canopy. Even when well thinned, the upland forests tend to be dark and shady, with little ground vegetation, and they are now mostly left unthinned and unbrashed, to minimise the risks of wind-throw. The resulting thicket forests that develop, and persist up to the time they are clear-felled at 30–45 years, are virtually impenetrable to large birds below the canopy and so have very little feeding value to species such as Ravens. Even more crucially, before the trees are planted, the sheep which were the Ravens' mainstay are removed and the ground is fenced against their re-entry. While the young developing forest and the now ungrazed ground vegetation often become good habitat for mammals and some birds before the young trees close into thickets, its food-value for birds which depended on sheep carrion is seriously reduced.

Where extensive forests begin to intrude into the feeding range of Ravens, they thus usually spell difficulty for the bird, unless there is some alternative or compensatory food source. Even by 1965, it was noticeable that, where blanket afforestation was spreading across the Cheviots and Southern Uplands, pairs of Ravens were beginning to drop out from territories which had always been regularly occupied. There seemed to be a conservation problem of some potential seriousness as afforestation spread apace across the uplands, as well as an interesting issue in land-use ecology. On discussing the matter with Ian Newton, we agreed that it was worth setting up a full-scale study to measure the scale of the impact on the whole Raven population across the Cheviots and Southern Uplands. Mick Marquiss became the main field-investigator, with the remit of visiting as many Raven nesting places as possible in these districts during the springs of 1974–76. I provided all my own previous field records back to the late 1940s and we put together all the other information available from local ornithologists in order to give earlier background for comparison with the new situation.

Mick had an energetic three seasons in the field and his survey soon made it clear that Raven breeding numbers across the whole study region had, indeed, declined considerably since the 1950s. The survey checked 123 previously known nesting areas, of which at least 81 had been used regularly. Of these, 44 of the once regular areas, one occasionally used area and three areas whose past use was uncertain were occupied in both 1974 and 1975. Of the total 48 pairs, 40 were proved to lay eggs but eight were non-breeders. This meant that at least 33 pairs out of an earlier minimum population of 81 pairs (41%) had dropped out. If an allowance was made for average former population (90 pairs seems reasonable), then 42 had been lost (47%), and actual breeders had dropped by 50 pairs (56%) (Marquiss, Newton & Ratcliffe, 1978).

Other factors besides afforestation had also been at work (see below) but, when the losses were examined one by one, it seemed likely that 22 once-regular areas and four irregular or uncertain areas had been deserted through forest encroachment within the Ravens' feeding range. When all nesting areas were matched against the extent of forestry encroachment, it was found that the greater the amount of afforestation within 3 km of the nesting place, the less likely was the area still to be occupied, the tendency

36. Managed grouse moor with a pattern of rotational heather-burning and suitable tree nesting habitat beyond the moor edge; Ravens are usually unwelcome on such ground and new arrivals seldom last long, although a pair returned to a tree-site in 1996. Langholm, Dumfriesshire (photo: D. A. Ratcliffe)

37. Blanket afforestation of sheepwalks with Sitka spruce; the three nesting pairs of Ravens in this area have declined to one. Southern Uplands, Kirkcudbrightshire (photo: D. A. Ratcliffe)

being highly significant. For 22 nesting areas, the exact year in which occupation ceased was known and, in all but one case, the year of desertion was immediately before, immediately after, or during, a year of tree-planting within 3 km of the nesting place. Ravens in these districts also showed more non-breeding and produced later, smaller broods in heavily afforested areas than in lightly afforested areas.

Mick's examination of food by pellet analysis supported these conclusions. Frequency of sheep's wool in pellets decreased as the extent of new forest increased in the vicinity of the nesting place and there were significant negative correlations when afforestation was measured within both 3 km and 5 km. Where afforestation levels were high, the castings showed the Ravens to be feeding especially on goat, deer, lagomorph and bird carrion, while, in the vicinity of recently planted areas, they were taking many voles (Table 2, column 4).

Richard Mearns repeated this Raven survey of the Cheviots and Southern Uplands in 1981 and found that at least 11 and possibly up to 16 further territories had become deserted since 1975 (Mearns, 1983). Of these new desertions, at least five could be attributed to forest encroachment, all of them previously regular breeding areas. By 1981, it therefore appeared that up to 31 pairs of Ravens had been lost from these two districts through this land-use change, representing 34% of a probable former population of 90 inland pairs.

Subsequently, there were reports that some of the vacant breeding areas were reoccupied and a full survey of the Southern Uplands Raven population in 1994 was organised by Chris Rollie through the Dumfries and Galloway Raptor Study Group. It showed that six of the areas where desertion was attributed to afforestation again held breeding pairs. How does this square with what I have said about the unfavourable conditions created by blanket afforestation? Clearly, where Ravens have returned to breed, food supply must again be adequate. One possibility is that birds less dependent on sheep carrion have appeared and have managed to subsist off a more varied diet that can be obtained where there is still a moderate amount of open ground left. Deer carrion may have increased to offset the loss of sheep carcasses and, in parts of the Galloway forests, Forest Enterprise staff have been leaving shot deer and goat carcasses to augment the food supply of Golden Eagles, which have been adversely affected in the same way as Ravens. Some of these carrion 'larders' are known to have been visited by Ravens. In some of the older forest areas, large clear-fells have recently re-expanded the extent of feeding range for open-ground birds, but this does not appear to have been a factor in any of the six reoccupations, where all the trees are still in their first generation.

In the Cheviots, however, Raven numbers have been at a low ebb, with usually no more than one or two pairs each year in areas with the least afforestation, though three pairs bred in 1996. The bird has long disappeared as a nester from the ground covered by the vast plantations of the north Tyne and has shown no inclination to return. This leaves a total of 25 Raven territories across the Southern Uplands and Cheviots which appear to have become permanently deserted through the encroachment of afforestation.

What of other districts? Afforestation has been too slight in the Pennines to have had any influence on Raven numbers. In the Lake District there has been rather more planting in some peripheral areas, and it is perhaps no coincidence that two of the few desertions of Raven territories known in the district occurred after extensive planting in their vicinity. The bird bred earlier in this century on the north front of the Pillar Mountain but, after the planting of Ennerdale in 1927, nesting ceased. Tree-nesting Ravens bred regularly on Great Mell Fell near Troutbeck up to 1974 but, after the planting of several large blocks of forest on adjoining sheepwalks around this time, the pair disappeared.

Argyll is another county extensively reafforested. In a recent study of Ravens in mid-Argyll, Chris Thomas (1993) concluded that blanket afforestation has rendered 43% of upland locations containing suitable breeding sites untenable for Ravens, but that afforested locations were still apparently suitable for occupancy where open ground remained, e.g. on the edge of forest blocks. Even locations in heavily afforested areas were regarded as suitable as long as the plantations were fragmented. The distribution of the forests was important; the amount of afforestation close to the nest, within about 1 km and closer, was more important than the total amount within 3 km. This implied that afforestation had caused a decline in breeding Raven population in Argyll, but lack of earlier data on distribution and numbers prevented an assessment of its scale. Thomas found that clutch size tended to decrease with increasing afforestation in the vicinity and breeding became less successful with increasing amounts of surrounding land under thicket forest.

A different situation was reported from central Wales, another fairly heavily afforested district of upland sheepwalks (Newton, Davis and Davis, 1982). In a study area of 475 km^2, where just under a quarter had been afforested with conifers between 1950 and 1970, Ravens bred at a higher density than in any other inland area studied in Britain. A total of 79 breeding territories were identified, including seven known during 1967–74 but vacant during the study period of 1975–79. The average density was 17 pairs/100 km^2, though this reduced to 15 pairs/100 km^2 if only the occupied territories were included. There were no census data for the preafforestation years, so that decline through tree-planting before the 1975 study began could not be ruled out. Nesting pairs were spaced farther apart in the more heavily afforested areas of the uplands, but the proportion of planted land within 1 km of the nest explained only 7% of the variation in nearest neighbour distances. The seven deserted territories were no more associated with planted areas than the occupied ones. Many pairs nested in conifers within the plantations and the only effect of increasing extent of closed forest around the nest site appeared to be a decrease in clutch size; occupancy and breeding success were unaffected.

The available evidence thus suggested that, in central Wales, afforestation did not have the adverse effect on Raven population that was found in the Cheviots and Southern Uplands. Newton *et al.* suggested several reasons for the difference. The first was that the blocks of forest in Wales were

38. A Cheviot Raven nest site on open moorland, pre-1924. Kielder, Northumberland (photo: Abel Chapman)

smaller and more scattered than in the massive blanket forests of the more northerly districts, so that open feeding range remained more widely available and all Raven pairs had some sheepwalk within easy reach. Most of the forest in Wales was younger than that in the north and still at a stage where Ravens could feed among the trees on voles and other prey. Sheep were also more numerous within plantations in Wales, where they had broken through fences and were allowed to live wild in some of the forest blocks.

39. The same Cheviot Raven site, under blanket afforestation, 1996; the Ravens ceased nesting at least 35 years ago (photo: D. A. Ratcliffe)

Finally, stocking density of sheep (and hence carrion supply, presumably) had also increased on the remaining sheepwalks during the previous decade, partly compensating for the loss of upland to forest. Comparing upland nests with different amounts of forest within 1 km, the proportion of pellets with sheep remains showed no appreciable decline with increase in forest, but the proportions with small mammal remains increased. In short, the feeding value of the Welsh uplands for the bird appeared to have been maintained, in contrast to the decline in carrying capacity experienced farther north.

Davis and Davis (1986) later thought that several instances of non-laying by pairs in the central Wales study area might be attributable to the high levels of afforestation in the surrounding areas, with the presumption of food shortage. A resurvey of this population by Cross and Davis in 1986 showed little change. Most of the young forest had passed into the thicket stage, increasing the area of closed canopy forest by 8% since 1978. Yet there was still no evidence that afforestation had caused any recent desertion of territories, though it may have played a part in earlier territory mergers. The total extent of forest had hardly increased and was $c.$ 25% of the study area, distributed as a patchwork of scattered plantations; it also retained a population of feral sheep.

Ravens tended to produce fewer eggs and young with increasing amounts of closed forest around the nests (measured at 1 km and 3 km radii), though those with lower levels of adjoining forest performed better than those with none at all, presumably due to differences in land productivity and exposure. Paradoxically, a few pairs with very high (60–80%) levels of forest around them performed remarkably well. Some pairs were also deriving benefit from new nest sites in forest. Cross and Davis (1986) nevertheless concluded that the extent of afforestation was probably close to, or at, a critical level, and that the future status of this central Wales Raven population could well be threatened by its further expansion.

Dare (1986a) in Snowdonia, Wells (in press) in Northern Ireland and Dixon (in press) in the Brecon Beacons concluded that Ravens were unaffected by the lesser amounts of afforestation in these districts.

Further blanket afforestation of the upland sheepwalks and deer forests can be expected to cause an additional decline in the number of breeding Ravens in some areas. Although the changes to the tax regime for private forestry in 1988 slowed the rate of new planting considerably, afforestation has continued to creep onwards year by year. Even in Galloway, one of the most heavily afforested regions of Britain, there is continual pressure for yet more planting, and further hill farms are lost to the conifers every year. Some of the largest expansions in recent years have been in areas where there were few Ravens, notably the Flow Country of Caithness and east Sutherland. There has also been little recent planting of new ground in Wales, where conversion of the present patchwork of plantations into extensive blankets has the potential to cause the biggest losses. It is a problem which needs keeping in mind and under review.

CHANGES IN FARMING PRACTICE

Hill sheep-farming has shown two related tendencies during recent decades. On the one hand, sheep-farmers have notably improved the management of their flocks. Veterinary attentions against parasites and pathogens have increased and mineral dietary supplements are routinely provided on many upland sheepwalks. Increasingly, flocks are wintered and then lambed on low 'inbye' ground or in large purpose-built sheds, and winter feeding is usual, including from 'big-bale' silage stored in the piles of large black plastic bags that are now a familiar sight on the upland farms. The result has been to lower mortality of both ewes and lambs, and this has tended to reduce the carrion supply for Ravens. The other marked tendency is the general increase in stocking levels of animals carried on the upland pastures. Such increase in sheep stocks has been made possible by improved husbandry but has also been fostered very directly by the headage subsidies for hill sheep available from the Government agriculture departments, which amount to an average of almost 60% of hill farmers' income (Egdell, Smith & Taylor, 1993). This has tended to increase mortality and, hence, carrion supply, again.

The effects on Ravens can thus show a contrast according to particular circumstances. Where the overall result is to reduce the amount of carrion available within the feeding range, the bird has serious problems, to the point where some pairs have dropped out, apparently because food supply has become critically low. This appears to have happened in at least five breeding territories in the Cheviots and Southern Uplands, where no other explanation for disappearance seems likely. Where, however, increased stocking has produced an enhancement of carrion supply, it has promoted an increase in Raven population. In the Lake District, as noted in Chapter 3, the mean breeding strength increased from 69 pairs during 1900–69 to 87 pairs during 1975–95 (a gain of 26%). The apparent dropping out of a few pairs from this population during the last 2 or 3 years may represent just a normal fluctuation, but could also result from locally reduced food supply as sheep mortality declines.

Snowdonia has shown the most startling change, over a 30-year period. When I looked at Ravens in this district in 1951–61, the numbers of crag-nesters appeared to be stable, at a virtual saturation level. Peter Dare made a more complete Raven survey of Snowdonia during 1978–85 (Dare, 1986a), and, when his figures were compared with precisely the same areas that I had covered, an increase of at least 40% was apparent – probably more in some parts of the district. Egdell, Smith and Taylor (1993) state that, since 1976, sheep numbers have increased by an overall 29% in Severely Disadvantaged Areas of Wales (which include Snowdonia).

More precise information has been gathered for the Carneddau massif of Snowdonia by Turner (1996) and this is summarised in Table 25. The 1992 total sheep numbers represent a 57% increase over those of 1952, and on a slightly reduced area. This closely parallels an increase of from 14 to 23 (64%) nesting pairs of Ravens in the same parts of this massif. The relationship between increase in sheep numbers and increased carrion supply may not, however, be linear or consistent, because mortality will always depend on several factors – of which increased competition for food is only one – and may be modified by management regime.

Rob Fuller has kindly allowed me to summarise findings and reproduce a relevant table from his internal BTO report on upland grazing and birds, which gives a more general countrywide picture. He finds that, during 1950–75, there was much variation in trends of sheep numbers between regions, some showing stability, others a steady increase, and yet others a drop after 1960. Since the mid-1970s, however, there have been large overall increases in sheep numbers in most regions, both upland and lowland, corresponding exactly in timing with the introduction of Hill Livestock Compensatory Payments. Wales stands out as showing a rise throughout the whole period since 1950, with a massive increase since 1969 to especially high stocking densities. Sheep numbers have also increased substantially in north Devon, northern England and the Southern Uplands. Rates of increase have varied regionally, with a virtual doubling of numbers in south Wales, Dyfed and Orkney, and a trebling in the Western Isles, while Highland Region, in particular, has shown little recent overall increase (Table 26). In several other regions showing a plateau or drop since 1980, sheep numbers remain far higher than in the mid-1970s. Notwithstanding the provisos about relating gross sheep numbers to Raven food supply, the geographical pattern of increase in sheep population identified by Fuller shows a remarkably good 'fit' with the pattern of variation in Raven breeding density and population across Britain.

In Northern Ireland, Jim Wells (1996, in press) has connected the increase in Ravens and their expansion into lowland areas (where they rest in quarries and trees) with the rise in sheep numbers to more than double the 1980 levels (Table 25).

Mearns (1983) surveyed the Galloway and Ayrshire coast in 1981 and found only 12–13 breeding pairs, compared with an estimated total of at least 37 pairs in 1961–62. He attributed the decline to changing agricultural practice on the farmed hinterland behind the nesting cliffs, with an increase in arable farming and improved sheep management. Cliff-top heathland and unimproved grassland had also decreased through reclamation, with cultivation pushed almost to the cliff edge in many places. Yet, in 1994, the Dumfries and Galloway Raptor Study Group found that there had been a remarkable recovery, with 24 pairs proved to nest and another 11 in occupation. This recovery is puzzling, but EU regulations no longer allow clearance of fallen stock by slaughterhouses. Farmers are now faced with removal charges as high as £15–40 per head. This may well have led both to an increase in carrion supply, and to greater tolerance to scavengers (Steven, 1996). The decline in poisoning of predators (by farmers as well as keepers) reported from some parts of Scotland could also have been a factor here. An increasing recovery of rabbit populations from myxomatosis has been noted in many areas and may have boosted food supply for some coastal Ravens.

The status of the Raven in Britain and Ireland will depend greatly on the future of hill farming. So long as the economically unviable upland sheep industry is propped up by public subsidies, Ravens will continue to occupy the sheepwalks. Their numbers will depend on the quality of the husbandry and, hence, the amount of carrion. Probably sheep mortality has an irreducible minimum on many sheepwalks, especially in the higher and more rugged mountains, so that the bird will always have a major food supply there. Population levels will, however, be affected by any changes in sheep-stocking rates and by the socio-economic factors that determine these.

There is an irony in the wish of many conservationists to obtain a reduction in the grazing pressure on many sheepwalks, to seek to arrest the

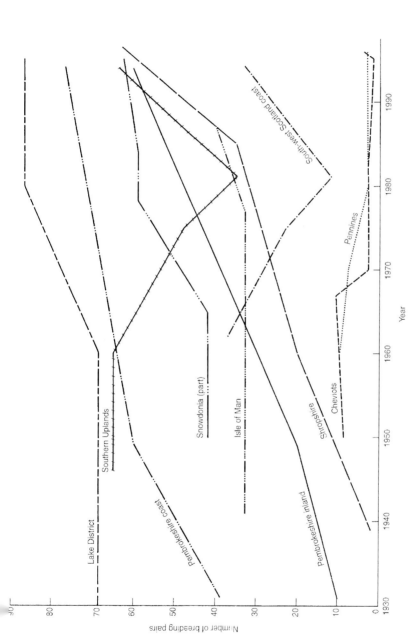

Figure 17. Trends in Raven breeding population in different districts of Britain.

NOTES: The districts shown are those for which figures are available. Increase in some parts of Wales appears to be more dramatic even than in Shropshire, while decrease in eastern Scotland has been appreciable; figures are insufficient to show these changes. Even during stable periods there is ususally slight year-to-year fluctuation in numbers, so that means have been used when a range of figures for breeding population is quoted by authors.

steady decline of heather and other dwarf shrubs, and to allow the recovery of such heath vegetation. The heather and bilberry moorlands are, on the whole, more productive and interesting places for birds in general, and other wildlife, than most grass- and bracken-dominated hills. The overgrazing of so much of our uplands can hardly be denied and measures are increasingly being promoted to compensate hill farmers for conceding desirable reductions in sheep-stocking densities on parts of the uplands. The Raven is one of the very few species to benefit from the present regime of maintaining maximum sheep stocks and the clear implication is that it must decline also if these are reduced. So, in the interests of achieving a more balanced management and diverse habitat, it may be necessary to accept local decline in Raven numbers from some of the high densities now found. This could easily be compensated by allowing the bird to reach its potential levels in those many areas where it is at present deliberately held down.

DISTURBANCE

Where persecuted, Ravens are shy and wary birds which keep a healthy distance from humans and their works. Yet, when left alone, they can lose their fear and become quite at ease with human presence and casual disturbance. Their occupation of medieval towns was proof enough of this familiarity and, at the present day, their habit of resorting in flocks to knackers' yards and rubbish dumps is further evidence of their unconcern about human presence. Ravens have long scavenged discarded food around the summit café of Snowdon and, in Lakeland, they are nowadays to be seen picking up scraps close to human visitors.

At the nest, they are rather more inclined to prefer privacy. When rock-climbing first became popular in Lakeland, several pairs of Ravens moved out of the most frequented crags. There was no decline in numbers because, in each case, they were able to occupy alternative cliffs within the territory where climbers seldom if ever went. In time, some pairs even became bold enough to move back to major climbing grounds and found nest sites away from the favourite routes, where they could rear their young in seclusion. Yet, on small crags which are constantly climbed, the frequency of disturbance close to the nest may simply be too much for the birds.

In Northumberland, several gritstone crags of 20 m height or less and rather limited extent have been permanently deserted through climbing disturbance, the Wanney Crags at the head of the Wansbeck being a case in point. At least five pairs of Ravens appear to have been lost in this way, and there is no evidence that they have resorted to alternative sites in trees – as they would seem well able to do. In the Pennines and Peak District also, even if grouse-keepering ceased to be a problem, many of the smaller scars and edges are so much climbed upon that Ravens are unlikely ever to return. Stanage Edge near Sheffield would offer plenty of scope for nesting sites, but it is scarcely conceivable that the bird would ever take up

residence there nowadays. On Dartmoor, there were at least 20 granite tors and hillside rocks where Ravens bred during the 1940s, but only three of these are still regularly occupied, all in rather secluded places. The rest are either favourite hikers' objectives or are close to popular tracks. Three tor sites on Bodmin Moor are also said now to be deserted.

By contrast, many pairs of Ravens have taken to nesting in quarries, including some that are still worked (see p. 151). Like Peregrines, they have adapted to the activity and noise, including blasting, because the nest with its sitting bird can often be placed far enough from the disturbance for this to be tolerated. It is when the incubating bird is flushed too often and kept off too long that Ravens soon give up a nesting place.

Hang-gliding and para-gliding are modern recreational activities that locally impinge adversely on the nesting of Ravens. At least two instances are reported in the north of England where broods of young Ravens died in the nest by becoming chilled when the parents were kept off for too long by gliders close by. John Mitchell believes that hang-gliders caused at least one pair of Ravens to desert during nest-building, and that this is becoming an increasing problem in the hills north of Glasgow. Occasional instances of nest desertion through inadvertent human disturbance are reported, usually when some unwonted activity chances to take place near to the nest site. Nest trees or woodlands may be felled and the sites destroyed. Such events do not necessarily cause desertion of the territory, provided there are alternative sites to which the birds can move. Even when permanent desertion follows, the effect on population is usually insignificant.

Rubbish dumps are much disturbed places, yet are important feeding sites for Ravens in some areas. They are used especially by non-breeders, but even induce a few pairs to nest close by and are probably behind the occasional return to city-nesting. The future of waste disposal and use of land-fill sites thus has a bearing on Raven ecology.

Casual disturbance is therefore not regarded as more than a purely local and limited threat to the existing Raven population, though it may be a factor acting against the spread of the bird as a breeder in certain areas.

PESTICIDES

There is little evidence that pesticides had any marked effects on Ravens at the population level, but there were pointers to individual instances of death and breeding failure during the period when organochlorine problems were at their height. In 1958 I found an incubating Raven dead on its nest in Lakeland. Ernest Blezard dissected the bird and could find no obvious cause of death, while subsequent analysis showed no trace of strychnine or phosphorus poisons; the tissues were not analysed for organochlorines, which had not then come under suspicion as secondary poisons. A second Raven was found dead below its nest in the Moffat Hills in 1961; it was not analysed chemically and, though its condition suggested death by poisoning, there were again no symptoms of strychnine or phosphorus compounds. In 1961 G. Trafford found two broods of young Ravens dead in the nest in Galloway and, in 1963, a Dumfriesshire nest had the unusually large number of four addled eggs with a single youngster. In the same area, a brood of two well-grown young Ravens was found dead in the nest in 1967, with only one adult still around; there was no physical injury and poisoning was suspected.

This is not much to go on and it is always possible that all these instances were the result of taking poisoned bait deliberately put down for 'vermin'. Some time after the highly toxic nature of dieldrin became known, a Deeside keeper told me that he had successfully used it in egg-baits against Crows, though 'it took longer than strychnine'. By 1961, there were indications that the organochlorine insecticides were causing the decline of raptors such as Peregrine and Sparrowhawk, through widespread secondary poisoning from the eating of contaminated prey. Symptoms of malaise in the Scottish Golden Eagle population, in the form of frequent egg-breaking and other breeding failure, seemed quite likely to be connected with the eating of carcasses of sheep which had been dipped with these same compounds. And if Eagles were thus affected, it was equally likely that other sheep-carrion-feeders, such as Common Buzzard and Raven, would show similar responses.

To test these ideas, it was necessary first to establish the extent and degree of contamination of these species by such residues. Samples of eggs were accordingly collected in the spring of 1963 and analysed by the staff of the Agricultural Scientific Services of the Department of Agriculture and Fisheries for Scotland. Neville Morgan undertook this work for the Nature Conservancy, and George Hamilton was responsible for the analyses. The results (Ratcliffe, 1965) were startling in showing that every egg of Golden Eagle, Buzzard and Raven which was analysed contained measurable residues of at least four organochlorine insecticides, namely DDE (the breakdown product of DDT), HEOD (the active principle of dieldrin), heptachlor epoxide (from heptachlor) and isomers of BHC (including lindane). The mean levels (parts per million, wet weight) are given in Table 27). The source of these chemicals was evidently mainly sheep-dips, the persistent residues being ingested by the birds when they fed on sheep carrion, and incidentally ate wool as well as fatty tissue.

It was later shown that low breeding success of Golden Eagles in western Scotland correlated with modest levels of contamination by these organochlorine residues, and that halving of these levels (through reduction in their use as sheep-dips) was followed by rapid recovery in breeding success (Lockie *et al.*, 1969). Ratcliffe (1970) showed that western Scottish Golden Eagles had undergone a 10% decrease in eggshell thickness index during 1951–65, an effect found in many birds of prey and attributable to contamination by DDT/DDE (Prestt & Ratcliffe, 1972). The associated frequent breaking of Golden Eagle eggs in the nest was an important part of the species' reduced breeding success (Lockie *et al.*, 1969). The lack of eggshell thinning in the Raven under similar DDT/DDE contamination levels to those in the Golden Eagle is discussed on p. 165. Nor was there any detectable reduction in breeding performance during the period when this effect became conspicuous in the Golden Eagle. The supposition was that the Raven was more resistant than most raptors and some corvids to the effects of organochlorine contamination, at least at the levels then observed.

To check on the situation, a sample of ten Raven eggs that I arranged to be collected, all from different nests, in the Lake District in 1970, was analysed at the Laboratory of the Government Chemist (D. C. Holmes). Apart from traces of PCBs (polychlorinated biphenyls; not analysed in the 1963 sample), the only residue found was ppDDE, at very low levels (Table 27), though present in every egg. In 1975, a further sample of single eggs from six Raven nests was collected from the Southern Uplands and analysed (Table 27); DDE remained at very low levels, dieldrin (HEOD) was still less, but PCBs were a little higher, all three compounds being present in each egg (Marquiss, Newton & Ratcliffe, 1978). These two sets of later results suggested that organochlorine risks to Ravens from sheep-dips had become still less of a possibility. However serious the pesticide threats may have been for some predators, there is thus no evidence that the toxic chemicals used in farming have *accidentally* posed a real danger to the species. It is possible that the most heavily contaminated individuals were affected, but no adverse population effect was detected. This should not divert attention from the fact that agricultural pesticides have been deliberately and widely misused to the detriment of this bird and other wildlife. Nor does it rule out adverse incidental effects of pesticides in other countries (see p. 242).

CONCLUSIONS

Virtually all recorded changes in Raven distribution and status in Britain and Ireland can be ascribed to human influence. The bird has declined and retreated under the onslaught of deliberate persecution, even up to the present day in some districts, and its range is limited by such direct hostility. It has both declined and increased in different areas through changes in land management that have altered its food supply. The Raven's resilience is such that it has made very substantial recovery in Wales, southwest England and Ireland during the last half-century or so, and shows signs of reacting quite rapidly to local improvements in conditions, including in Scotland. Its breeding performance gives it the potential to extend its present range considerably. A watchful eye should be kept on its status, even so, especially in the light of the BTO finding of a long-term decline in brood size. Perhaps the most serious threat is the wish of game-shooting interests to put the clock back into the nineteenth century, by amending the law to allow legalised killing of predatory birds which they dislike. The Raven could become more widespread if hostility from game-preservers were relaxed, but the pressure seems likely to be in the opposite direction.

The issue belongs to the politics of wildlife conservation and must be treated accordingly. Raven enthusiasts must be prepared to champion their bird and challenge the opposition vigorously. There is no need to retreat defensively into tedious arguments about 'scientific evidence', either. It is enough to state with conviction and passion that a great many people enjoy watching birds, not shooting them, and especially birds as majestic and inspiring as the Raven.

CHAPTER 13

Ravens elsewhere in the world

Voous (1960) suggests that *Corvus corax* is second only to the Peregrine in its ecological adaptability and Cramp and Perrins (1994) remark that it is so wide-ranging that the concept of habitat is hardly applicable. It is one of the most widespread of all bird species in the northern hemisphere, though south of the equator it is replaced by other species of Raven. Our species breeds from the high arctic to the tropics, in an extremely wide range of habitats. It is reported from Grinnell Land on Ellesmere Island at over 80°N and occurs widely around the coasts of Greenland up to 78°N. Apart from an absence from Novaya Zemlya and a central region of the Siberian sector, it is widespread throughout the circumpolar Arctic and extends southwards through the whole boreal and temperate zones. In the New World the Raven occurs as far south as Honduras, but in the Old World, to the south of the African Mediterranean, Persian Gulf and northern India, it is replaced by other forms with species rank. Australia has evolved its own small group of distinctive Ravens.

Raven habitats thus include barren grounds of fell-field and tundra, coastal terrain, both rocky and 'soft', coniferous and broad-leaved forests, scrub, savannah, steppe, semi-desert and desert (cold and hot). In some regions the poorer kinds of agricultural grassland and heath are much favoured. Habitat varies from the truly natural to the completely man-made. The bird tolerates a range of temperatures, from the numbing cold at the edge of the polar wilderness and alpine Tibet at 5,800 m, to the searing desert heat of Death Valley, California. It also exists between extremes of rainfall and aridity. Whether this has involved a range of physiological adaptations between different populations across this climatic spectrum can only be a matter for conjecture.

The history and status of the Raven on mainland Europe tend to mirror the picture in Britain and Ireland, although on a much larger scale. The map in Cramp and Perrins (1994) shows that the species has largely disappeared from much of the agricultural lowlands of France, the Netherlands, Germany and Czechoslovakia. It remains numerous on the rocky coast of Brittany, in the main mountain ranges of central and southern Europe, and through most of the Iberian peninsula. Though absent as a breeder from parts of northern Denmark, it is more or less continuously distributed eastwards and northwards from northern Germany, Poland and the Ukraine (i.e. through Fennoscandia, Russia and other European states of the former USSR), and through Turkey. In many regions where it had declined or disappeared, the Raven has increased again during the last half century or so. Cramp and Perrins (1994) quote H. Shirihai as saying that the species was drastically reduced by pesticides in Israel, and that poisoning for pest control caused a major decline in Hungary at this time.

The various Raven studies reported for other parts of the world suggest that the ecology of *Corvus corax* has much similarity throughout. Regional and local diet varies according to what is available, but the Raven is everywhere essentially an opportunist scavenger and omnivore. In wilderness areas, both coastal and inland, it subsists on large mammal remains that become available naturally, e.g. by following the reindeer herds with their attendant wolf-packs, gathering to gorge on stranded cetaceans, or joining with vultures to scavenge the leavings of large carnivores in the tropics. The carcasses of domesticated or semi-domesticated animals are a major food source in many regions but, where small ground mammals are abundant, they form an important element of diet. Abundance of large insects (e.g. grasshoppers and locusts) also provides a major seasonal food item, while a vegetable component is sometimes significant.

Throughout its range, the Raven is a solitary and territorial nester, with a breeding density that appears generally to vary in parallel with food supply. In many regions it is sparingly distributed, but where there is an abundant food supply it can reach a high breeding density over a considerable area, as in the Canaries and parts of the USA. Though still widely persecuted as a suspected predator of domestic stock and game, or a nuisance-robber of vegetable crops, it assembles wherever people discard food in quantity and lives on familiar terms with humans where its value as a scavenger is appreciated. Most populations are evidently relatively sedentary, and flocking and communal roosting of non-breeding birds are general. Numbers of some of these gatherings far exceed those recorded in Britain and Ireland, however. In Idaho, USA, Engel *et al.* (1992) counted a mid-July peak of 2,103 Ravens at one communal roost on a 6-km section of power transmission line with 15 towers, and another maximum of 2,289 birds on a longer section of the line on 9 August. Heinrich (1990) mentions daytime aerial congregations of about 2,000 Ravens in New Mexico and 1,000 in Utah, and Cotterman and Heinrich (1993) record a roost of 1,500 birds in low scrub in the Mojave Desert, California.

Ravens maintain their nesting associations with Peregrines in some other parts of the world, sharing the same cliffs and tolerating each other's close company. In northern regions, unoccupied Raven cliff eyries are a favourite nest site for Gyr Falcons, and Rough-legged Buzzards are often in close proximity (e.g. White & Cade, 1971). The avoidance of too close an acquaintance with eagles of various species is general as regards nesting locations.

SUBSPECIES OF *CORVUS CORAX*

Hartert recognised no fewer than 13 subspecies of the Northern Raven. Ten different subspecies of *Corvus corax* (Fig. 18) are accepted nowadays and distinguished by trinomials (Cramp & Perrins, 1994), though the validity of these distinctions must inevitably remain matters of opinion. They are:

1. *corax* Linnaeus 1758, the nominate subspecies, occurring throughout north-west Europe (British Isles, France and Fennoscandia to western Siberia as far as the Yenisei). It grades into form *kamtschaticus* from central Siberia through western Russia to southern Europe and east to Transcaucasia, northern Iran and Kazakhstan.

2. *hispanus* Hartert and Kleinschmidt 1901, found in Iberia, Balearic Islands, Sardinia, perhaps Corsica and Italy. It has shorter wings and a more arched bill.

3. *laurencei* Hume 1873, ranging from the eastern Mediterranean (eastern Greece and Cyprus) through the Middle East to eastern Kazakhstan and northern regions of the Indian subcontinent and reaching China. Slightly larger than nominate subspecies, birds with worn plumage become browner on nape, mantle and throat (cf. *C. ruficollis* below).

4. *tingitanus* Irby 1874, found in Mediterranean coastal regions of north Africa. It is small, with shorter, stouter bill, less elongated throat hackles and relatively long wings and short tail.

5. *canariensis* Hartert and Kleinschmidt 1901, confined to the Canary Islands. Another small Raven similar to *tingitanus*.

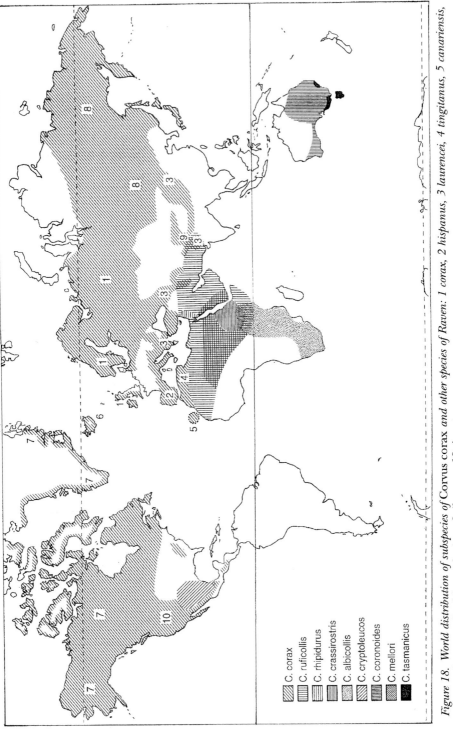

Figure 18. World distribution of subspecies of Corvus corax and other species of Raven: 1 corax, 2 hispanus, 3 laurencei, 4 tingitanus, 5 canariensis, 6 varius, 7 principalis, 8 kamtschaticus, 9 tibetanus, 10 sinuatus. NOTE: In north-eastern Africa, C. ruficollis and C. rhipidurus overlap the whole range of C. crassirostris, and C. albicollis overlaps its southern range.

6. *varius* Brünnich 1764, confined to the Faeroes and Iceland. It is slightly larger than the nominate subspecies, which it replaces here.

7. *principalis* Ridgway 1887, found in northern and eastern North America and Greenland. One of the largest subspecies, it is known as the Common or Northern Raven in North America.

8. *kamtschaticus* Dybowski 1883, occurring through eastern Asia from Siberia to Mongolia and Japan. It is not always well defined from *C. corax*.

9. *tibetanus* Hodgson 1849, found in the mountains of central Asia from Tien Shan and Pamirs to Himalayas and Tibet. This is the largest subspecies of *C. corax*, recorded at 6,400 m on Everest and normally living at 4,000–5,000 m.

10. *sinuatus* Wagler 1829, occurring through mountainous country of western North America, including the Rockies, but extending also through Central America to Nicaragua. This, the Western Raven, is a smaller form than *principalis*.

The smallest subspecies are found in the warmest and most southerly regions, while the largest are in the coldest northern and montane regions, conforming to Bergman's Rule.

After the nominate subspecies, the subspecies *hispanus, canariensis, varius, principalis* and, perhaps, *sinuatus* could be regarded as well studied, but the others are mostly rather little known.

OTHER SPECIES OF RAVEN

More divergent forms are given specific status, though all kinds of Raven across the world share certain features in common. They have predominantly black plumage, heavy bills and a croaking voice; they build similar nests, mostly in rocks and/or trees, and lay much the same type of egg. All are given to aerial displays and show flocking of juveniles and non-breeders. They favour carrion for food and become bold as scavengers, associating with people wherever human activities provide good food sources. Most are relatively sedentary and movements are related especially to marked seasonal climatic change. (For further details see Goodwin, 1986; Wilmore, 1977; Cramp & Perrins, 1994.)

African Ravens

South of the Mediterranean regions of Africa, *Corvus corax* gives way to the Brown-necked Raven, *Corvus ruficollis*, a smaller species (weighing only about half as much) with a brown, coppery tinge over its head, neck and undersides. It has a preference for desert habitats and occurs through north Africa from the Cape Verde Islands and the whole Saharan region to Ethiopia, Somalia and northern Kenya. Voous (1960) notes that it occurs also through Arabia and Iran to Pakistan, north-west India and Turkestan, but he casts doubt on its specific distinctness on the grounds that intermediates between *ruficollis* and *corax* occur in Iran and India.

The Brown-necked Raven is the form most generally adapted to hot arid conditions. Abel Chapman (1921) described a day-long train-crossing of the Nubian Desert in the Sudan, during which the only birds he saw were three Ravens, presumably of this species. His vivid description of this 'howling wilderness' leaves no doubt that it is one of the most inhospitable places on Earth, shunned by most other vertebrate life. The ecology and habits of this species appear to be similar to those of our Raven, though the sources of carrion are usually different and the diet includes locusts, crickets, termites, lizards, small snakes, and fruit of date palm (Cramp & Perrins, 1994). In north-east Africa is the smaller subspecies distinguished as the Dwarf Raven, *Corvus ruficollis edithae*, which overlaps geographically with the Brown-necked form. It nests in loose colonies, in smaller trees and bushes (especially acacias and palms) than the other, and not in rocks. Some ornithologists believe it merits full species rank.

Also in arid or semi-arid regions of north Africa is the Fan-tailed Raven, *Corvus rhipidurus*. It occurs mainly in the tropical zone between the equator and 20°N, but also along the Red Sea hinterland of Saudi Arabia and in Israel. Chapman (1921) described the Fan-tailed Ravens in the Sudan as 'eerie, wraith-like creatures with long necks and an immense crinoline-like expansion of the secondaries, so exaggerated as almost to coalesce with the short and rounded tail'. He drew sketches that matched this description, but a photograph of the bird in the Middle East shows merely a broad-winged, short-tailed bird. There are variations, however, and, in the north of their range, birds are distinguished as the race *stanleyi*, named in honour of the late Stanley Cramp. Somewhat vulturine in flight profile, the broad wings no doubt give the bird special soaring ability and it is said often to be seen rising in thermals. It is little bigger than the Carrion Crow but has longer wings. The voice is high-pitched and it gives falsetto croaks. When walking the bird has a habit of keeping its rather short bill open, as though panting. Nesting is in rocks, with separate pairs markedly territorial, but its biology and relationships with Brown-necked Ravens (with which it overlaps) are little known (Wilmore, 1977; Cramp & Perrins, 1994).

The White-necked Raven, *Corvus albicollis*, is widespread across the southern half of Africa, with an eastern distribution in the north, but becoming increasingly western with distance south. A bird with a thick, heavy bill, it is between our Raven and the Carrion Crow in size. It inhabits open mountainous country, including areas with some woodland, but occurs commonly around villages and encampments, and is often seen boldly strutting about safari lodges and camps. Wilmore (1977) says it has a good sense of smell and is the first to find dead animals, often leading vultures to them. It also travels long distances in search of food. Breeding period depends on location; it begins in February in Ruwenzori, but extends from September to November in Malawi, and October to December in Kenya.

The Thick-billed Raven, *Corvus crassirostris*, has, like the Dwarf Raven, a highly localised distribution, mainly in Eritrea, Ethiopia, south Somalia and the south-east Sudan. With its huge, almost bulbous bill and a white patch on the back of the head, it has recognisable affinities with the previous species, but is larger and about the size of our Raven. It is described as frequenting mountainous country at 1,500–2,400 m (Goodwin, 1986), but is often another camp-familiar and said also to be good at 'smelling out' dead animals (Wilmore, 1977). Flocks are sometimes alleged to cause crop damage, but it is generally regarded as a beneficial scavenger. It indulges in 'grand aerobatics' during the breeding season.

Wilmore says that, 'The true representative of the Raven in Africa is the Pied Crow *Corvus albus*', which is slightly larger than the Carrion/Hooded Crow and has a bill as massive as a Raven's. It occurs southwards from a line from Senegal to the Sudan. Pied Crows depend on human refuse and so favour populated areas. Their breeding is triggered by rain and so the nesting period varies between different parts of Africa, but after breeding they move from wet to dry areas, and so have irregular migratory habits. The species has spectacular morning and evening flights, gathering in flocks of 100 or more to circle high in the air, avoiding the extreme mid-day heat. It is parasitised by the Great Spotted Cuckoo, which lays up to three eggs in host nests.

North American and Australian Ravens

The New World has the American White-necked Raven, *Corvus cryptoleucos*, a smaller Crow-type bird than *C. corax*, with a stubbier bill. It is localised in the hot south and south-west (but not the west coast) of the USA and Mexico, and inhabits deserts, open plains, arid farmlands and foothill country.

The last group of Ravens is in Australia, which has three species. The Australian Raven, *Corvus coronoides*, is widespread across the eastern half of the country and along a south coastal belt to the south-west. It is partly a woodland species, along gum creeks, but occurs especially in open, savannah-type woodland and fairly open country, including farmland. Though mainly a tree-nester, in eucalypts, it has adapted to nesting on power-line towers and occurs on the edge of cities. A small Raven, little bigger than our Carrion Crow, its high-pitched wailing call sounds like a baby crying. In eastern Australia, but not the west, the distribution of this species corresponds with that of sheep. The close-grazed sheep pastures give easy availability of insects and lizards. It is regarded as having economic importance and has been much studied, especially in relation to alleged predation on lambs, particularly by flocks. Yet, as with Crows in Britain, there is relatively little evidence that these Ravens kill healthy lambs but quite a lot to suggest that farmers' husbandry contributes appreciably to the losses that are incurred. Control of numbers, which includes blasting of roosts, besides all the more usual methods, affects mainly the mobile flocks and leaves most of the territorial breeders unharmed. Banding has shown that 64% of youngsters die in their first year (Wilmore, 1977).

The so-called Little Raven, *Corvus mellori*, is almost the same size and overlaps in range with the previous species, though it is restricted to a much smaller area, in south-eastern Australia. In plumage, measurements, voice, behaviour (including mating and territory) and movements, it is different from the previous species. It nests in rather low trees, at up to 1,524 m in the southern Alps, and is semi-colonial, with five pairs per hectare, though nest territories are still maintained. This species breeds at the end of its second year (as distinct from the third for the Australian Raven) and occupies the breeding territory for only 3 months, after which family parties join up and travel around nomadically in search of food. It does not attack lambs but is less wary and more easily shot than the larger species.

The rarest species is the Forest Raven, *Corvus tasmanicus*, confined to Tasmania and two limited coastal areas of south-eastern mainland Australia. In size it equals *Corvus coronoides* and it is the heaviest Australian corvid, with the stoutest bill, and a deep voice. Though it nests in trees, especially in the wet sclerophyll forests and pinewoods, it is not only a woodland bird, but also occurs in a wide range of country, from open savannah woodland to farmland. Established pairs defend territories throughout the year, but juveniles and non-breeders form flocks with the other two species. A form with longer tail and wings but less massive bill in New South Wales is distinguished as subspecies *boreus*, but is thought to be losing ground through forest clearance on the New England Tablelands.

There are only three other Australian corvids; two of them – the Torresian Crow, *Corvus orru*, and the Little Crow, *Corvus bennetti* – are widespread Crows hardly less than the Little Raven in size. The third – the much smaller House Crow, *Corvus splendens* – is an unwelcome immigrant from India and Sri Lanka. All five of the native Australian corvids are notable for the white eyes of the adults.

CHAPTER 14

Intelligence in Ravens

The Raven has long been held in popular regard as the most intelligent of birds. Its ready adaptation to captivity, ability to mimic human speech, mischievousness and tendency to play tricks on other animals, and the general knowingness of its manner, have all helped to give the bird an aura of canniness and what the Victorians called 'sagacity'. It was regarded by some of the earlier ornithological luminaries, such as Charles Waterton, William Macgillivray and Alfred Newton, as the most intelligent member of the bird kingdom. And, when the pioneer of modern ethology, Konrad Lorenz, who kept many species of birds in captivity and studied their behaviour, described the Raven as 'that species [of bird] which I consider to have the highest mental development of all', we have to take special note.

Yet what do these seeming characteristics of intelligence amount to in practice? Heinrich (1990) has pointed out that, despite the general assumptions (including his own) of intelligence in Ravens, there is precious little objective evidence in support. Lorenz himself, despite repeated references to the Raven's high intelligence, has disappointingly little to say about it, and that mainly anecdotal as a result of chance insights provided by his few tame birds. Wilmore (1977) mentions work on brain structure which suggests that Raven brain development, with high hemispheric indices, shows evolution to higher degrees of complexity than in most birds – consistent with the possession of higher intelligence.

The activities of birds which add up to behaviour appear to be largely the result of a programmed inheritance of response to particular stimuli. When, without any prior conditioning by exposure to a certain stimulus and knowledge of the results of its response, a bird reacts in a way that is good for it, we call the process 'instinct'. The knowledge of the right way to react appears to be part of its constitutional make-up from birth, and is akin to the programmed instructions which determine the actions of an automaton or robot. Instincts are in-built across a whole species, or even wider groupings. Examples are chicks which beg for food on hatching, birds which undertake migrations without any parental guidance, and nest-building which conforms to a precise pattern of materials and structure for the species.

Birds and other animals also have a considerable ability to learn behaviour as individuals through a trial-and-error process involving rewards and disincentives: the conditioned reflexes which lead to habit formation. Some measure of intelligence may be involved at the outset as the creature 'learns' the best response for the circumstances but, once the habit is formed, it becomes automatic and is not under the mediation of conscious control. It might be argued that learning ability is a condition of intelligence, and the Raven's skill in acquiring tricks has an appealing 'smartness' to many people. An example is the tame bird which was able to throw coins for the reward of pieces of meat (Nelson, 1907).

The more critical issue is whether a bird can use its knowledge to adjust its responses to new, perhaps once-off, situations to give a better outcome than if blind instinct or habit were applied. This is the ability to *reason* which is the defining aspect of intelligence. As Krushinskii *et al.* (1982) have put it, elementary reasoning capacity is 'the ability of an animal to grasp simple causal relationships between objects and phenomena in its environment and to apply this knowledge in new situations'. It amounts to deductive thinking, and is to be divined especially in spur-of-the-moment behaviour. Our concern is not only to understand how well Ravens are able to do this, but to know whether they can do it better than other birds.

Some of the critical evidence comes from formal experiments with Hooded or Carrion Crows, cousins of the Raven and commonly assumed to be in the same high avian intelligence class, though Lorenz (1970) points out that Crows and Ravens differ in certain aspects of behaviour. Joe Crocker (1985, 1987), who worked under L. V. Krushinskii in Moscow University, has described the results of testing a group of six Hooded Crows with problems in which motivation was the finding of hidden food. Each Crow had an 'arbitrary' and a 'causal' problem to solve in finding the food. In the arbitrary task, there were two features of the receptacle containing the food – colour and shape – of which one was no more likely than the other to have any reason to be connected with food. The experimenter chose to place the food always in a green, apple-shaped container, as distinct from a red, hourglass-shaped one. The birds soon learned to pick the correct container, but could have been responding either to colour or shape, or both. Crocker then arbitrarily decided that shape should be the relevant cue and, in a second stage of the experiment, swapped the colours, noting how long it took to learn that red, apple-shaped containers now held the food.

The containers in the causal problem were also different colours (blue and yellow), but the second difference this time was that one was opaque and the other transparent; shape was kept identical for both containers. Food was always hidden in an opaque container. This introduced an element of rationality into the choice: food in an opaque container was hidden, whereas in a transparent one it obviously was not. Once the birds had learned to choose the opaque container, opacity should become the more powerful cue, and they should be less confused than before when the colours were swapped over in the second stage of the later experiment. This is what happened, suggesting that the Crows recognised and acted upon cues which had a cause-effect connection, instead of an unrelated one, with the food reward.

Crocker notes that, in other tests in which animals had to locate food which they had been shown, Crows, along with wolves, turtles and lizards, did well, whereas pigeons, voles, frogs and fish did badly. And when exposed to a dimensional problem, with flat and 3-D figures (rationally, food could only be under the latter), the Crow family scored better than wolves and dogs, but less well than monkeys, dolphins and bears. It is a moot point whether contrived tests such as these, involving such a limited range of features, are a fair measure of an animal's all-round reasoning ability – rather as human intelligence tests are a doubtful means of measuring a person's overall intellectual powers.

For other evidence of reasoning ability, we have to look at anecdotal cases of behaviour in the field that appeared suggestive but were isolated examples not subjected to replication or control. They have to exclude instances of habits which have evidently been learned by one individual from another, such as the frequent observation that Ravens in the deer forests of the Scottish Highlands will rapidly home-in on the report of a rifle, in anticipation of a feast from the 'gralloch' of shot deer left on the hill. Yet the first Ravens which learned this trick must have been associating cause and effect, and perhaps a number of Ravens in quite different places all learned the connection independently. An analogous situation was when researchers on Red Grouse began marking nests they had found with canes placed a short distance away. Crows in the area rapidly learned what the canes signified and the marked nests were systematically predated. The Crows were not smart enough to know when the canes began to mark the location of poisoned egg-baits instead, but the human observers learned a lesson all the same.

Rather on the same theme, a Ptarmigan nest on the high fjells of Norway was robbed by Ravens soon after I found and photographed it, even though the sitting bird had not been flushed. The two Ravens – seen hanging about at a distance – had probably decided that my interest in the spot was worth investigating. This was an isolated occurrence and hardly one from which to draw any conclusions, but Ravens have evidently learned to watch people in case they discard food, and might find nests by following their trail. They could then begin to make the connection that human interest in a certain place = nest discovered, and to observe the possibilities still more closely. An observant Galloway shepherd suggested to me that Carrion Crows had learned to watch humans examining Curlew nests, and predated these as a result – although they also seem well able to find Curlew nests without assistance. Konrad Lorenz observed that, in contrast to Jackdaws, a Raven will slip away to conceal objects (including food) when its companions are not watching and will seek places out of sight and not normally visited by them. He also noted that Ravens learned to store food out of *reach* of humans who had robbed them. Concealment behaviour was thus not wholly determined by instinct.

The Corvidae in general are especially alert to new food opportunities. At a petrol filling-station in Lapland, a Hooded Crow was perched on the roof just above the pumps. A local car pulled in, towing an open trailer on which was a sealed brown-paper sack. Within seconds, the Crow was down on the sack, had ripped it open and was bolting the grain that it held. I could not have guessed what the sack contained, but the Crow had evidently learned this at some point – at least to the extent of knowing that such objects were likely to contain edible things.

Much of the evidence for intelligence in birds, both experimental and anecdotal, concerns their responses to food, but sometimes there are other needs that appear to exercise their minds. My wife and I once saw an occupied Raven nest in a hole right under the fourteenth arch of a viaduct with 20 arches. It was quite inaccessible and we merely viewed it from ground-level far below, before retiring to the car beyond the first arch at the other end of the viaduct. Our position was at an acute angle to the line of the viaduct, which thus appeared as a solid wall, with only the entrances to the arches on our side visible, and we could not see the nest. When the parent Raven, which had been flying around just ahead, decided it was safe to return, it flew under the third arch at our end and we saw it no more, i.e. it returned to the nest under the cover of the hidden side of the viaduct and so avoided giving the nest position away. This could have been chance but, at the time, it seemed like a clever weighing of the circumstances in order to mislead us, for the bird had no means of knowing that we had already seen the whereabouts of its nest. The particular point of interest was that the value of the bird's manoeuvre depended entirely on the position, relative to the viaduct, that we had chanced to take.

A rather similar event befell me when I discovered the tree-nest shown on p. 149. I was walking up a small valley and making my way along the slope opposite to a little side-gill, which later proved to hold the nest, when a pair of Ravens appeared ahead of me. It was in the days before I had any binoculars, so I failed to spot the nest and headed up the valley in the hope of finding hidden rocks higher up. The Ravens kept ahead of me for some time, but eventually faded out over the head of the valley. I saw there were no rocks and returned towards the little tree-lined branch-gill, about 1 km down the glen. When I neared it, a Raven appeared in the air again and, when I reached the foot of the gill, the second bird flew out and gave the position of the nest away. After disappearing over the valley head in front of me, the two Ravens had clearly flown over into a parallel valley and doubled back down this before crossing over again to reach their nest, remaining hidden from me all the time. If I had not decided to return the same way, they would have fooled me into thinking there was no nest.

Verner (1909) described finding three Raven nest sites in Spain, situated in deep, vertical fissures of limestone tors, where the nests were quite hidden but holes in the rocks gave a front and rear entrance. It was never possible to see the birds leave their nests, unless two people walked on either side of these rocks to watch both exits, for the sitting bird always left by the opening hidden to a single passer-by. Although birds were seen around these places, it was some time before Verner realised exactly where the nests were, so good was their concealment and their owners' ability to avoid giving the game away. While the Raven's eye for nest sites giving maximum difficulty of access could be innate, Verner's experience suggests a little more than this.

Smith (1905) related a number of incidents involving tame Ravens which might be regarded as evidence of their knowing ways. Perhaps the most convincing was the following:

> Another Raven, kept in a yard, in which a big basket sparrow-trap was sometimes set, watched narrowly the process from his favourite corner, and managed, when the trap fell, to lift it up, hoping to get at the sparrows within. They, of course, escaped before he could drop the trap. But, taught by experience, he opened communications with another tame Raven in an adjoining yard, and the next time the trap fell, while one of them lifted it up, the other pounced upon the quarry.

Wilmore (1977) mentions an instance where a Raven in a wire-netting cage was freed by some of its wild fellows, which dug under the cage from the outside, while the captive dug from the inside, until there was a hole big enough for it to escape through. At least two cases are recorded where Ravens in wooden-sided cages next to other captive birds have dug holes in the wood big enough for them to grab their neighbours. A Kestrel was caught and killed in this way (Smith, 1905). In the other instance, the Raven baited the hole with its own food; chicks of a domestic fowl in the next cage came to the bait and were promptly caught (D'Urban & Mathew, 1892).

Morris (1895) knew a tame Raven which habitually went hunting with a dog that it had grown up with. The dog entered the cover and drove out hares and rabbits, which the Raven seized, and the dog then came to its assistance so that the victim never escaped. The bird was said to be more use than a ferret and would enter a barn with several dogs, to enjoy rat-hunting. Morris also described how sparrows were baited with corn in a yard and then shot from a window open just wide enough to take the gun muzzle, neither gun nor shooter being visible from outside. A tame Raven in the yard had never seen a normal shot fired, yet, when anyone appeared with a gun heading for the place of concealment, the bird showed great alarm and hid itself, but watched and pounced out on any birds that were shot.

Ravens trained by Otto Koehler in Germany in the late 1940s could count up to six or seven. One Crow studied by the University of Mississippi learned to watch a clock face, wait until the hands pointed to 12, and then operate a lever to obtain a reward of grain (Wilmore, 1977). Kilham (1989) had a tame Raven which was fond of sweeping sand or dirt with swings of its bill and, on occasion, was seen to pick up a small flat piece of wood to use as a tool in this exercise.

The propensity of tame Ravens and other corvids for taking and hiding bright and metallic objects, such as jewellery, coins and spoons, is well known. Golf balls are another such trophy. The concealment part of this behaviour is clearly related to the well-developed caching activity and, although these objects have no food-value, they are sometimes mixed with items of food. Charles Dickens, replacing one tame Raven that had died with a second bird, said that, 'The first act of this Sage was to administer to the effects of his predecessor, by disinterring all the cheese and halfpence he had buried in the garden'. Presumably, the birds have some intuition that the uneatable objects have some value or meaning, and store them for a purpose – which, at present, we hardly understand.

The functional meaning of voice mimicry in birds is not well understood, but may have little to do with reasoning ability. When captive animals are rewarded with food on imitating their owners, there must be a strong inducement to continue and expand the copying of voices and words. Some captive Ravens have been described as having not only an aptitude for learning human speech, but a diligence in extending their repertoire and polishing their performance. Lorenz (1952) described how his tame Raven, Roah, had a particular set of flight movements and an innate call which it gave when wanting other Ravens to follow it; yet, when the bird wished its master to follow, it would give the same movements but call its own name, with human intonation, instead of the innate call. Lorenz had not trained the bird in this context, and said that, 'The old Raven must, then, have possessed a sort of insight that "Roah" was my call-note!'

Some observers have regarded the wild Raven as an especially wary bird and one that it is difficult to trap. They seem to recognise jaw-traps as engines of destruction and to avoid them at all costs, though they will not hesitate to attack victims already caught in such traps. Lorenz (1970) noted that his tame birds reacted with extreme fright to any alteration of *known* objects. Tony Cross found that two pairs deserted nests with young because too obvious signs of human presence had been left close-by. Yet Heinrich (1990) had no difficulty in inducing Ravens to enter baited cage-traps and catching them there. And, once caught, many of the birds seemed to become tame quite rapidly. Several writers have remarked on the bird's apparent ability to recognise people carrying guns, and to stay well out of range, while coming close to shepherds and farm-workers. Although where persecuted the Raven usually remains the wildest of the wild, it soon becomes confiding where it is unmolested and food is on offer. Des Thompson saw one walking about and picking up scraps beside a throng of hikers on Helvellyn summit and, when Roderick Corner was eating his lunch on the Thirlmere fells, a Raven flew down and sat on a rock about 3 m away. This ability to make fairly rapid adjustments of behaviour suggests that the bird is a good learner.

Doubtless many of the stories about Raven intelligence have been coloured by projections and wishes to endow the bird with human qualities – and even more with magical ones – and this must be especially true the further back in history one goes. The earlier associations of the species with humankind were so mixed up with myth and superstition that any accounts of them have to be discounted as credible evidence of intelligence. There is the tale from Ancient Rome of the Raven which, in a time of great heat, dropped pebbles into an urn to raise its water level until this became reachable to drink – but this same story has had at least one other animal as the operator. Many of the time-honoured tales of Ravens are the purest whimsy that say far more about their human inventors than about the bird. Only further careful experimentation of a rigorous, scientific kind is likely to throw more light on the subject of Raven intelligence.

APPENDIX 1

Distribution of Raven flocks and roosts in Britain and Ireland

SOUTH-WEST ENGLAND

Cornwall

A flock of 34 birds flying north-west at Bude, 1 September 1936 (H. G. Hurrell). About 150 counted in a communal roost in the Four Burrows wood (also a Rook roost) near Carvinack Farm; mostly fed in a field near Chacewater where butcher's offal was spread, 4 October 1945. During the breeding season numbers dwindled to 16 on 14 March. The last recorded use of this feeding area was on 12 September 1949 when 25 birds were present (Ryves, 1948; Penhallurick, 1978). In 1948, about 100 Ravens were seen from early autumn until the year end flying over Altarnun, with 112 on 20 September. The roost was in a small wood at Clitters on the edge of Bodmin Moor; H. K. L. Whitehouse counted 170 here on 17 September 1955 and there were still 120 on 27 January 1956. Up to 1978 there were no further reports of such large gatherings, 'though sizeable flocks are still seen most winters on the higher moors', e.g. at Dozmary Pool there were 47 on 14 January 1969, 61 on 30 November 1972 and 40 on 13 November 1975 (Penhallurick, 1978). More recent counts are of 17 on Rough Tor on 7 March 1993, 14 at Bosistow on 14 October 1993, and 52 at Crowdy Reservoir on 21 November 1993.

Devon

A flock of 42 on Dartmoor in 1951. A roost developed in January 1955 at Wrangaton, the birds being attracted by the carcasses of sheep trapped in snow-drifts, though they moved between two woods; it grew to a maximum of 86 on 11 April, but dwindled through the summer and was gone by September (Hurrell, 1956). At Winkleigh: 95 in 1960 to 130 in 1962, and 100+ on several subsequent occasions (Moore, 1969). Much larger assemblies have since been reported (Dare, in Sitters, 1988), with a roost of 345 near Torrington in April 1977 – an unusual number for the breeding season. G. Kaczanow also reports a regular roost in a Dartmoor wood where a breeding season total of approximately 100 birds was recorded in 1994. As well as large summer gatherings at a few communal feeding places, autumn and winter roosts of 50–100 occur elsewhere; up to 25 on Lundy.

Somerset

A gathering of 110 on dead sheep, Exmoor, 4 March 1978 (Somerset Ornithological Society, 1988). More recent Bird Reports for the county are for small numbers (3–6), seen widely, and, in 1991, 12–18 at Vinnicombe from August, and c. 20 at Elsworthy on 9 September.

WALES

Lovegrove, Williams and Williams (1994) state that regular roosts of 20–30 birds are commonplace and up to 50 not infrequent.

Glamorgan

In the mid-1960s, up to 60 were seen in several roosts in the Swansea valley, near Neath, in the Morriston and Llansamlet areas, and at Llanharan. There were fewer in succeeding years, but 57 were seen at Llanwonno on 4 April 1977 – the largest recent assembly. In the 1980s and early 1990s, up to 20 were widely reported in most winters, the largest flock being 50 at Mynydd Eglwysilan on 30 September 1990 (Hurford & Lansdown, 1995). A land-fill site near Merthyr Tydfil was used by over 30 birds, even during the breeding season in 1993–95 (A. Dixon).

Pembrokeshire

R. M. Lockley saw a flock of 126 roosting on Skomer in September 1953. Ridpath (1953) saw 31 at a knacker's yard on Cuffern Mountain, and Mylne (1961) recorded 150 in this locality and another 50 near Haverfordwest, 24 km away, in April 1959. More recently, up to 80 Ravens have been seen on Skomer in September when the fledgling Manx Shearwaters are evidently an attraction, since the flock disperses in October. Up to 250 occurred at Dudwell, drawn by discarded offal from a slaughterhouse, and 170 were seen here as late as April. There are regular communal roosts in winter, with up to 150 at Walwyn's Castle in 1954 and up to 263 at Roch in 1964, but they are mostly smaller, at 30–50 birds (Donovan & Rees, 1994).

Carmarthenshire

D. K. Bryson saw 40–50 near Cwmanne on 27 March 1946; most were paired and displaying, with continual croaking.

Breconshire

Every autumn for several years, H. A. Gilbert saw a flock over a hill near Brecon; they arrived in September and, after November (when most were flying in pairs), numbers decreased and they disappeared after Christmas; maximum number 32 in November 1944. Peers and Shrubb (1990) mention flocks/roosts of 66 near Pontsticill in February 1982 and 43 on Mynydd Eppynt in early 1989.

Radnorshire

Peers (1985) records a post-breeding flock of 30–40 birds together in the Elan Valley or Radnor Forest.

Cardiganshire

Ingram, Salmon and Condry (1966) said that communal roosts and flocks were occasionally seen in all months, though the numbers they mentioned were not large, e.g. 21 at Aber-mad in mid-September 1955. Well over 100 were seen on sheep carrion in the Tywi valley in March 1975 (Lovegrove, Williams & Williams, 1994). Davis and Davis (1986) found that, in the Cardiganshire uplands, Raven flocks were especially associated with a refuse tip and nearby abattoir, where often 50 and sometimes up to 100 birds were seen during 1975–80. Most of the birds roosted in trees near the tip and were often seen flying towards the tip in the late afternoon from points up to 15 km away. In spring the flocks became widely dispersed over the sheepwalks in smaller groups, pairs and singles.

Merioneth

'Some years ago, a flock of 50 or 60 ravens used to roost together, regularly, at Craig-yr-Ogof on Bwlch-y-Groes [between Llanuwchllyn and Dinas Mawddwy]; but from having been so much shot at there, they have now deserted it for a quieter retreat across the valley. There, I have sometimes seen considerably more than that number come in to roost at night, a body which must have represented the collected birds from a large area of country' (Bolam, 1913). A roost in a pine clump near the Dovey estuary was observed by W. A. Cadman; numbers rose from 30–40 in 1937 to 70 in 1939 but fell to 20–30 by 1946–47. It was used all year, with a peak in September and fewer in mid-winter (when most arrived to roost in pairs), but no further decrease during February–April and smallest numbers in May–August.

Caernarvonshire

Orton (1948) noted that Raven flocks were unusual in the mountains of Snowdonia during the early years of the present century and the largest group he saw was of only 8 birds.

Only small parties, of 10–12 birds, were occasionally seen on the Carneddau during October–February 1950–53; no roost was located here or elsewhere in Snowdonia (D. A. Ratcliffe). Snowdon summit has long been a regular haunt of non-breeding Ravens, but usually in small numbers, and a flock of 30 there on 22 April 1969 was unusual (Dare, 1986a). Dare estimated that probably fewer than 40–50 unattached Ravens frequented Snowdonia during 1978–85. In May 1977, I saw 30+ birds at a rubbish tip near Pwllheli in the Lleyn peninsula. Alex Turner also saw 30 Ravens circling around the cliff below Snowdon summit on 24 January 1987.

Anglesey

The largest British–Irish roost in recent times occurs in the Newborough Forest, planted on the sand-dunes of south-west Anglesey. It is said to have existed for at least 10 years, but counts are available only for 1995–96, as follows (Iolo Williams and Nigel and Daniel Brown):

1995   3 January – 189; 5 January – 226; 21 February – 267; 27 October – 250; 19 November – 351; 30 December – 330.

1996   21 January – 510; 27 January – 431; 25 February – 583; 1 April – 310; 25 April – 325; 24 May – 759, within 1 ha; September – 856; 30 November – 1384.

This large number may explain why only small flocks/roosts are known in adjoining inland districts of north Wales.

Denbighshire

A flock of 150 was seen on Mynydd Hiraethog in May 1992 (Lovegrove, Williams & Williams, 1994). A gathering of 30 birds, loafing and feeding on sheep pastures on the Migneint-Hiraethog area on 30 April 1982, consisted wholly of distinct pairs (Dare, 1986a).

MIDLANDS

Herefordshire

A flock of approximately 100 birds in the winter of 1995–96 on Garway Hill (K. Mason).

NORTHERN ENGLAND

Cumberland and Westmorland

Lake District: In Kentmere, on 1 March 1931, Coombes (1948) watched a somewhat dispersed flock of probably 50–60 Ravens, some working in groups along the hill 'flying round and round each other with much croaking', and one group which perched as 7 distinct pairs on a particular rock. Coombes saw another flock of 30–50 drifting along the peaks from Crinkle Crags to Great End in the Scafell range on 6 and 11 April 1947, never attempting to feed, despite the abundance of sheep carcasses. A flock of 16 Ravens flew from the crags of Deepdale, under Fairfield, on 3 September 1949 and 30+ from the same place on 21 September 1950 (D. A. Ratcliffe). On 11 August 1957, E. Blezard watched 50 Ravens in a loose flock over the fell tops at Mardale Head, and saw at least 46 fly into Blea Water Crags at dusk, evidently to roost, on 2 November 1958.

In 1985, Geoffrey Fryer first noticed a regular evening flight line of Ravens over his home in Windermere to a roost on low wooded hills west of the lake. The birds pass as singles, pairs or small groups of up to 8 or 9, but once 18, and mostly follow a narrow corridor only *c.* 200 m wide. The onset of flight time varies between 19.00–19.50 h BST, and lasts from 10 minutes to over 1 hour. The flight is continued through the breeding season, with numbers ranging from 29 on 13 April 1985 to 62 on 26 April 1985, and 71 on 27 April 1996 to 53 on 2 May 1996, but varying widely from one day to the next. D. Hayward counted approximately 80 birds in this Claife Heights roost during the winter of 1995–96.

Pennines: The following observations cover the period 1909–64.

1909	Rundale: 25 April – 19; 7 May – 21; 5 August – 24; 26 October – 23; 3 November – 57; 4 November – 48 (E. B. Dunlop and A. Mason, in Blezard, 1954).
1910	Rundale: 21 January – 30, nearly all in pairs; 4 February – 27; 1 March – 30; 4 March – 24; 22 March – 6; 23 March to 30 July (5 visits) – 0; 30 July – 2; 1 August – 7; 2 August – 7; 5 August – 6; 19 August – 8; 27 August – 26; 29 August – 34; 10 September – 42; 7 October – 40; 11 November – 27; 4 December – 34; 14 December – 38 (E. B. Dunlop and A. Mason).
1911	Rundale: 3 February – 4; 23 and 27 February – 0; 26 April (and several previous visits) – 0; 4 May – 6; 9 May – 10, mostly in pairs; 28 May – 0; 12 June – 0; early August – up to 5; 12 September – 15; 1 October – *c.* 30; 4 October – *c.* 43; 5 October – 52; 25 December – 17; early spring: small numbers attributed to mild weather by Dunlop, but more likely birds were in alternative roost site (E. B. Dunlop and A. Mason).
1912	Rundale: 19 January – 36, flying in pairs; 1 March – 6; 3 March – 0; 14 May – 18, in pairs; 3 November – 50 (E. B. Dunlop and A. Mason)
1913	Rundale: 26 January – 30, nearly all in pairs; 2 March – 35; 25 March – one pair; 30 October – 40 (E. B. Dunlop and A .Mason).
1925	Greencastle, Knock Ore Gill: 28 June – 3; 27 September – 27 (E. Blezard and R. Graham).
1931	High Cup Dale: 4 June – numerous recent droppings and feathers, birds not present that day (E. Blezard).
1951	Locality not stated, but either Scordale or High Cup Dale: 18 November – 31, noticeably in pairs (R. W. Robson).

1957–58	Scordale: between October and February – up to 20 (W. Atkinson).
1959–60	Wild Boar Fell: winter – up to 20 (W. Thompson).
1964	High Cup Dale; 1 January – 27 (G. Horne).

Yorkshire

Although, to the south of the Lakeland Pennines, small numbers of non-breeding Ravens are still widely seen, large groups appear to be absent. A flock of 12 in Arkengarthdale in January 1972 is one of the more notable records.

Northumberland

There are again few records of sizeable flocks. Brian Harle found records of a winter group of 17 on the Wark moors in 1937 and 24 at Uswayford, Coquetdale, in 1955. The 1982 Ornithological Report for Northumberland and Durham mentioned a flock of 12 on the border with Cumbria near Alston on 24 April.

Isle of Man

'After the breeding season Ravens are seen in family parties and from early September occur in small flocks of up to 30 – recently a flock of 46 was seen on Ayres heath in early November, probably attracted by the rubbish tip where smaller flocks have previously been noted'. A quarry roost in Patrick held up to 40 birds. Flocks of non-breeding Ravens are sometimes seen in March and April (Cullen & Jennings, 1986).

SOUTHERN SCOTLAND

During many years of bird recording in Galloway, Donald Watson had records of only modest Raven flocks. A party of 20 was seen in the upper Ken Valley by L. A. Urquhart on 4 September 1952, but no more were reported until 1979, when 10 were noted by R. N. Hogg on moorlands west of New Galloway. The largest number, 21, was seen near Talnotry, a breeding area in the central Stewartry hills, on 31 January 1981 by K. C. R. and H. S. C. Halliday. Since then, groups of 9–14 have been recorded on eight occasions in widely scattered upland localities. The largest known roost has only just been discovered, by J. Adair, at a woodland near Kirkcowan, Wigtownshire: at least 27 and probably over 30 birds were present on 6 April 1996. Chris Rollins saw 23 Ravens mobbing a juvenile Golden Eagle in southern Ayrshire in September 1996. No records of flocks for the eastern part of the region have come to light.

SCOTTISH HIGHLANDS

Arran

On 7 December 1991, an assembly of 40 Ravens was seen mobbing a Golden Eagle in Glen Chalmadale, evidently contesting a deer gralloch (A. R. Church, *per* J. Mitchell).

Perthshire

Near Pitlochry, crag roost: 'during winter' 1941 – 120; 18 April 1943 – 80; 1 May 1943 – 40; By 3 May 1943 – all gone (E. J. Ferguson).

Glen Lednock, Comrie, crag roost: 12 December 1974 – 51; 5 October 1975 – 34; 7 December 1980 – 50+; 11 January 1981 – 50+ (Scottish Ornithologists' Club).

Glen Lochan, Amulree, crag roost: April 1982 – 90 (E. Day); 3 May 1982 – 12, in ones and twos mainly (P. Stirling-Aird).

Glen east of Ardtalnaig, Glen Almond: April 1990 – 12.

Wendy Mattingley has followed Raven roosts in central Perthshire during 1990–95, with the following results:

1990	Glen Quaich, Amulree, woodland roost: February – c. 60; 4 March – 70; 11 March – 25–30; 24 March – c. 8; 11 April – 3–4; August–September – 0; 28 October – c. 15; 4 November – 55; 1 December – 56+.
1991	Glen Quaich: 6 January – 67–82; 20 January – 50+; 3 February – 20; 24 February – c. 20; 11 March – c. 20; 6 October – small numbers; 26 October – 0; 9 November – 1; 1 December – 15; 8 December – 60; 24 December – 0.
1992	Glen Quaich: 12 January – 0; 24 October – 69+; 14 November – c. 50; 12 December – 68.
1993	Glen Quaich: 30 January – c. 15; 7 and 21 February – 0; 24 October – 0. Amulree: 6 February – 25–30; 20 February – c. 20; 5 March – 20+; 21 March – < 20; 2nd week: September – 0; 24 October:21+; 13 November – 0. Logiealmond: 4 December – 122; 10 December – 166.
1994	Glen Quaich: 6 February – 0; 25 September – 0; 26 November – < 5; 9 December – c. 20; 17 December – 0; 28 December – 2–3. Amulree: 11 March, 26 November, 9 December – all 0. Logiealmond: 4 January – 160; 22 January – 193; 30 January – 155; 5 February – 164; 20 February – 53; 20 March – c. 40; September – occupied; 22 October – 79+; 6 November – 73; 27 November – 53; 17 December – 52–67; 24 December – 75.
1995	Glen Quaich: 15 January, 18 February, 11 March – all 0; 10 October – 56; 29 October – 51–61; 31 October – 38; 19 November – few or none; 3 December – 5. Logiealmond: 7 January – 84; 31 January – 60–70; 12 February – 121; 20 February – 67; 6 March – 76; 19 March – 44; 3rd week April – 0; 10 September – c. 10; 10 October – few or none; 22 October – 26; 29 October – 29; 25 November – 57+; 3 December – 20–50. Kenmore, Loch Tay: 12 March – 31–37; 4 April – 12; 8 September – c. 20; 15 September – a few; 9 October – 36+; 12 October – 80–100; 19 October – 111; 29 October – 25; 13 November – 20+; 19 November – 50+; 26 November – 35+; 3 December – 140–150. Kenmore, 2nd site: 8 September – c. 80; 15 September – 49; 24 September – 69–80. Kenmore, 3rd site: 15 September – 53; 24 September – 23.

Overall numbers in mid-winter for these mid-Perth roosts are evidently c. 200, and the different flocks are believed to represent a single population. Co-ordinated counts at all roosts gave 135–145 birds on 29 October and 165–205 on 3 December (W. Mattingley, L. Steele, E. Cameron, K. Thomson).

On 14 January 1996, Wendy Mattingley discovered a new roost with 307 Ravens in northern Perthshire. Local comments suggested that it may have been in existence for several years previously. It is sufficiently far from the mid-Perth sites to hold a completely separate population, though this cannot be proved.

Angus

Mostly small numbers (up to 7) non-breeding birds have been seen in recent years, but numbers have slowly increased (R. Downing). A flock of 24 was about in Glen Ogil in March 1983 (A. Rollo, per A. G. Payne) and peaks of 24 Ravens were seen on 7 January 1993, and 23 on 9 September 1994, both parties heading south to south-east (K. Slater)

Aberdeenshire, Banffshire, Kincardineshire

Ravens are seen mainly in the stalking season, feeding on gralloch – small groups of up to 8 birds and occasionally up to 30. A party of 12 was about the main corrie of Lochnagar on 4 September 1985 (A. G. Payne) and groups of 29 and 7 birds were near Glas Maol, on the Angus border, in October 1987.

Argyll

In the south, Kintyre has sizeable flocks, with over 100 at a roost on the Largie estate in 1989. In mid-Argyll, 245 birds were reported at one site on 31 August 1994. The rubbish tip at Moleigh near Oban became a well-known resort of non-breeding Ravens, with up to 100 in 1991 and a spring peak of 64 on 17 March 1994 (data from Argyll Bird Reports). Despite these assemblies, Chris Thomas found little evidence of 'floaters' during the breeding season in mid-Argyll in 1989–91. There were occasional singles or pairs, but no more than four were seen together.

North of Loch Linnhe, Jeff Watson reports small spring–summer flocks: near Drimnin, Morvern – 11 on 15 April 1982; near Acharacle, Sunart – 15+ on 13 May 1982; Claish Moss, Sunart – 8 on 27 March 1983; east of Kilchoan, Ardnamurchan – 20+ on 12 July 1984. Hancox (1985) recorded flocks of 30 and 20 during September to mid-October 1977 and 1978, and 25 in February 1979 at Drimnin in the north of Morvern, but only smaller groups of up to 11 birds in the spring and summer.

Inverness-shire

On the small crags of Beinn a' Chlaidheimh, a grouse-moor area on the border with Morayshire, I watched a flock of around 100 Ravens gather to roost at dusk on 22 August 1959.

Sutherland

Harvie-Brown and Buckley (1887) said that over 100 roosted in a rock near Balnacoil up to 1878. No recent roosts of this size are reported, but Angus (1983) says that flocks of up to 50 occur in the straths in winter. A. Vittery saw 52 near Kinbrace on 10 September 1996.

INNER HEBRIDES

Islay

Roost in Islay estate woodlands: December 1973 – 95; early in 1975 – 100+; December 1977 – 72; 1985/86 – 105 (Elliott, 1989); 20 October 1994 – over 100 (Argyll Bird Report).

Mull

Tobermory rubbish tip:. 23 November 1985 – 100; 19 February 1989 – 35 (Argyll Bird Report); 1995 –50–60 regularly; May 1996 – 160 (P. Haworth).

Tiree

Rubbish tip: 1987 – 32; 23 June 1989 – 28; 25 April 1993 – 27 (Argyll Bird Report).

Skye

Parties of varying size, to a total of at least 100, assembling before flying into a roost on coastal cliffs at Tarvaig, north of Portree, 20 June 1962 (D. A. Ratcliffe).

OUTER HEBRIDES

Ravens are especially abundant on Lewis during winter near Stornoway, where large numbers roost in the woodlands at night and forage by day for carrion and on the rubbish tips. A count of 200 on a rubbish tip near Stornoway was made on 29 June 1981 (Cunningham, 1983). On North Uist, A. B. Duncan counted 276 at a roost near Newton in December, 1938. Also on North Uist, small groups of 7–9, and one of 20, added up to 104 on 13 February 1996 (W. Neill). At a roost near the southern end of South Uist, numbers were 64 on 30 January, 92 on 5 February and 80 on 22 February 1996. By April 1996, numbers at the North Uist roost had risen to about 300 birds, and may have included all the non-breeders from South Uist, as none remained there (W. Neill). On St Kilda, several Ravens were constantly foraging in the rubbish tip of the military encampment in 1959, but it was unclear whether they were breeders or non-breeders. The nearest nest, with young, was only 800 m away.

ORKNEY

Flocks of 20–25 birds were frequent where food was plentiful, and a regular roost of 100–150 was known on a Mainland cliff in 1974–77. The largest roost in 1981–82 was of *c.* 60 (Booth, Cuthbert & Reynolds, 1984). The roost has moved to a place difficult to observe and count, but at least 89 Ravens were present on 26 April 1993 and at least 80 on 16 May 1994 – more probably 100+ on both dates (C. J. Booth).

Fair Isle

Flocks of 40–45 reach the island occasionally between late March and May, and of 25–30 also during September to October (Dymond, 1991).

SHETLAND

Nineteenth-century observations on Raven flocks are noted above (pp. 104, 120). A list compiled by Peter Ellis from the Shetland Bird Reports shows that flocks of 40–70 birds were frequently seen at well-scattered localities across the archipelago during 1975–92. Favourite gathering places, with plentiful food, were Fitful Head, south Mainland (abundant rabbits) – 40–79; Boddam, south Mainland (slaughterhouse) – 35–56; Lerwick area, central Mainland (rubbish tip) – 50–130+; various localities, north Mainland (rubbish tips) – 19–150; mid- and north-east Yell (rubbish tip) – 15–80; Lambaness area, north Unst (rubbish tip) – 43–150. Flocks often roosted close to their feeding places. During construction of the Sullom Voe oil terminal large rubbish dumps attracted over 300 Ravens which roosted nearby and a roost of over 300 was noted on cliffs south of Lerwick in August 1968 (Ewins, Dymond and Marquiss, 1986; P. K. Kinnear). Despite the earlier observations that Shetland Raven numbers increased in autumn and declined again in the spring, in recent years there was little relation between flock size and time of year. Flocks of up to 130 have been seen during the breeding season and a total of at least 400 flock birds was known during the early 1980s. Numbers have tended to decrease subsequently, with 70 birds as the largest flock recorded since 1988.

IRELAND

I am indebted to Declan McGrath for collating recent records from the Irish Bird Reports and other sources. Praeger (1937) reported flocks of 27 on Connor's Pass and 17 on Brandon Mountain, both in the Dingle peninsula of Kerry, and 56 near Malinmore, Donegal, in the immediately preceding years. Connor's Pass appears to be a long-term roost, for over 50 birds were counted there on 27 April 1967, and 70 on 28 December 1992. In May 1961, 20 were roosting in a clump of ash trees at Tallaght, near Dublin; up to 25 were at a roost at Aghavarragh, Wicklow in August 1975; and 80 at Rockwell, Tipperary on 31 October 1991. In 1991, a flock of 40–45 birds was seen throughout the year at Benduff slate quarry and rubbish dump in Cork. The largest reported roost for Ireland is of 150 Ravens in woodlands near Ashford Castle, Galway, in the early 1980s, but this had declined to only 35 birds in September 1988. In the Glen Veagh National Park, Donegal, Mac Lochlainn (1984) recorded flocks of up to 16 birds in summer. On the coast, a flock of 80 was seen at the Cliffs of Moher, Clare, on 30 August 1979, while at Cape Clear, Cork, the highest day count was 61 on 2 August 1980. Other Cork coast records were 76 at Dursey Island on 9 October 1986, 42 at Cape Clear on 16 September 1987, and 32 at Mizen Head on 14 September 1988.

APPENDIX 2

Calls of the Raven

The calls of the Raven are difficult to deal with, partly because there is so much variation, but also because different people render them differently in making phonetic descriptions. Some of the interpretation of function also seems to be subject to disagreement and is sometimes little more than guesswork. Although it is about the species in North America, the account by Heinrich (1990) seems admirable for its justified caution in drawing conclusions about the functions of different Raven calls. He says that Ravens probably have a greater variety of calls than any other bird. The oft-repeated statement that the species utters no less than 64 recognisably different calls appears to date from the days of the Roman augurs, to whom they had immense significance (Forbes, 1905). Chapman (1924) commented that, 'The repertoire of the Raven is surprisingly varied. Perched on some jutting crag – possessing the earth – he will go on soliloquising in a dozen different keys and cadences'. Heinrich comments that, despite the Raven's voice receiving probably more attention than any other aspect of the bird, he is convinced that there is nothing that we know less about. He finds that trying to match published descriptions of calls with his own experience is a frustrating experience, and that even the objectivity of sonagrams has been of little help. I could not agree more.

For Europe, Goodwin (1986) summarised the available information, drawing on the studies of captive Ravens by Gwinner (1964). Cramp and Perrins (1994) have made a fuller synthesis of published work, integrating different descriptions and reproducing sonagrams of the main calls; they recognise 12 adult calls, with a few variations, and provide statements of function. I shall try to provide a brief account, based on my own field experience, supplemented by published work where necessary.

The most familiar Raven note in Britain and Ireland is that described by Cramp and Perrins (1994) as the 'pruk-call', but which my ears receive as 'crok', usually uttered as a rather measured double 'crok-crok', especially when the bird is flying, but also from a perch. Sometimes it seems closer to 'rok', and can sound almost like a hoarse bark of a dog. The note may be repeated several times, with short intervals between. It is far-carrying and, in still air, audible up to at least 2.5 km.

This call appears to be the same as the 'quork' of Heinrich (1990), which I recognise as a less usual variant, and the 'kra' of Gwinner (1964), which I do not recognise at all (but see below). I also have difficulty with the interpretation of function given by Cramp and Perrins (1994), that it 'almost always appears to involve real or imagined threat from non-conspecific enemy'. It appears to me most usually to be a kind of advertisement call, and is often given by distant or high-flying birds, well away from a nest and at any time of the year. Perhaps it advertises presence to other Ravens, though it is also given by birds in flocks. It may well have a territorial significance at times, as Heinrich suggests, though he remarks that there are many variations and they undoubtedly convey various meanings. In situations where mild alarm might be appropriate, as when humans approach a nesting haunt, much the same call may be given, but it tends to take on a more urgent tone, sometimes rising in pitch and increasing in frequency. It also changes into a distinct 'tork-tork', and a male with a sitting mate will give a rapid 'tork-te-tork-tork'. Another variant is a distinct, rather drawn-out and resonant 'glonk', sometimes repeated.

Some birds give a distinctive alarm call when humans are close to or at the nest – a rapidly repeated, staccato 'crok-crok-crok-crok', given with a characteristic winnowing flight of rapid, short wing-beats, head pulled in and bill pointed downwards. Each note is sharper than the normal 'crok' and becomes almost 'crak' at times. Not all nesting Ravens make this call and some voice alarm with angry-sounding 'raark-raark-raark' cries, frequently delivered from some perch close by and rising to the intensity of a rasping screech. A shriller variant of this note is given when a Raven is attacked by a stooping Peregrine, as well as during apparently territorial chases of one Raven by another. This call is usually heard at a distance and sounds more like a snarl than the typical alarm directed at humans – a rapidly repeated 'aark-aark-aark' given on an ascending scale as a falcon hurtles down in attack. Although Gwinner (1964) describes his 'kra' note as the typical Raven call, he then says it is particularly given during 'wild flight chases', which leads me to wonder if he is referring to this snarling call.

There is great variation in the calls to be heard between one bird and the next. Ravens idling together in flight in their nesting haunts, and not obviously alarmed, often make a liquid gurgling sound, ending with a distinct 'plop' – as though drawing a cork from a bottle, but in the wrong sequence! Its significance is unclear. Some birds give a resonant 'goingg', like the twanging of a stringed instrument, as humans approach the nesting place. The same individual can ring the changes on the standard 'crok', with 'craark', 'crork', 'quork', 'cork', 'glock', 'glank', 'clonk' and a metallic-sounding 'clink'. Individuals also vary enormously in their demonstrativeness to human intruders, some keeping up a great deal of noise while others remain almost silent.

Heinrich (1990) describes a distinctive 'yell' given by juvenile Ravens when they wish to summon other non-breeders in defence of a food store against the more dominant territorial adults. He notes that it is similar to the call juveniles make when they want to be fed: higher-pitched than the typical 'crok', and evidently correponding to the 'rüh' call of Gwinner (1964), regarded by him as a location call. Recently fledged juvenile Ravens call a good deal, in evident imitation of the parental 'crok' but with a higher pitch and sounding more like a squawk. Heinrich often heard a distinctive 'knocking' sound and found that other students of Raven voice recorded a similar call. It was described by Zirrer (1945) as 'a loud, metallic, bell-like note, well imitated by the striking of a light hammer on a heavy piece of tin'. Apparently this note is given by females, and Goodwin comments that similar mechanical sounding calls are commonly given by females in many species of corvid, in sexual or self-assertive situations. A 'trill' reported by Heinrich was associated with food, but also with other excitement.

There is then a variety of softer calls given more as conversational exchanges between birds, mostly quiet mutterings that are inaudible at any distance. Gwinner (1964) described a deep, throaty 'gro' call used for contact between paired birds at close hand, between parents and flying young, by parents offering to feed young, and by a male offering to feed his mate. A variety of 'cooing', 'crooning' and 'whining' calls between birds are recorded. Goodwin (1986) observes that, 'Established pairs may use personal variants of various innate calls and displays. These may function to facilitate immediate recognition at a distance as well as, presumably, to enhance the emotional bond between them'. Heinrich noted a soft parental 'korr' at the nest, which caused young to gape, and an oft-repeated 'rrack' that made them duck into the nest bowl and evidently served as a warning. The nestlings in turn have a husky 'kraah' note which serves as a food call. Heinrich also speaks of the Raven's song, quoting Bent ('a curious gargling, strongly inflected talk') and Zirrer ('a soft musical warble' with 'lisps, croaks, buzzing sounds and gulps'). This is no doubt the 'soliloquising' referred to by Chapman (1924). It is said to be heard most often in late summer.

The Raven's propensity for mimicking the human voice has long been seen as part of the bird's special charisma. This may introduce further complications in studying the voice of captive birds. Some writers have said that this does not increase the repertoire of wild Ravens since, unlike some species such as the Starling, they do not imitate other creatures outside captivity. Bolam (1913) nevertheless heard one wild bird which did so while perched in a tree just above his hide:

> There he sat cawing and chuckling to himself for a long time, sometimes changing his perch a little, the variety of his repertoire being extraordinary; no sound uttered beyond a half-tone, but almost every voice of the mountain being imitated in turn, and some of the notes defying all similes. He yapped like a dog, reproduced the call of a grouse, and a crow, to perfection, bleated like a sheep, or a goat, and made all sorts of metallic noises, from the click of a blacksmith's anvil to the creaking of a rusty hinge.

Voice mimicry is one of the most frequently remarked of the Raven's traits in captivity and writers often speak of the bird's habit of practising its lines. Many amusing tales have been told of the Raven's speech abilities. Smith (1887) recounted the story of a tame Raven which had learned certain lines and quirks of behaviour. This bird strayed beyond its master's grounds and was espied on the ground by a passing commercial traveller (they had them even then), who descended from his gig intent upon its capture. The Raven turned round and said in a clear voice, 'Good-bye, old fellow, good-bye old cock', and then hopped briskly away. The startled salesman proceeded into the village, where – doubtless fortified in the nearest tavern – he related his alarming encounter with Satan! Then there was the bird which, living near the guard-room at Chatham barracks, more than once turned out the guard by the authenticity of its commands. Smith (1905) said of one of his tame Ravens that, 'A bad cough, which I had, he managed to imitate so well that people who passed down an adjoining lane, thought it inconsiderate of me to expose a gardener who had such a hacking cough to all weathers in my garden'.

Charles Dickens was another Raven-fancier, who kept at least three of these birds, and commented on their varying individual personality and aptitude for learning to mimic speech. The two brightest were the inspiration for his depiction of the pet Raven, Grip, in *Barnaby Rudge.* Housed in a stable, one of these birds 'applied himself to the acquisition of stable language, in which he became such an adept, that he would perch outside my window and drive imaginary horses with great skill, all day'.

APPENDIX 3

Appearance of the Raven

Cramp and Perrins (1994) give a detailed account, from which I have summarised salient features of the nominate subspecies of the Raven. *Corvus corax* is the largest of the Crow tribe, and the largest of all passerines, while *C. corax corax* is one of the medium-sized forms, the subspecies *varius, laurencei, principalis, kamtschaticus* and *tibetanus* being larger.

THE NOMINATE SUBSPECIES

Size (mm)

Adult:	There is moderate sexual size dimorphism, with the male as the larger sex. Some observers state that this is clear in the field when the two birds are seen together, but in some pairs it is by no means obvious, especially at a distance.
Male:	Wing 410–470; wing-span 1,250–1,500; tail 220–250; bill to tail 6,300–6,700
Female:	Wing 390–440; wing-span 1,200–1,350; tail 210–245; bill to tail 6,000–6,400
Juvenile	
Male:	Wing 400–430; tail 207–243
Female:	Wing 390–417; tail 214–233

The Raven has a massive bill, with a length (to skull) of 73–83.5 mm in males and 70.9–80.3 mm in females, for European birds.

Weight (g)

Adult:	
Male:	1,080–1,588 (lower, Cramp & Perrins, 1994; upper, G. Bolam, 1913)
Female:	993–1,475 (E. Blezard, unpublished data); 798–1,315 (Cramp & Perrins, 1994)
Juvenile:	
Sex not given:	885 (G. Bolam, 1913, a single February bird)

Heinrich (1990), who caught and measured numerous wild *Corvus corax corax* in North America, said that adults and juveniles cannot be distinguished by size, and his graph of bill length (75–93 mm) and body weight (1,040–1,525 g) shows this very well. Bolam (1913), who examined many Ravens killed by keepers in north Wales, took the contrary view, saying that juveniles were usually distinctly smaller than adults, and that the difference was especially marked in the size of bills. He also claimed that the tendency for the tip of the bill to have a hook increased with age. Yet, apart from a few body weights, Bolam gave no measurements in support of his views. Kilham (1989) noted that a tame bird which broke 1.5 cm off the tip of its lower mandible had restored this loss in 2 months, so that the bill continues growth to cope with wear.

Plumage

The apparently completely black, glossy plumage of Ravens is enlivened in close view by a colourful glossy iridescence, predominantly of purple, but varying from a blue to red tint, and appearing also to be shot with bronze and green in places. The primaries and tail feathers are the least strongly tinted. The colour becomes bluer on worn feathers, while abrasion of wing and tail plumage produces a browning effect. The bases of body feathers tend to light greyness. The sexes are similar, though females are on average less strongly purple than males.

The five outer primaries have markedly emarginated webs, giving the 'splayed-finger' effect in the outstretched wing so characteristic of buzzards, eagles and vultures – all birds in which soaring is an especially notable feature of flight (p. 24). Wing-loading is likely to be low, in common with the broad-winged raptors, though I have not seen figures. The tail feathers shorten from the centre to the edge, giving the tail its conspicuous wedge or diamond shape; more prominent in the adult than the juvenile (pp. 24 and 192).

Juveniles lack the distinctive 'beard' feathers on the throat of the adult and are much sleeker about the head, an effect which may make them appear smaller than old birds. Juveniles have a less glossy plumage, mostly lacking the colourful sheen, especially on the upper parts, which tend to black-brown, and the brownness increases as feathers become worn. In flight, juveniles appear broader in the wing than adults, but their outer primaries are less deeply emarginated. Both adults and juveniles have the base of the bill and nostrils covered with long bristles (p. 191). The plumage has a musty smell typical of the Corvidae, though not as pronounced as in the petrels.

Complete albinos occur rarely, though, in an early record, Dr John Caius referred to two in Cumberland in August 1548. Ussher and Warren (1900) mention an 1857 record of two pure-white young found in the same nest in Ireland and Macgillivray (1837) reported an albino on Pabbay in the Outer Hebrides. Partial albinos appear to be almost equally rare in Britain or Ireland (one on Harris in the Outer Hebrides; Macgillivray, 1837) and even birds with a few white feathers are seldom seen. Nelson (1907) gave three records of pied birds for Yorkshire. Sage (1962) notes that partial albinism in Faeroe Ravens was once common and that a white-speckled mutant was known there since the Middle Ages, once forming a considerable proportion of the population. It led Brünnich to name this Faeroe race *Corvus varius* in 1764. A rapid decline of this mutant began around 1850 and it was last seen in 1902 (Williamson, 1965). The same pied variety was occasionally seen on Iceland, but was always at very low frequency there.

Moult

Juveniles begin to moult their first plumage about 2 months after fledging and take another 2 months to complete this. In Britain and Ireland, depending on fledging time, the time of onset thus varies from June–August and of completion from August–October. They then have more of the appearance of adults, with plume-like hackle feathers on the throat and sides of the neck, and glossy plumage. The tail, wings and wing-coverts are browner than in the adult and the iridescent colours are less fully developed. Adults usually begin their complete annual moult when the breeding season is well advanced or over, from mid-April to mid-June, starting with the innermost primary, nearest the carpal joint (p 1), and working outwards (descendent). The outermost primary (p 10) is shed on average 122 days after the first. The inner tail feathers (t 1) are cast first (usually with p 4) and the outermost (t 6) about 60 days later. Outer secondaries (s 1) are lost first with p 5, and tertials from p 6 onwards. Moult is usually completed after 140 days, when the last primary has regrown, between early September and early November.

Bare parts

Apart from the eyes, and the rare cases of albinism, an adult Raven is black all over. The bill, legs, feet and claws are black, and even the inside of the throat and the tongue are so coloured in the adult – the startling carmine-red mouth and yellow gape of the nestlings having been replaced, first by an intermediate greyish-flesh colour. Heinrich and Marzluff (1992) state that the age at which the mouth turns fully black depends on social status. In highly dominant paired birds it can be as early as 8 months, but in subordinate unpaired birds not until 28 months or even more. Kilham (1989) noted that two yearlings raised by hand had black mouths. The iris is blue-grey in the juvenile but changes to dark brown as adult plumage is acquired. The eye is relatively small for the size of the bird.

APPENDIX 4

Scientific names of animal and plant species in the text

ANIMALS

Birds

Blackbird *Turdus merula*
Black Crow *Corvus capensis*
Black Grouse *Lyrurus tetrix*
Black-headed Gull *Larus ridibundus*
Black-throated Diver *Gavia arctica*
Brünnich's Guillemot *Uria lomvia*
Buzzard *Buteo buteo*
Carrion Crow *Corvus corone corone*
Chaffinch *Fringilla coelebs*
Chough *Pyrrhocorax pyrrhocorax*
Common Gull *Larus canus*
Cormorant *Phalacrocorax carbo*
Curlew *Numenius arquata*
Domestic Fowl *Gallus domesticus*
Domestic Pigeon *Columba livia*
Dotterel *Eudromias morinellus*
Dunlin *Calidris alpina*
Eagle Owl *Bubo bubo*
Eider *Somateria mollissima*
Fieldfare *Turdus pilaris*
Fulmar *Fulmarus glacialis*
Gannet *Sula bassana*
Golden Eagle *Aquila chrysaetos*
Golden Plover *Pluvialis apricaria*
Great Spotted Cuckoo *Clamator glendarius*
Greenshank *Tringa nebularia*
Great Black-backed Gull *Larus marinus*
Great Northern Diver *Gavia immer*
Guillemot *Uria aalge*
Gyr Falcon *Falco rusticolus*
Hen Harrier *Circus cyaneus*
Herring Gull *Larus argentatus*
Heron *Ardea cinerea*
Hobby *Falco subbuteo*
Hooded Crow *Corvus corone cornix*
Jackdaw *Corvus monedula*
Kestrel *Falco tinnunculus*
Kittiwake *Rissa tridactyla*
Lesser Black-backed Gull *Larus fuscus*
Magpie *Pica pica*
Mallard *Anas platyrhynchos*
Meadow Pipit *Anthus pratensis*
Merlin *Falco columbarius*
Mute Swan *Cygnus olor*
Partridge *Perdix perdix*
Peregrine *Falco peregrinus*
Pheasant *Phasianus colchicus*
Ptarmigan *Lagopus mutus*
Puffin *Fratercula arctica*
Razorbill *Alca torda*
Redwing *Turdus iliacus*
Red Grouse *Lagopus lagopus*
Red Kite *Milvus milvus*
Red-legged Partridge *Alectoris rufa*
Reed Bunting *Emberiza schoeniclus*
Ring Ouzel *Turdus torquatus*
Rock Dove *Columba livia*
Rook *Corvus frugilegus*
Rough-legged Buzzard *Buteo lagopus*
Sandhill Crane *Grus canadensis*
Shag *Phalacrocorax aristotelis*
Short-eared Owl *Asio flammeus*
Sparrowhawk *Accipiter nisus*
Starling *Sturnus vulgaris*
Stonechat *Saxicola torquata*
Turkey vulture *Cathartes aura*
Wheatear *Oenanthe oenanthe*
Whimbrel *Numenius phaeopus*
Wren *Troglodytes troglodytes*

Mammals

Brown hare *Lepus europaeus*
Brown rat *Rattus norvegicus*
Common shrew *Sorex araneus*
Coyote *Canis latrans*
Grampus (Killer whale) *Grampus griseus*
Hedgehog *Erinaceus europaeus*
Lemmings *Lemmus* spp., *Myopus* spp., *Dicrostonyx* spp.
Mink *Mustela vison*
Mole *Talpa europaea*

Appendix 4

Mountain hare *Lepus timidus*
Orkney vole *Microtus orcadensis*
Otter *Lutra lutra*
Pine marten *Martes martes*
Polar bear *Thalassarctos maritimus*
Polecat *Mustela putorius*
Pygmy shrew *Sorex minutus*
Rabbit *Oryctolagus cuniculus*
Red deer *Cervus elaphus*
Red fox *Vulpes vulpes*
Red squirrel *Sciurus vulgaris*
Reindeer (Caribou) *Rangifer tarandus*
Short-tailed field vole *Microtus agrestis*
Soay sheep *Ovis aries*
Stoat *Mustela erminea*
Water shrew *Neomys fodiens*
Water vole *Arvicola amphibius*
Weasel *Mustela nivalis*
Wild cat *Felis silvestris*
Wolf *Canis lupus*
Wood mouse *Apodemus sylvaticus*

Reptiles, amphibians, fish

Adder *Vipera berus*
Brown trout *Salmo trutta*
Common lizard *Lacerta vivipara*
Frog *Rana temporaria*
Toad *Bufo bufo*

Invertebrates

Antler moth *Cerapteryx graminis*
Ants Formicidae
Barnacles *Balanus balanoides, Chthamalus stellatus*
Black slug *Arion ater*
Brackish water snail *Hydrobia ulvae*
Clam *Pecten maximus*
Cockle *Cardium edule*
Craneflies Tipulidae
Dor beetle *Geotrupes sylvaticus*
Emperor moth *Pavonia pavonia*
Fox moth *Macrothylacia rubi*
Grasshoppers Orthoptera
Ground beetles *Carabus* spp., *Pterostichus* spp.
Horse mussel *Modiolus modiolus*
Limpet *Patella* spp.
Lugworm *Arenicola marina*
Mussel *Mytilus edulis*
Northern eggar moth *Lasiocampa quercus callunae*
Oyster *Ostrea edulis*
Periwinkle *Littorina littoralis, L. saxatilis*
Pill beetle *Byrrhus pilula*
Rove beetle *Staphylinus* spp.
Sea-urchin *Echinus esculentus*
Serpulid tube-worm *Spirorbis borealis*
Shorecrab *Carcinus maenas*
Skipjack beetle *Agriotes lineatus*
Spider crab *Hyas* spp.
Venus shell *Venus gallina*
Whelk *Nucella lapillus*

PLANTS

Trees and tall shrubs

Alder *Alnus glutinosa*
Ash *Fraxinus excelsior*
Beech *Fagus sylvatica*
Birch *Betula verrucosa, B. pubescens*
Bird cherry *Prunus padus*
Cedar *Cedrus libani*
Corsican pine *Pinus nigra* var. *maritima*
Douglas fir *Pseudotsuga menziesii*
Elder *Sambucus nigra*
Elm *Ulmus procera, U. glabra*
Grand fir *Abies grandis*
Hawthorn *Crataegus monogyna*
Holly *Ilex aquifolium*
Horse chestnut *Aesculus hippocastanum*
Juniper *Juniperus communis*
Larch *Larix kaempferi*
Lodgepole pine *Pinus contorta*
Noble fir *Abies procera*
Norway spruce *Picea abies*
Oak *Quercus robur, Q. petraea*
Rowan *Sorbus aucuparia*
Scots pine *Pinus sylvestris*
Silver fir *Abies* spp.
Sitka spruce *Picea sitchensis*
Sweet chestnut *Castanea sativa*
Sycamore *Acer pseudoplatanus*

Wellingtonia *Sequoia gigantea*
Yew *Taxus baccata*

Medium shrubs

Blackthorn *Prunus spinosa*
Bramble *Rubus fruticosus* agg.
Common gorse *Ulex europaeus*
Western gorse *Ulex gallii*

Dwarf shrubs

Bilberry *Vaccinium myrtillus*
Cowberry *Vaccinium vitis-idaea*
Crowberry *Empetrum nigrum, E. hermaphroditum*
Heather (Ling) *Calluna vulgaris*
Least willow *Salix herbacea*

Ferns, mosses and herbs

Bent *Agrostis canina, A. tenuis, A. stolonifera*
Black bog-rush *Schoenus nigricans*
Bog moss *Sphagnum* spp.
Bracken *Pteridium aquilinum*
Cotton grass *Eriophorum vaginatum, E. angustifolium*
Deer sedge *Scirpus cespitosus*
Fescue *Festuca ovina, F. vivipara, F. rubra*
Flying bent *Molinia caerulea*
Great woodrush *Luzula sylvatica*
Heath rush *Juncus squarrosus*
Mat grass *Nardus stricta*
Mountain sedge *Carex bigelowii*

APPENDIX 5

Names of the Raven

Hraefen, Hraefn	England (Anglo-Saxon)
Hremu	England (Old English)
Ralph, Ravvin, Rewin, Rauin	England
Croupy Craw, Corby, Croaker	Northern England
Corbie, Corbie Craw	Scotland (Carrion Crows are also sometimes called Corbies, so that there is scope for confusion)
Fhitheach, Biadhtach, Bran	Scotland (Gaelic) (for others see Forbes, 1905)
Cigfran, Gigfran, Fran	Wales
Feeagh	Isle of Man
Grand Corbeau	France
Corbo, Corv, Crovo	Italy
Cuervo	Spain
Kolkrabe	Germany
Raaf	Netherlands
Ravn	Denmark, Norway
Korp	Sweden
Korppi	Finland
Hrafu	Iceland
Voron	Russia
Oreb	Hebrew
Common Raven	North America

An olden-time and now obsolete term for a gathering of the species was an 'unkindness of Ravens'.

Bibliography

ABRAHAM, G. D. 1919. *On Alpine Heights and British Crags*. Methuen, London.
ALLEN, E. F. & NAISH, J. M. 1935. Raven's nest with ten eggs. *Br. Birds* 28, 27.
ALLIN, E. K. 1968. Breeding notes on Ravens in north Wales. *Br. Birds* 61, 541–5.
ANDREWS, I. J. 1986. *The Birds of the Lothians*. Scottish Ornithologists Club, Edinburgh.
ANDRIESCU, C. & CORDUNEANU, V. 1972. Raspindirea corbuli (*Corvus corax* L.) in Judetal Botosani. *Muzeul de Stüntele Naturii Dorohoi, Botsani, Studii si comunicari 1972*, 199–204.
ANGUS, S. 1983. *Sutherland Birds*. The Northern Times, Golspie.
ARMSTRONG, E. 1958. *The Folklore of Birds*. The New Naturalist. Collins, London.
ASPDEN, W. 1928–29. Further notes on Puffin Island 1928. *Br. Birds* 22, 103–6.
BANNERMAN, D. A. 1953. *The Birds of the British Isles*, vol I. Oliver & Boyd, Edinburgh.
BANNERMAN, D. A. 1961. *The Birds of the British Isles*, vol IX. Oliver & Boyd, Edinburgh.
BARTHOLOMEW, J. Undated. *The Survey Gazetteer of the British Isles* Ninth Edition. Bartholomew, Edinburgh.
BATTEN, H. M. 1923. *Inland Birds*. Hutchinson, London.
BAXTER, E. V. & RINTOUL, L. J. 1953. *The Birds of Scotland*, vol I. Oliver & Boyd, Edinburgh.
BELL, T. H. 1962. *The Birds of Cheshire*. John Sherratt, Altrincham.
BENT, A. C. 1946. Life histories of North American Jays, Crows and Titmice. Bulletin No. 191, Smithsonian Institution, United States National Museum.
BERROW, S. D. 1992. The diet of coastal breeding Ravens in Co. Cork. *Ir. Birds* 4, 555–8.
BIRCHAM, P. M. M. 1989. *The Birds of Cambridgeshire*. Cambridge University Press, Cambridge.
BLEZARD, E. 1928. On the Raven. *Trans. Carlisle Nat. Hist. Soc.* 4, 16–22.
BLEZARD, E. 1943. The Birds of Lakeland. *Trans. Carlisle Nat Hist. Soc.* Vol. VI. Published by the Society.
BLEZARD, E. 1946. The Lakeland Pennines and their birds. *Trans. Carlisle Nat. Hist. Soc.* VII, 100–15. Published by the Society.
BLEZARD, E. 1954. Lakeland ornithology. *Trans. Carlisle Nat. Hist. Soc.* Vol. VIII. Published by the Society.
BOLAM, G. 1912. *The Birds of Northumberland and the Eastern Borders*. Blair, Alnwick.
BOLAM, G. 1913. *Wildlife in Wales*. Frank Palmer, London.
BOLAM, G. 1932. A catalogue of the Birds of Northumberland. *Trans. Nat. Hist. Soc. Northumberland, Durham & Newcastle-upon-Tyne*. VIII. 1–165.
BOOBYER, G. 1995. *Breeding Birds of the Cairngorms. An Analysis of Change 1968–91*. Unpublished Joint Nature Conservation Committee Report to Scottish Natural Heritage. JNCC, Peterborough.
BOOTH, C. J. 1979. A study of Ravens in Orkney. *Scott. Birds* 10, 261–7.
BOOTH, C. J. 1985. The breeding success of Ravens on Mainland Orkney, 1983–85. *Orkney Bird Rep.* 1985, 59–62.
BOOTH, C. J. 1986. Raven breeding for the first time at 6 years old. *Scott. Birds* 14, 51.
BOOTH, C. J. 1996. Ravens nesting on buildings in Orkney. *Scottish Birds*, 18, 159–65.
BOOTH, C. J., CUTHBERT, M. & REYNOLDS, P. 1984. *The Birds of Orkney*. The Orkney Press, Kirkwall.
BORRER, W. 1891. *The Birds of Sussex*. R. H. Porter, London.
BRAMWELL, D. 1959–60. Some research into bird distribution in Britain during the late-glacial and post-glacial periods. *Bird Rep. Merseyside Naturalists Ass.* 1959–60.
BROWN, L. H. & WATSON, A. 1964. The Golden Eagle in relation to its food supply. *Ibis* 106, 78–100.
BUCKLAND, S. T., BELL, M. V. & PICOZZI, N. 1990. *The Birds of North-east Scotland*. North-East Scotland Bird Club, Aberdeen.

BUCKLEY, T. E. & HARVIE-BROWN, J. A.. 1891. *A Vertebrate Fauna of the Orkney Isles*. David Douglas, Edinburgh.
BUCKNILL, J. A. 1900. *The Birds of Surrey*. Porter, London.
BULL, H. G. 1888. *Notes on the Birds of Herefordshire*. Hamilton, Adams & Co., London.
BURNS, R. 1939. *Scots Ballads*. Seeley Service, London.
BUXTON, J. (ed.) 1981. *The Birds of Wiltshire*. Wiltshire Library and Museum Service, Trowbridge.
BYRKJEDAL, I. 1980. Nest predation in relation to snow-cover – a possible factor influencing the start of breeding in shorebirds. *Ornis Scand.* 11, 249–52.
BYRKJEDAL, I. 1987. Antipredator behavior and breeding success in Greater Golden Plover and Eurasian Dotterel. *Condor* 89, 40–7.
CHAPMAN, A. 1907. *The Birdlife of the Borders*. 2nd edition. Gurney and Jackson, London.
CHAPMAN, A. 1921. *Savage Sudan. Its Wild Tribes, Big-game and Bird-life*. Gurney and Jackson, London.
CHAPMAN, A. 1924. *The Borders and Beyond*. Gurney and Jackson, London.
CHEVERTON, J. M. 1989. *Breeding Birds of the Isle of Wight*. Isle of Wight Natural History and Archaeological Society, Newport.
COOK, M. 1992. *The Birds of Moray and Nairn*. The Mercat Press, Edinburgh.
COOMBES, R. A. H. 1948. The flocking of the Raven. *Br. Birds* 41, 290–5.
COOMBS, C. J. F. 1978. *The Crows*. Batsford, London.
Cornwall Bird-watching and Preservation Society. 1991. *Birds in Cornwall*. Annual Report.
COTTERMAN, V. & HEINRICH, B. 1993. A large temporary roost of Ravens. *Auk* 110, 395.
COWARD, T. A. 1910. *The Vertebrate Fauna of Cheshire and Liverpool Bay*, vol. I. Witherby, London.
COWIN, W. S. 1941. A census of breeding Ravens. *Yn Shirrag ny Ree (Peregrine)* 1, 3–6.
COX, S. 1984. *A New Guide to the Birds of Essex*. Essex Birdwatching and Preservation Society.
CRAIGHEAD, J. J. & CRAIGHEAD, F. C. 1956. *Hawks, Owls and Wildife*. Stackpole Company and Wildlife Management Institute, Harrisburg, PA.
CRAMP, S. & PERRINS, C. M. (eds.) 1994. *Handbook of the Birds of Europe, the Middle East and North Africa. The Birds of the Western Palearctic. Vol. VIII Crows to Finches*. Oxford University Press, Oxford.
CRIGHTON, G. M. 1976. *The Birds of Angus*. Published by author, Brechin.
CROCKER, D. R. 1987. *Foraging Behaviour in Bullfinches* (Pyrrhula pyrrhula). Ph.D. thesis, University of London.
CROCKER, J. 1985. Respect your feathered friends. *New Scientist* 10 October, 47–50.
CROSS, A. V. 1995. Colour-marked Ravens. *BTO News* No. 196, 20.
CROSS A. V. & DAVIS, P. E. 1986 Monitoring of Ravens and land use in central Wales. Unpublished report to the Nature Conservancy Council.
CULLEN, J. P. 1978. A census of breeding Ravens in the Isle of Man. *Peregrine* 46, 264–73.
CULLEN, J. P. & JENNINGS, P. P. 1986. *Birds of the Isle of Man*. Bridgeen Publications, Douglas.
CUNNINGHAM, P. 1983. *Birds of the Outer Hebrides*. The Melven Press, Perth.
DARE, P. J. 1986a. Raven *Corvus corax* populations in two upland regions of north Wales. *Bird Study* 33, 179–89.
DARE, P. J. 1986b. Aspects of the breeding biology of Ravens in two upland districts of north Wales. *Naturalist* 111, 129–37.
DARE, P. J. 1988. Ravens in Devon: the coastal breeding population. *Devon Birds* 41, 15–17.
DAVIS, P. E. & DAVIS, J. E. 1986. The breeding biology of a Raven population in central Wales. *Nature in Wales* 3, 44–54.
DAY, J. C., HODGSON, M. S. & ROSSITER, N. 1995. *The Atlas of Breeding Birds in Northumbria*. Northumberland & Tyneside Bird Club.
DEANS, P., SANKEY, J., SMITH, L., TUCKER, J., WHITTLES, C. & WRIGHT, C. (eds.) 1992. *An Atlas of the Breeding Birds of Shropshire*. Shropshire Ornithological Society.
DICK, C. H. 1916. *Highways and Byways in Galloway and Carrick*. Macmillan, London.

DICKEY, D. R. & VAN ROSSEM, A. J. 1938. The Birds of El Salvador. *Publ. Field Mus. Nat. Hist. Zool. Ser.* 23, 1–635.
DICKSON, R.C. 1992. *Birds in Wigtownshire.* G. C. Book Publishers, Wigtown.
DINGLE, T. J. 1980. Ravens nesting on electricity pylons. *British Birds* 73, 479.
DIXON, A. (in press) Breeding biology of a Raven *Corvus corax* population in the Brecon Beacons. *Welsh Birds.*
DOBSON, R. 1952. *The Birds of the Channel Islands.* Staples Press, London.
DONOVAN, J. & REES, G. 1994. *Birds of Pembrokeshire.* Dyfed Wildlife Trust.
D'URBAN, W. S. M. & MATHEW, M. A. 1892. *Birds of Devon.* R.H. Porter, London.
DYMOND, J. N. 1991. *The Birds of Fair Isle.* Published by J. N. Dymond.
EGDELL, J., SMITH, D. & TAYLOR, J. 1993. Agricultural policy in the Uplands: the move to extensification. *RSPB Conservation Rev.* 7, 27–34.
ELLIOT, R. E. 1989. *Birds of Islay.* Helm, London.
ELLIS, P. M., OKILL, J. D., PETRIE, G. W. & SUDDABY, D. 1994. The breeding performance of Ravens from a sample of nesting territories in Shetland during 1984–1993. *Scott. Birds* 17, 21–34.
ENGEL, K. A. & YOUNG, L. S. 1989. Spatial and temporal patterns in the diet of Common Ravens in southwestern Idaho. *The Condor,* 91, 372–78.
ENGEL, K. A. & YOUNG, L. S. 1992a. Daily and seasonal activity patterns of Common Ravens in southwestern Idaho. *Wilson Bull.,* 104, 462–71.
ENGEL, K. A. & YOUNG, L. S. 1992b. Movements and habitat use by Common Ravens from roost sites in south-western Idaho. *J. Wildl. Manage.* 56, 596–602.
ENGEL, K. A., YOUNG, L. S., STEENHOF, K., ROPPE, J. A. & KOCHERT, M. N. 1992. Communal roosting of Common Ravens in southwestern Idaho. *Wilson Bull.* 104, 105–21.
ETCHÉCOPAR, R-D. & HÜE, F. 1967. *The Birds of North Africa.* Oliver & Boyd, Edinburgh.
EVANS, A. H. 1911. *A Fauna of the Tweed Area.* David Douglas, Edinburgh.
EWINS, P. J., DYMOND, J. N. & MARQUISS, M. 1986. The distribution, breeding and diet of Ravens *Corvus corax* in Shetland. *Bird Study* 33, 110–16.
FIELDING, A. H. & HAWORTH, P. F. 1995. Testing the generality of bird-habitat models. *Conservation Biol.* 9, 1466–81.
FISHER, J. 1966. *The Shell Bird Book.* Ebury Press and Michael Joseph.
FORBES, A. R. 1905. *Gaelic Names of Beasts (Mammalia), Birds, Fishes, Insects, Reptiles, etc.* Oliver and Boyd, Edinburgh.
FORREST, H.E. 1907. *The Vertebrate Fauna of North Wales,* Witherby, London.
FOX, G. A., YONGE, K. S. & SEALEY, S. G. 1980. Breeding performance, pollutant burden and eggshell thinning in Common Loons *Gavia inimer* nesting on a Boreal Forest Lake. *Ornis Scandinavica* 11, 243–48.
FREUCHEN, P. & SALOMONSEN, F. 1960. *The Arctic Year.* Jonathan Cape, London.
FROST, R. A. 1978. *Birds of Derbyshire.* Moorland Publishing Company, Buxton.
FRYER, G. 1986. Notes on the breeding biology of the Buzzard. *British Birds* 79, 18–28.
FULLER, R. J. 1996. Relationships between grazing and birds, with particular reference to sheep in the British uplands. *BTO Res. Rep.* No. 164. Unpublished.
GALLOWAY, B. & MEEK, E. R. 1983. Northumberland's birds. *Trans. Nat. Hist. Soc. Northumb.* 1978–83, 44.
GASTON, A. J. 1985. Murres on the menu. *Nature Canada* 14, 40–43.
GIBBONS, D. W., REID, J. B. & CHAPMAN, R. A. 1993. *The New Atlas of Breeding Birds in Britain and Ireland: 1988–1991.* Poyser, London. (Compiled for the British Trust for Ornithology, Scottish Ornithologists' Club and Irish Wildbird Conservancy.)
GIBBONS, D., GATES, S., GREEN, R. E., FULLER, R. J. & FULLER, R. M. 1995. Buzzards *Buteo buteo* and Ravens *Corvus corax* in the uplands of Britain: limits to distribution and abundance. *Ibis* 137, 575–84.
GIBSON, J. A. 1953. The status of the Raven on the Clyde islands. *Glasgow & W. Scotland Bird Bull.* 2, 5.
GILBERT. 1946. Raven flocks in Brecon. *British Birds,* 29, 52–3.
GILBERT, H. A. & BROOK, A. 1924. *Secrets of Bird Life.* Arrowsmith, London.

GILBERT, H. A. & BROOK, A. 1931. *Watchings and Wanderings among Birds.* Arrowsmith, London.
GLADSTONE, H. S. 1910. *The Birds of Dumfriesshire,* Witherby, London.
GLEGG, W. E. 1935. *A History of the Birds of Middlesex.* Witherby, London.
GOODWIN, D. 1986. *Crows of the World.* 2nd edition. British Museum (Natural History), London.
GORDON, S. 1912. *The Charm of the Hills.* Cassell, London.
GORDON, S. 1915. *Hill Birds of Scotland.* Cassell, London.
GORDON, S. 1938. *Wild Birds in Britain.* Batsford, London.
GOTHE, J. 1961. Zur Ausbreitung und zum Fortpflanzungsverhalten des Kolkraben (*Corvus corax* L.) unter besonderer Berüksichtigung der Verhältnisse in Mecklenburg. Pp. 63–129 in H. Schildmacher, *Beiträge zur Kenntnis deutscher Vögel.* Fischer, Jena.
GURNEY, J. H. 1921. *Early Annals of Ornithology.* H. F. & G. Witherby, London.
GWINNER, E. 1964. Untersuchungen über das Ausdrucks und Sozialverhalten des Kolkraben (*Corvus corax corax* L.). *Z. Tierpsychol.* 21, 657–748.
GWINNER, E. 1965a. Beobachtungen über Nestbau und Brutpflege des Kolkraben (*Corvus corax*) in Gefangenschaft. *J. Ornithol.* 106, 145–78.
GWINNER, E. 1965b. Uber den Einfluss des Hunger und anderer Faktoren auf die Versteck – Activität des Kolkraben (*Corvus corax*). *Vögelwarte* 23, 1–4.
GWINNER, E. 1966. Uber einige Bewegungsspiele des Kolkraben (*Corvus corax* L.). *Z. Tierpsychol.* 23, 28–36.
HAARTMAN, L. VON 1969. *Comm. Biol. Soc. Sci. Fenn.* 32.
HANCOX, M. 1985. Some observations on the flocking and foraging behaviour of Ravens at Drimnin, Argyll. *Scott. Naturalist* 37–40.
HARLOW, R. C. 1922. The breeding habits of northern Ravens in Pennsylvania. *Auk* 39, 399–410.
HARRIS, M. P. & MURRAY, S. 1978. *Birds of St. Kilda.* Institute of Terrestrial Ecology, Banchory.
HARRISON, C. J. O. 1966. Variation in the distribution of pigment within the shell structure of birds' eggs. *J. Zool., Lond.* 148, 526–39.
HARRISON, C. J. O. 1987. Pleistocene and prehistoric birds of south-west Britain. *Proc. Univ. Bristol Spelaeol. Soc.* 18, 81–104.
HARRISON, G. R. (ed.), DEAN, A. R., RICHARDS, A. J. & SMALLSHIRE, D. 1982. *The Birds of the West Midlands.* West Midland Bird Club, Warwickshire.
HARVEY, L. A. & ST. LEGER-GORDON, D. 1953. *Dartmoor.* The New Naturalist, No. 27. Collins, London.
HARVIE-BROWN, J. A. 1906. *A Fauna of the Tay Basin and Strathmore.* David Douglas, Edinburgh.
HARVIE-BROWN, J. A. & BUCKLEY, T. E. 1887. *A Vertebrate Fauna of Sutherland, Caithness and West Cromarty.* David Douglas, Edinburgh.
HARVIE-BROWN, J. A. & BUCKLEY, T. E. 1888. *A Vertebrate Fauna of the Outer Hebrides,* David Douglas, Edinburgh.
HARVIE-BROWN, J. A. & MACPHERSON, H. A. 1904. *A Fauna of the North-west Highlands and Skye.* David Douglas, Edinburgh.
HAURI, R. 1956. Beitrage zur Biologie des Kolkraben (*Corvus corax*). *Ornith. Beob.* 53, 28–35.
HEATHCOTE, A., GRIFFIN, D. & SALMON, H. M. (eds.) 1967. *The Birds of Glamorgan.* D. Brown, Cowbridge.
HEINRICH, B. 1988. Winter foraging at carcasses by three sympatric corvids, with emphasis on recruitment by the raven *Corvus corax. Behav. Ecol. Sociobiol.* 23, 141–156.
HEINRICH, B. 1990. *Ravens in Winter.* Barrie & Jenkins, London.
HEINRICH, B. & MARZLUFF, J. 1992. Age and mouth color in the Common Raven. *Condor* 94, 549–50.
HEWSON, R. 1981. Scavenging of mammal carcasses by birds in West Scotland. *J. Zool., Lond.* 194, 525–37.

HICKEY, J. J. 1942. Eastern population of the Duck Hawk. *Auk* 59 (2), 176–204.
HICKLING, R. 1978. *Birds in Leicestershire and Rutland*. Leicestershire and Rutland Ornithological Society, Leicester.
HOLMES, R. T. 1970. Differences in population density, territoriality and food supply of Dunlin on arctic and sub-arctic tundra. Pp 303–19 in A. Watson (ed.), *Animal Populations in Relation to their Food Resources*. Blackwell, Oxford.
HOLYOAK, D. T. 1967. Breeding biology of the Corvidae. *Bird Study* 14, 153–168.
HOPE JONES, P. 1980. Bird scavengers on Orkney roads. *Br. Birds* 73, 561–68.
HOUSTON, D. 1977. The effect of Hooded Crows on sheep-farming in Argyll, Scotland. 1. The food supply of Hooded Crows, 2. Hooded Crow damage to hill sheep. *J. Appl. Ecol.* 14, 1–15, 17–29.
HOWARD, E. 1920. *Territory in Bird Life*. John Murray, London.
HUDSON, W. H. 1915. *Birds and Man*. Duckworth, London.
HUDSON, P. J. 1992. *Grouse in space and time: The population biology of a managed gamebird*. The Game Conservancy Trust, Fordingbridge.
HURFORD, C. & LANSDOWN, P. 1995. *The Birds of Glamorgan*. Published by the authors and sponsored by Cardiff Naturalists' Society and the Countryside Council for Wales.
HURRELL, A. G. 1951. Ravens using thermals. *Br. Birds* 44, 88–9.
HURRELL, H. G. 1956. A Raven roost in Devon. *Br. Birds* 49, 28–31.
HUTCHINSON, C. D. 1989. *Birds in Ireland*. Poyser, Calton.
INGRAM, G. C. S., SALMON, H. M. & CONDRY, W. M. 1966. *The Birds of Cardiganshire*. West Wales Naturalists' Trust.
JAEGER, E. C. 1963. Aerial bathing of Ravens. *Condor* 65, 246.
KELSALL, J. E. & MUNN, P. W. 1905. *The Birds of Hampshire and the Isle of Wight*. Witherby, London.
KENNEDY, D. & WALKER, A.B. 1988. The great transformer. *Nature Canada* 17, 34–9.
KILHAM, L. 1985. Sustained robbing of American crows by Common Ravens at a feeding station. *J. Field Ornithol.* 56, 425–6.
KILHAM, L. 1989. *The American Crow and the Common Raven*. Texas A. and M. University Press, College Station.
KING, S. 1995. *The Nesting Success and Breeding Density of Ravens in the South-west Lake District, Cumbria, with Particular Reference to Food Supply*. Unpublished M.Sc. thesis, Manchester Metropolitan University.
KNIGHT, R. L. 1984. Responses of nesting Ravens to people in areas of different human densities. *Condor* 86, 345–6.
KOCHERT, M. N., STEENHOF, K., ROPPE, J. & MULROONEY, M. 1984. Raptor and Raven nesting in the PP&L Malin to Midpoint 500kV Transmission Line. Pp 20–39 in *Snake River Birds of Prey Research Project, Annual Report 1984*. U.S. Department of the Interior Bureau of Land Management, Boise District, Idaho.
KRUSHINSKII, L. V., ZORINA, Z. A., DOBROKHOTOVA, L. P., BONDERCHUK, L. S. & FEDOTOVA, I. B. 1982. [The experimental study of reasoning abilities in birds.] *Ornitologiya*, 17, 22–35. (In Russian)
LACK, P. 1986. *The Atlas of Wintering Birds in Britain and Ireland*. Poyser, Calton.
LAWRENCE, R. D. 1986. *In Praise of Wolves*. Henry Holt, New York.
LEWIS, E. 1938. *In Search of the Gyr-falcon*. Constable, London.
LLOYD, L. R. W. *c.* 1925. *Bird facts and fallacies*. London.
LIKHACHEV, G. N. 1951. Multiplication and food of *Corvus corax* in protected Tula forests. *Bull. Soc. Nat. Moscov* 56, 45–53.
LILFORD, LORD 1895. *Birds of Northamptonshire and Neighbourhood*. Porter, London.
LITTLEFIELD, C. D. 1986. Predator control to enhance greater sandhill crane productivity on Malheur NWR, Oregon. *1986. U.S. Dep. Inter. Fish Wildl. Serv., Malheur Natl. Wildl. Refuge*. Princeton, Oreg., 35 pp.
LOCKIE, J. D. 1955. The breeding and feeding of Jackdaws and Rooks, with notes on Carrion Crows and other Corvidae. *Ibis* 97, 341–69.
LOCKIE, J. D., RATCLIFFE, D. A. & BALHARRY, R. 1969. Breeding success and organo-chlorine residues in Golden Eagles in west Scotland. *J. Appl. Ecol.* 6, 381–89.

LOCKLEY, R. M. 1949. *The Birds of Pembrokeshire*. West Wales Field Society, Cardiff.
LOCKLEY, R. M. 1953. Aerial assembly of Ravens in December. *Br. Birds* 46, 347–8.
LOOFT, V. 1983. Die Bestandsentwicklung des Kolkraben in Schleswig–Holstein. *Corax* 9, 227–32.
LORAND, S. & ATKIN, K. 1989. *The Birds of Lincolnshire and South Humberside*. Leading Edge, Hawes.
LORD, J. & MUNNS, D. J. 1970. *Atlas of Breeding Birds of the West Midlands*. Collins, London.
LORENZ, K. Z. 1952. *King Solomon's Ring*. Methuen, London.
LORENZ, K. 1940. Die Paarbildung beim Kolkraben. *Z. Tierpsychol.*, 3, 278–92.
LORENZ, K. 1970. *Studies in Animal and Human Behaviour*, vol. 1. Methuen, London.
LORENZ, K. 1971. Pair-formation in Ravens. Pp. 17-36 in H. Friedrich (ed.), *Man and Animal: Studies in Behaviour*. Paladin, London.
LOVEGROVE, R., WILLIAMS, G. & WILLIAMS, I. 1994. *Birds in Wales*. Poyser, London.
MACGILLIVRAY, W. 1837. *A History of British Birds, Indigenous and Migratory*, vol. 1. Scott, Webster and Geary, London.
MACKENZIE, O. 1924. *A Hundred Years in the Highlands*. Arnold, London.
MAC LOCHLAINN, C. 1984. Breeding and wintering bird communities of Glenveagh National Park, Co. Donegal. *Ir. Birds* 2, 482–500.
MACPHERSON, H. A. 1892. *A Vertebrate Fauna of Lakeland*. David Douglas, Edinburgh.
MACPHERSON, H. A. & Duckworth, W. 1886. *The Birds of Cumberland*. Charles Thurnam, Carlisle.
MCWILLIAM, J. M. 1936. *The Birds of the Firth of Clyde*. Witherby, London.
MARQUISS, M. & BOOTH, C. J. 1986. The diet of Ravens *Corvus corax* in Orkney. *Bird Study* 33, 190–5.
MARQUISS, M., NEWTON, I. & RATCLIFFE, D. A. 1978. The decline of the Raven *Corvus corax*, in relation to afforestation in southern Scotland and northern England. *J. Appl. Ecol.* 15, 129–44.
MARTIN, A. 1987. *Atlas de las aves indificantes en la Isla de Tenerife*. Instituto de Estudios Canarios. Monografia XXXII.
MARZLUFF, J. M., HEINRICH, B. & MARZLUFF, C. S. 1996. Raven roosts are mobile information centres. *Anim. Behav.* 51, 89–103.
MATTINGLEY, W. A. 1995. Winter diet of Ravens in Perthshire. *Scott. Birds* 18, 71–7.
MEARNS, R. 1983. The status of the Raven in southern Scotland and Northumbria. *Scott. Birds* 12, 211–18.
MITCHELL, F. S. 1892. *The Birds of Lancashire*. Gurney and Jackson, London.
MITCHELL, J. 1981. The decline of the Raven as a breeding species in central Scotland. *Forth Naturalist & Historian*, 6, 34–42.
MOFFETT, A. T. 1984. Ravens sliding in snow. *Br. Birds* 77, 321–2.
MOORE, N.W. 1957. The past and present status of the Buzzard in the British Isles. *Br. Birds* 50, 173–97.
MOORE, R. 1969. *The Birds of Devon*. David & Charles, Newton Abbot.
MORRIS, F. O. 1895. *A History of British Birds*, vol. 1. John Nimmo, London.
MOSS, R. 1969. A comparison of Red Grouse (*Lagopus l. scoticus*) stocks with the production and nutritive value of heather (*Calluna vulgaris*). *J. Anim. Ecol.* 38, 103–12.
MUIRHEAD, G. 1889. *The Birds of Berwickshire*, vol. I. David Douglas, Edinburgh.
MUNRO, I. 1988. *The Birds of the Pentland Hills*. Scottish Academic Press, Edinburgh.
MURTON, R. K. 1971. *Man and Birds*. The New Naturalist, Collins, London.
MYLNE, C. K. 1961. Large flocks of Ravens at food. *Br. Birds* 54, 206–7.
NASH, J. KIRKE 1935. *The Birds of Midlothian*. Witherby, London.
NEALE, J. J. 1901. The Raven at home and in captivity. *Rep. Trans. Cardiff Nat. Soc.* 32, 49–53.
NELSON, T. H. 1907. *The Birds of Yorkshire*. 2 vols. A. Brown, London.
NETHERSOLE-THOMPSON, D. 1932. Observations on the field habits, haunts and nesting of the Raven in 1932. *Ool. Rec.* 12, 67–72, 79–82.
NETHERSOLE-THOMPSON, D. & NETHERSOLE-THOMPSON, M. 1979. *Greenshanks*. Poyser, Berkhamsted.

NETHERSOLE-THOMPSON, D. & WATSON, A. 1981. *The Cairngorms. Their Natural History and Scenery.* The Melven Press, Perth.

NEWTON, A. (ed.) 1864–1902. *Ootheca Wolleyana: An Illustrated Catalogue of the Collection of Birds' Eggs,* vol. I. R.H. Porter, London.

NEWTON, I. 1986. *The Sparrowhawk.* Poyser, Calton.

NEWTON, I., DAVIS, P. E. & DAVIS, J.E. 1982. Ravens and Buzzards in relation to sheep-farming and forestry in Wales. *J. Appl. Ecol.* 19, 681–706.

NOGALES, M. 1994. High density and distribution patterns of a Raven *Corvus corax* population on an oceanic island (El Hierro, Canary Islands). *J. Avian Biol.* 25, 80–4.

NOONAN, G. 1971. The decline and recovery of the Raven in Dublin and Wicklow. *Dublin & N. Wicklow Bird Rep.* 2, 16–21.

OAKES, C. 1953. *The Birds of Lancashire.* Oliver & Boyd, Edinburgh.

OGGIER, P. A. 1986. Siedlungsdichte und Sozialverhalten des Kolkraben *Corvus corax* im Wallis. *Orn. Beob.* 83, 295–9.

ORTON, K. 1948. Bird life in the mountains. Chapter 8 in H. R. C. Carr & G. A. Lister (eds) *The Mountains of Snowdonia.* 2nd edition. Crosby Lockwood, London. Originally published in 1924.

PALMER, E. M. & BALLANCE, D. K. 1968. *The Birds of Somerset.* Longmans, Green & Co, London.

PATON, W. & PIKE, O. G. 1929. *The Birds of Ayrshire.* Witherby, London.

PEARSALL, W. H. 1950. *Mountains and Moorlands.* The New Naturalist, Collins, London.

PEERS, M. 1985. *Birds of Radnorshire and Mid-Powys.* Peers, Llangammarch Wells.

PEERS, M. & SHRUBB, M. 1990. *Birds of Breconshire.* Brecknock Wildlife Trust, Brecon.

PENHALLURICK, R. D. 1978. *Birds of Cornwall and the Scilly Isles.* Headland Publications, Penzance.

PICOZZI, N. & WEIR, D. 1976. Dispersal and causes of death in Buzzards. *British Birds* 69, 193–201.

POOLE, K. G. & BROMLEY, R. G. 1988. Interrelationships within a raptor guild in the central Canadian Arctic. *Canadian J. Zool.* 66, 2275–82.

PRAEGER, R. L. 1937. *The Way that I Went.* Hodges, Figgis and Co, Dublin.

PRENDERGAST, E. D. V. & BOYS, J. V. 1983. *The Birds of Dorset.* David & Charles, Newton Abbot.

PRESTT, I. & RATCLIFFE, D. A. 1972. Effects of organochlorine insecticides on European birdlife. Pp 486–513 of *Proceedings of the XVth International Ornithological Congress,* E. J. Brill, Leiden.

PRILL, H. 1982. Einige ökologische Aspekte beim Kolkraben im Verlauf seiner Ausbreitung. *Ornithol. Rundbrief Mecklenburgs* 25, 24–9.

RADFORD, M. C. 1966. *The Birds of Berkshire and Oxfordshire.* Longmans, London.

RATCLIFFE, D. A. 1962. Breeding density in the Peregrine *Falco peregrinus* and Raven *Corvus corax. Ibis* 104, 13–39.

RATCLIFFE, D. A. 1965. Organo-chlorine residues in some raptor and corvid eggs from northern Britain. *Br. Birds* 58, 65–81.

RATCLIFFE, D. A. 1970. Changes attributable to pesticides in egg breakage frequency and eggshell thickness in some British Birds. *J. Appl. Ecol.* 7, 67–115.

RATCLIFFE, D. A. 1993. *The Peregrine Falcon.* 2nd edition. Poyser, London.

RAWNSLEY, H. D. 1899. *Life and Nature at the English Lakes.* James MacLehose, Glasgow.

RIDPATH, M. G. 1953. Damage by the Raven in Pembrokeshire. Unpublished Research Report No. 36, of the Ministry of Agriculture and Fisheries, Infestation Control Division, Tolworth.

RITCHIE, J. 1920. *The Influence of Man on Animal Life in Scotland.* Cambridge University Press, Cambridge.

ROYAL SOCIETY FOR THE PROTECTION OF BIRDS. 1980. *Silent Death. The Destruction of Birds and Mammals through the Deliberate Misuse of Poisons in Britain.* RSPB, Sandy.

ROYAL SOCIETY FOR THE PROTECTION OF BIRDS & NATURE CONSERVANCY COUNCIL. 1991. *Death by Design. The Persecution of Birds of Prey and Owls in the UK 1979–89.* RSPB and NCC, Sandy and Peterborough.

RYVES, B. H. 1948. *Bird Life in Cornwall.* Collins, London.
SAGE, B. L. 1959. *A History of the Birds of Hertfordshire.* Barrie and Rockliff, London.
SAGE, B. L. 1962. Albinism and melanism in birds. *Br. Birds* 55, 201–20.
SAXBY, H. L. 1874. *The Birds of Shetland.* Maclachan & Stewart, Edinburgh.
SCOTT, W. 1806. *Minstrelsy of the Scottish Border,* vol. II. Longman, Hurst, Rees and Orme, London; Constable, Edinburgh.
SCHUFELDT, R. W. 1890. *The mythology of the Raven.* Macmillan, London and New York.
SEEBOHM, H. 1883. *A History of British Birds.* R. H. Porter, London.
SELLIN, D. 1987. Zu Bestand, Ökologie und Ethologie des Kolkraben (*Corvus corax*) in Nordosten des Bezirkes Rostock. *Vogelwelt* 108, 13–27.
SHARROCK, J. T. R. 1976. *The Atlas of Breeding Birds in Britain and Ireland.* T. & A. D. Poyser, Berkhamsted. (Compiled for British Trust for Ornithology and Irish Wildbird Conservancy.)
SHRUBB, M. 1979. *The Birds of Sussex: Their Present Position.* Phillimore, London.
SIMSON, C. 1966. *A Bird Overhead.* Witherby, London.
SITTERS, H. P. 1988. *Tetrad Atlas of the Breeding Birds of Devon.* Devon Bird Watching and Preservation Society, Yelverton.
SKARPHEDINSSON, K. H., NIELSEN, O. K., THORISSON, S., THORSTENSEN, S. & TEMPLE, S. A. 1990. Breeding biology, movements and persecution of Ravens in Iceland. *Acta Nat. Islandica,* 33, 1–45.
SMITH, A. C. 1887 *The Birds of Wiltshire.* R. H. Porter and H. F. Bull, London and Devizes.
SMITH, D. G. & MURPHY, J. R. 1973. Breeding ecology of raptors in the eastern Great Basin of Utah. *Brigham Young Univ. Sci. Bull.* (*Biol. Series*) 18 (3), 76 pp.
SMITH, R. BOSWORTH 1905. *Bird Life and Bird Lore.* John Murray, London.
SOLER, J. J. & SOLER, M. 1991. Analisis comparado del regimen alimenticio durante el periodo otoño-invierno de tres especies de corvidos en un area de simpatria. *Ardeola* 38, 69–89.
SOLER, J. J., SOLER, M. & MARTINEZ, J. G. 1993. Grit ingestion and cereal consumption in five corvid species. *Ardea* 81, 143–9.
SOMERSET ORNITHOLOGICAL SOCIETY. 1988. *Birds of Somerset.* Allan Sutton Publishing, Gloucester.
SPEEDY, T. 1920. *The Natural History of Sport in Scotland with Rod and Gun.* Blackwood, Edinburgh.
ST JOHN, CHARLES 1884. *A tour in Sutherlandshire.* David Douglas, Edinburgh.
STEENHOF, K. & KOCHERT, M. N. 1982. Nest attentiveness and feeding rates of Common Ravens in Idaho. *The Murrelet,* Spring 1982, 30–32.
STEENHOF, K., KOCHERT, M. N. & ROPPE, J. A. 1993. Nesting by raptors and Common Ravens on electrical transmission line towers. *J. Wildl. Manage.,* 57, 271–81.
STEVEN, G. 1996. Letter about disposal of fallen farmstock and effect on scavengers. *Scottish Bird News,* No 42, June 1996.
STEVENSON, H. 1866. *The Birds of Norfolk,* vol. I. Van Voorst, London; Matchett & Stevenson, Norwich.
STIEHL, R. B. 1985. Brood chronology of the Common Raven. *Wilson Bull.* 97, 78–87.
STROUD, D. A. (ed.). 1989. *The Birds of Coll and Tiree: Status, Habitats and Conservation.* Scottish Ornithologists' Club/Nature Conservancy Council, Edinburgh.
SWAINE, C. M. 1982. *Birds of Gloucestershire.* Allan Sutton Publishing, Gloucester.
SWAINSON, C. 1886. *The Folk Lore and Provincial Names of British Birds.* Elliot Stock, London.
TEMPERLEY, G. W. 1951. A history of the birds of Durham. *Trans. Nat. Hist. Soc. Northumb.* vol. 9.
THOM, V. M., 1986. *Birds in Scotland.* Poyser, Calton.
THOMAS, C. J. 1993. *Modelling the Distribution and Breeding Performance of the Raven* Corvus corax *in Relation to Habitat.* Ph.D. thesis, University of Glasgow.
THOMAS, D. K. 1992. *An Atlas of Breeding Birds in West Glamorgan.* Gower Ornithological Society, Neath.
THOMPSON, D. B. A., WATSON, A., RAE, S. & BOOBYER, G. 1995. Recent changes in breeding bird populations in the Cairngorms. *Bot. J. Scotland* 48, 99–110.

TICEHURST, N. F. 1909. *A History of the Birds of Kent.* Witherby, London.
TINBERGEN, N. 1953. Carrion Crow striking Lapwing in the air. *Br. Birds* 46, 377.
TURNER, A. 1996. *A Habitat Management Survey of the Unenclosed Portion of the Carneddau SSSI.* Unpublished report to the Countryside Council for Wales.
TYLER, S., LEWIS, J., VENABLES, A. & WALTON, J. 1987. *The Gwent Atlas of Breeding Birds.* Gwent Ornithological Society, Cardiff.
USSHER, R. J. & WARREN, R. 1900. *The Birds of Ireland.* Gurney and Jackson, London.
VAN VUREN, D. 1984. Aerobatic rolls by Ravens on Santa Cruz Island, California. *Auk* 101, 620–21.
VERNER, W. 1909. *My Life among the Wild Birds in Spain.* John Bale and Danielsson, London.
VOOUS, K.H. 1960. *Atlas of European Birds.* Nelson, Amsterdam.
WALPOLE-BOND, J. 1914. *Field Studies of Some Rarer British Birds.* Witherby, London.
WALPOLE-BOND, J. 1938. *A History of Sussex Birds,* vol. I. Witherby, London.
WALSH, P. H. & MCGRATH, D. 1988. *Waterford Bird Report, 1976–1986.* Irish Wildbird Conservancy.
WATSON, A. 1965. A population study of Ptarmigan (*Lagopus mutus*) in Scotland. *J. Anim. Ecol.* 34, 135–72.
WEIR, D. N. 1978. Effects of poisoning on Raven and raptor populations. *Scott. Birds* 10, 31.
WELLS, C. 1995. Review of the status of breeding Ravens in Cheshire. *Cheshire & Wirral Bird Rep. 1994*, p. 100.
WELLS, J. H. 1996. The status of the Raven in Northern Ireland. *Irish Birds,* in press.
WHILDE, T. 1994. *The Natural History of Connemara.* Immel Publishing, London.
WHITAKER, J. 1907. *Notes on the Birds of Nottinghamshire.* Walter Black, Nottingham.
WHITE, C. M. & CADE, T. J. 1971. Cliff-nesting raptors and Ravens along the Colville River in Arctic Alaska. Pp. 107–50 in *The Living Bird,* 10th annual. Cornell Laboratory of Ornithology.
WHITE, G. 1789. *Natural History and Antiquities of Selborne.* White, London.
WHITLOCK, F. B. 1893. *The Birds of Derbyshire.* Bemrose, London.
WILLIAMSON, K. 1965. *Fair Isle and its Birds.* Oliver and Boyd, Edinburgh.
WILMORE, S. B. 1977. *Crows, Jays and Ravens.* David and Charles, Newton Abbot.
WITHERBY, H. F., JOURDAIN, F. C. R., TICEHURST, N. F. & TUCKER, B. W. 1938. *The Handbook of British Birds,* vol. 1. Witherby, London.
ZIRRER, F. 1945. The Raven. *Passenger Pigeon* 7, 61–7.
ZUNIGA, J. M., SOLER, M. & COMACHO, I. 1982. *Status de la avifauna terrestre de la Hoya de Guadix: Aspectos ecologicos.* Trab. Mongr. Dep. Zool. Univ. Granada. (N. S.) 5, 17–51.

TABLE 1: *Raven breeding population of Britain and Ireland by county*

County	No. of breeding pairs			Most recent date
	Total	Coastal	Inland	
ENGLAND				
Channel Isles	10	10	–	1988–91
Isle of Wight	1	1	–	1988–91
Dorset	8	7	1	1996
Devon*	350	55	295	1988
Cornwall*	100	60	40	1978
Somerset*	50	10	40	1988
Wiltshire	2	–	2	1995–96
Gloucester	5	–	5	1982
Hereford*	30	–	30	1996
Shropshire*	75	–	75	1996
Derbyshire	1	–	1	1996
Cheshire	1+	–	1+	1995
Yorkshire	6–7	–	6–7	1995
Lancashire	8	–	8	1995
Westmorland	39	–	39	1996
Cumberland	46	2	44	1996
Durham	1–2	–	1–2	1995
Northumberland	3	–	3	1996
Isle of Man	40	15–18	22–25	1986
Total	776–778	160–163	613–618	
WALES				
Monmouth (Gwent)	75	–	75	1995
Glamorgan	110	10	100	1995
Carmarthen	100	3–4	95+	1994
Pembroke	140	77	60+	1994
Cardigan	165–190	20+	145–170	1994
Brecon	160	–	160	1995
Radnor	30–50	–	30–50	1985
Montgomery	120–140	–	120–140	1994
Merioneth*	100	–	100	1994
Caernarvon	115	15	100	1995
Anglesey	27	22	5	1994
Denbigh	50–55	–	50–55	1994
Flint	10–12	–	10–12	1994
Total	1202–1274	147–148	1050–1122	
SCOTLAND				
Berwick	2	2	–	1988–91
Dumfries	20	–	20	1996
Kirkcudbright	32	5	27	1994
Wigtown	24	21	3	1994
Ayr	20	7	13	1994
Renfrew	1	–	1	1994
Peebles	3	–	3	1994
Roxburgh	3	–	3	1996
Midlothian	1–2	–	1–2	1994
Dunbarton	9–13	–	9–13	1995
Stirling	7–10	–	7–10	1995
Bute	20–23	13–16	7	1991

TABLE 1 (continued)

County	No. of breeding pairs			Most recent date
	Total	Coastal	Inland	
SCOTLAND (continued)				
Perth	50	–	50	1996
Argyll*	335	170	165	1995
Angus	4	–	4	1995
Aberdeen	1–2	–	1–2	1995
Inverness*	180–200	90–110	90	1978
Ross & Cromarty*	85	35	50	1988–91
Sutherland*	70	30	40	1988–91
Caithness*	10	6	4	1988–91
Orkney	90	80	10	1995
Shetland	156	146	10	1995
Total	1123–1155	605–628	518–527	
Total for Britain	3101–3207	912–939	2181–2267	
IRELAND, NORTHERN				
Antrim*	69	33	36	1996
Down*	34	5	29	1996
Tyrone*	27	–	27	1996
Londonderry*	27	–	27	1996
Fermanagh*	23	–	23	1996
Armagh*	21	–	21	1996
IRELAND, REPUBLIC				
Dublin	⎫	–	–	⎫
Wicklow	⎪	–	–	⎪
Kildare	⎬ 80	–	–	⎬ 1987
Carlow	⎪	–	–	⎪
Wexford	⎭	–	–	⎭
Kilkenny	10	–	–	1990s
Waterford	50	20	30	1976–86
Tipperary*	25	–	–	– ⎫
Roscommon*	10	–	–	– ⎪
Laois*	5	–	–	– ⎪
Westmeath*	5	–	–	– ⎪
Cavan*	5	–	–	– ⎪
Cork*	100	–	–	– ⎪
Kerry*	100	–	–	– ⎬ 1990s
Clare*	50	–	–	– ⎪
Galway*	50	–	–	– ⎪
Mayo*	70	–	–	– ⎪
Sligo*	50	–	–	– ⎪
Leitrim*	50	–	–	– ⎪
Donegal*	100	–	–	– ⎭
Total for Ireland	961	–	–	–

NOTES:
* Counties where numbers are estimated.
In addition, there has been sporadic nesting since 1950 in the following counties:
  England: Stafford, Worcester, Warwick.
  Scotland: Selkirk, Clackmannan, Moray, Nairn.
  Ireland: Offaly, Monaghan, Meath, Longford.

286 Table 2

TABLE 2: Food of the Raven in various regions of Britain and Ireland

1	2		3		4		5		6		7		8		9		10		11	
Region	N Wales (Merioneth)		N England & S Scotland		N England & S Scotland		Central Wales		Central Wales		Central Perthshire		Orkney		Shetland		Shetland		Cork SW Ireland	
Date	c. 1910		1935–70		1974–76		1976–79		1976–79		1990–93		1982–83		1982–83		1982–83		1989	
Pellet source	WR		BT		BT		BT		WR		WR		BT+WR		BT		WR		BT	
No. & % total of pellets	No.	%	No.	%	No.	%	No.	%	No.	%	No.	%	No.	%	No.	%	No.	%	No.	%
Food item																				
MAMMALS																				
Sheep	170	39	76	60	389	56	534	90	60	60	70	10	121	13	261	48	21	28	} 33	} 85
Other large mammals	28	6	–	–	159	23	–	–	–	–	5	<1	–	–	–	Present	–	Present		
Lagomorph	37	9	30	24	189	27	35	6	22	11	398	94	758	80	173	32	9	12	2	5
Vole/Mouse	49	11	25	20	114	16	72	12	7	4	67	9	70	7	–	–	–	–	–	–
Rat	48	11	–	–	–	–	9	1	–	–	–	–	61	6	–	Present	–	Present	–	–
Mole	54	12	1	1	–	–	22	4	2	1	9	1	–	–	–	–	–	–	–	–
Stoat/Weasel	2	<1	3	2	16	2	–	–	–	–	–	–	–	–	–	–	–	–	–	–
Other mammals	6	1	3	2	33	5	15	2	15	2	15	2	2	<1	27	5	4	5	–	–
BIRDS																				
Geese & ducks	–	–	–	–	–	–	–	–	–	–	–	–	60	6	–	–	–	–	–	–
Gulls & waders	–	–	–	–	8	1	–	–	–	–	–	–	26	3	11	2	6	8	–	–
Sea-birds	–	–	–	–	2	<1	–	–	–	–	–	–	20	2	58	11	6	8	–	–
Red Grouse	Present		6	5	36	5	–	–	–	–	–	–	–	–	–	–	–	–	–	–
Pigeon	–	–	2	2	28	4	–	–	–	–	–	–	9	1	–	–	–	–	–	–
Passerine	–	–	3	2	–	–	–	–	} 1	} <1	–	–	–	–	24	4	–	1	1	3
Domestic Fowl	–	–	–	–	–	–	–	–			–	–	27	3	5	1	1	1	–	–
Unidentified	25	6	–	–	84	12	13	2	7	4	118	17	80	8	105	19	11	14	4	10
Eggshell	1	<1	8	6	195	28	10	2	–	–	6	<1	196	21	76	14	11	14	5	13
FISH																				
Freshwater	} 2	} <1	–	–	–	–	–	–	–	–	–	–	–	–	–	–	–	–	–	–
Marine			–	–	–	–	–	–	–	–	–	–	21	2	40	7	6	8	1	3
REPTILES & AMPHIBIA	–	–	–	–	24	3	–	–	–	–	–	–	–	–	–	–	–	–	–	–

Table 2 (continued)

INVERTEBRATES													
Beetles	31	7	16	13	}125	}18	61	10	7	4	29	4	31
Other insects/spiders	1	<1	–	–	–	–	–	–	–	–	74	11	}9
Other terrestrial	–	–	–	–	12	2	1	<1	–	–	–	–	–
Marine	47	11	–	–	–	–	–	–	–	–	–	–	–
VEGETABLE													
Grass, moss, debris	79	18	67	53	651	93	12	2	–	–	483	69	–
Cereal grains	21	5	1	1	16	2	3	<1	–	–	4	<1	94
Fruits & seeds	76	18	–	–	–	–	13	2	–	–	76	11	14
Seaweeds	Present												
OTHER													
Stone & grit	17	4	24	19	300	43	–	–	–	–	115	16	–
Soil	–	–	17	13	–	–	–	–	–	–	1	<1	2
Refuse	–	–	–	–	7	1	10	5	161	81	6	<1	–
Total no. of pellets	433		127		697		591		198		700		945

–	–	–	–	–	–	11	1	34	–	–
–	10	3	1	1	–	1	1	5	30	77
–	–	–	–	–	–	–	–	–	–	–
–	–	–	–	–	–	13	13	21	54	
132	–	Present	Present	34	87					
–	24	6	1	4	5	23	59			
8	1	9	12							
231	43	9	12							
–	–	–	–	–	–					
35	6	51	67							
540		76		39						

NOTES:
WR: Pellet sample from winter communal roost.
BT: Pellet sample from pairs in breeding territories.
Column 2. From Bolam (1913). Hills around Llanuwchllyn, Merioneth. Large mammals: cattle. Lagomorphs: entirely rabbits. Other mammals: dog, common shrew (3), water shrew (1), hedgehog. Birds: mainly Red Grouse, Starling.
Column 3. From E. Blezard (unpubl.). Hills of Lakeland, N Pennines, Cheviots and Southern Uplands. Other mammals: water vole (1), common shrew (1). Eggshells: Red Grouse (4 from 1 locality), Mallard (1), Domestic Fowl (1). Includes two stomachs from nestlings.
Column 4. From Marquiss, Newton & Ratcliffe (1978). Hills of Southern Uplands and Northumberland. Large mammals: deer (mainly), cattle. Other mammals: fox (4 sites), red squirrel (1), hedgehog (2), shrews. Birds: Mallard (8 from 5 localities). Reptiles: common lizard (3 localities). Amphibians: frog and toad (7 localities). Invertebrates: earthworms (3 localities).
Columns 5 & 6. From Newton, Davis & Davis (1982). Hills and farmland in Cardigan, Radnor, Brecon and Carmarthen. Large mammals: horse. Other mammals: dog (8), fox (2), cat (1), pygmy shrew (3), wood mouse (1). Birds: Fieldfare (2), Chaffinch (2), Starling (1), Meadow Pipit (1), Wren (1), corvid (1), Domestic Fowl (4). Eggshells: Black-headed Gull, Domestic Fowl (4) and duck (1).
Column 7. From Mattingley (1995). Hills and marginal land in central Perthshire. Large mammals: red deer. Lagomorphs: predominantly rabbit (31% frequency in identified remains), with mountain hare forming the rest (6%). Other mammals: voles/mice – mainly short-tailed field vole, common shrew (4), wood mouse (1). Birds: especially Red Grouse. Insects: mainly moth cocoons. Fruits and seeds: mainly rowan berries (69), with a few bird cherries (30) and birch seeds (4). Vegetable matter was mostly not identifiable.
Column 8. From Marquiss & Booth (1986). Coasts and maritime moorlands of Orkney. Large mammals: cattle, horse, goat. Other mammals: seal, otter, wood mouse. Sea-birds: mainly auks, Fulmar, Shag. Eggshells: mainly sea-birds and waders but also wildfowl. Berries: crowberry, bilberry.
Columns 9 & 10. From Ewins, Dymond & Marquiss (1986). Column 9 from coasts and maritime moorlands of Shetland, Column 10 from communal roost near rubbish tips. Large mammals: pony, cattle and goat. Other mammals: dog, cat, rat, mouse. Birds: Fulmar, Shag, gull, Blackbird, Fieldfare, Redwing, Starling, corvids. Eggshells: Shag, Guillemot, Curlew, gulls, waders. Berries: cowberry.
Column 11. From Berrow (1992). Coast of Co. Cork. Mammal remains: mostly sheep, with rabbit identified twice. Bird and insect remains were not identified to species, but 12 genera or species of marine invertebrate were named. Seaweeds were predominantly *Corallina officinalis*, but also unidentified fucoids.

TABLE 3: Long-distance recoveries of ringed Ravens

Place ringed	Date ringed	Place recovered	Date recovered	Distance (km)	Direction (°)	Age (days)
Tresta, Shetland	13 May 1993	North Sea oil-rig	16 Aug 1993	229	39	120
Ballycastle, Antrim	11 May 1987	West Stow, Suffolk	25 Jan 1988	551	125	284
Tregaron, Dyfed	13 Apr 1983	Blidworth, Nottingham	22 Oct 1983	210	62	217
Amlwch, Anglesey	15 Apr 1960	Tantallon Castle, Berwick	25 Mar 1962	317	20	734

NOTES:
1. West Stow is on the edge of the Breckland and close to a mid-nineteenth-century breeding place of Ravens.
2. The sea-cliffs at Tantallon Castle may well have been a former nesting haunt.

TABLE 4: Distance of Raven movements, from ringing recoveries

Region	Distance (km)							Total
	0	1–10	11–25	26–50	51–100	101–200	>200	
SW England	2	1	1	2	1	–	–	7
Wales	22	27	47	33	21	7	2	159
N England	6	6	15	40	18	2	–	87
Isle of Man	2	10	10	2	–	1	–	24
S Scotland	3	1	4	18	12	1	–	39
Highlands	2	3	1	2	8	1	–	17
Orkney & Shetland	4	10	5	4	5	–	1	29
Northern Ireland	5	10	6	3	6	2	1	33
S Ireland	19	17	43	38	12	9	–	138
Total	65	85	132	142	83	22	4	533

NOTES:
1. Source: BTO Ringing Scheme.
2. All recoveries: 1924–93.

TABLE 5: *Distance of Raven movements, according to age, from ringing recoveries*

Year	Distance (km)							Total
	0	<10	11–25	26–50	51–100	101–200	>200	
1st*	31	13	2	1	–	–	–	47
1st**	21	34	39	55	32	8	3	192
2nd	2	8	36	29	27	7	–	109
3rd	3	4	18	24	8	2	1	60
4th	1	7	12	14	4	2	–	40
5th	–	5	5	6	3	2	–	21
6th	1	2	4	3	2	–	–	12
7th	1	–	1	2	1	–	–	5
8th	–	1	–	4	–	–	–	5
9th	–	–	1	1	–	–	–	2
10th	–	–	–	–	–	–	–	–
11th	–	–	–	–	–	–	–	–
12th	–	1	1	–	–	–	–	2
13th	–	3	1	–	–	–	–	4
Totals	60	78	120	139	77	21	4	499

NOTES:
1. 1st year* Up to 75 days.
2. 1st year ** 76–365 days.
3. Source: BTO Ringing Scheme.
4. The total of 499 recoveries excludes:
    20 recoveries recorded as long dead, leg and ring only, ring only.
    13 recoveries of birds ringed as adults and subadults (age unknown).
    1 record of a bird recovered a second time, 33 days after release from first recovery.

TABLE 6: *Direction of Raven dispersal, from ringing recoveries*

Region	Direction																
	N	NNE	ENE	E	ESE	SSE	S	SSW	WSW	W	WNW	NNW	O
SW England	1	–	–	2	–	1	1	–	–	–	–	–	2
Wales & W Midlands	5	17	11	31	13	14	8	13	7	5	6	7	22
N England	2	5	5	17	21	6	5	2	4	7	2	4	6
Isle of Man	2	9	1	2	–	–	–	1	5	–	–	2	2
S Scotland	2	5	9	5	3	2	2	2	3	2	3	2	3
Highlands & Islands	1	–	3	1	1	1	2	–	1	–	–	1	2
Orkney & Shetland	3	3	–	1	1	2	5	1	1	2	1	4	6
Northern Ireland	2	2	1	3	3	1	1	2	5	1	5	2	5
S Ireland	13	11	3	6	11	10	18	15	12	13	3	4	19
Total	31	52	33	68	53	37	42	36	38	30	20	26	67

NOTES:
1. Source: BTO Ringing Scheme.
2. All 533 recoveries are included.
3. Category O is for birds recovered at the ringing locality.

TABLE 7: *Raven ringing recoveries according to month and age*

Year	Jan	Feb	Mar	Apr	May	Jun	Jul	Aug	Sep	Oct	Nov	Dec	Total
1st*	–	–	–	4	33	10	–	–	–	–	–	–	47
1st**	18	18	19	4	–	12	24	22	23	21	18	13	192
2nd	8	6	18	33	22	2	2	4	1	6	3	4	109
3rd	2	3	7	15	9	5	5	1	4	4	2	3	60
4th	3	1	6	10	8	1	1	2	1	4	–	3	40
5th	–	3	2	6	2	3	2	1	1	–	1	–	21
6th	–	–	1	3	2	1	–	–	–	3	–	2	12
7th	–	–	2	–	–	–	1	–	–	2	–	–	5
8th	–	–	1	2	1	1	–	–	–	–	–	–	5
9th	–	–	–	–	1	1	–	–	–	–	–	–	2
10th	–	–	–	–	–	–	–	–	–	–	–	–	–
11th	–	–	–	–	–	–	–	–	–	–	–	–	–
12th	–	–	–	–	–	–	2	–	–	–	–	–	2
13th	–	–	2	1	–	–	–	–	1	–	–	–	4
Totals	31	31	58	78	78	36	37	30	31	40	24	25	499

NOTES:
1. 1st year* Up to 75 days.
2. 1st year** 75–365 days.
3. Source: BTO Ringing Scheme.
4. The total of 499 recoveries excludes:
   20 recoveries recorded as long dead, leg and ring only, ring only.
   13 recoveries of birds ringed as adults and subadults (age unknown).
   1 record of a bird recovered a second time, 33 days after release from first recovery.

TABLE 8: Raven tree-nests: choice of tree species by region

Tree	SW England	Wales & Midlands	N England	Isle of Man	S Scotland	Highlands & Islands	Ireland
CONIFERS							
Scots pine	59	155	1	3	1	2	31
Corsican pine	–	9	–	–	–	–	–
Lodgepole pine	1	–	–	–	–	–	–
Spruce (unspecified)	–	56	–	3	–	1	–
Norway spruce	–	6	–	–	–	–	1
Sitka spruce	1	1	–	–	–	–	5
Larch	3	44	–	–	1	–	–
Douglas fir	2	4	–	–	–	–	–
Grand fir	3	–	–	–	–	–	–
Silver fir	1	–	–	–	–	–	–
Cedar	1	–	–	–	–	–	–
Conifer (unspecified)	5	19	–	1	–	–	–
Yew	–	1	–	–	–	–	–
Total	76	295	1	7	2	3	37

Table 8 (continued)   293

BROAD-LEAVES

Oak	12	76	–	–	–	8
Beech	28	48	–	1	–	1
Ash	1	21	–	–	–	–
Birch	–	20	–	–	1	1
Sycamore	4	21	–	1	1	–
Alder	–	6	–	1	–	–
Elm	1	–	–	–	–	–
Hawthorn	1	14	–	–	1	–
Rowan	2	5	–	–	–	–
Elder	–	2	–	–	–	–
Horse chestnut	–	7	–	–	–	–
Sweet chestnut	–	1	–	–	–	–
Total	49	221	–	2	3	10
Total records	125	516	1	9	5	47

NOTES:

Data are all from records submitted to the BTO Nest Records Scheme – 705 out of a total of *c*. 2600. Some records are for the same nest in different years but, taken as a whole, they give a broad indication both of the relative frequency of tree-nests in the different regions and of the preferences for different tree species. In addition, I have other records of the use of trees in two regions (figures are for different territories).

Northern England: Scots pine (4), birch (3), larch (2), beech (1), alder (1), rowan (1).
Southern Scotland: Scots pine (7), Norway spruce (5), rowan (5), ash (2), larch (1), silver fir (1), sycamore (1), alder (1).

TABLE 9: *Raven nest sites on buildings in Britain and Ireland*

Location	Date	Source
OLD BUILDINGS		
Louth Church, Lincolnshire	1693	Lorand & Atkin (1989)
St Nicholas Cathedral, Newcastle	Late 1700s	J. Hancock
Tower on Glastonbury Tor, Somerset	1840s	Palmer & Ballance (1968)
Willet Hill Folly, Somerset	Pre-1850	Palmer & Ballance (1968)
Corfe Castle, Dorset	Early/Mid-1800s	Smith (1905)
Mausoleum at Castle Howard, Yorkshire	Up to 1856	Nelson (1907)
Beverley Minster, Yorkshire	Up to 1840	Nelson (1907)
Kirkwall Cathedral and Bishop's Palace, Orkney	c. 1850	J. Wolley
Tower of Magee College, Londonderry	1868	Ussher & Warren (1900)
Guildhall clock tower, Swansea	1973, 1974	Thomas (1992)
Ruined mansion, Ballintemple House, Carlow	1973–75	G. Noonan
Ruined castle in Co. Limerick	1972–81	P. Brennan & P. Jones
Clonea Castle, Co. Waterford	1977, 1984, 1986	Walsh & McGrath (1988)
Coolnamuck Castle, Co. Waterford	1983	Walsh & McGrath (1988)
Ballynatray ruin, Co. Waterford	1982–83	Walsh & McGrath (1988)
Kilcoe Castle, Co. Cork	1989	Berrow (1992)
Rossbrin Cove Castle, Co. Cork	1989	Berrow (1992)
Ruined chapel, Mainland Orkney	1989-95	Booth (1996)
Bell tower of old church, Mainland Orkney	1977–95	Booth (1996)
Tower of ruined church, Egilsay, Orkney	1987	Booth (1996)
Old lighthouse, North Ronaldsay, Orkney	1966–95 (not every year)	Booth (1996)
The Town Hall, Chester	1996	TV news item
MODERN CONSTRUCTIONS		
Two disused buildings on wartime airfield, Mainland Orkney	1976–87	Booth (1996)
Man-made tower, Shetland	Early 1980s	Ewins *et al.* (1986)
Derelict industrial stacks, Baglan Bay oil refinery, Glamorgan	1980s	Thomas (1992)
Glascarnoch Dam, Ross-shire	1970–73	E. Bartlett
Cooling tower of Yelland power station, Barnstaple, Devon	1987	C. F. Snook
OLD MINE BUILDINGS		
Lead-mine pit-head gear, Plynlimon, Cardigan	1930s	A. Brook
Wheel case below Beinn y Phott, Isle of Man	1943	Cullen & Jennings (1986)
Bleak House, Rattlebrook, Dartmoor, Devon	1940s	W. Robinson
	1959	P. Dare
Old engine house at Red Mires Lake, Dartmoor	Pre-1953	Harvey & St Leger-Gordon (1953)
Old engine houses and stacks at tin mines, Cornwall	Pre-1978	Penhallurick (1978)
Derelict mine chimney, SW Cambrian Mountains	Pre-1975	Davis & Davis (1986)
Old engine house, Pibble Hill mine, Galloway	1982	R. Mearns
Passmore's engine house, Treskillard, Cornwall	1991	Cornwall Bird-watching and Preservation Society, 1991

Table 9 (continued)

DESERTED FARMHOUSES AND COTTAGES

Location	Date	Reference
Ruined shooting box south of Listonshiels, Pentland Hills	c. 1924	Munro (1988)
Burn Divot cottage (derelict), Wark Moors, Northumberland	1930s	H. M. S. Blair
Ruined building, Yarner Wood, Devon	1954	B. Campbell
Ruined farmhouse, Valley, North Uist	1981	D. J. R. Counsell
Ruined cottage, Birsay, Orkney	1983	E. R. Meek
Eleven different ruined houses and cottages, Mainland, Sanday, Stronsay and Damsay (seven territories), Orkney	1973–95	Booth (1996)
Unmanned lighthouse outbuilding, Sanday, Orkney	1994	Booth (1996)
Ruined building, Tiree, Argyll	1989	R. A. Broad
Ruined cottage, Elan Valley, Radnorshire	1990	A. V. Cross

RAILWAY VIADUCTS AND BRIDGES

Location	Date	Reference
St Pinnock railway viaduct, Cornwall	1940–60	Penhallurick (1978)
Glynn Valley railway viaduct, Cornwall	1960	C. J. Stevens
Big Water of Fleet railway viaduct (disused), Galloway	1968, 1975–78	A. D. Watson, G. Horne, D. A. Ratcliffe
Bridge over gorge, Pollaphuca, Kildare	1973	G. Nooman
Keswick line railway viaduct (disused), Cumberland	1986	S. Richardson
Tower of Tamar suspension bridge, Devon	Pre-1988	Sitters (1988)
Cornwood viaduct (disused), Devon	1966 – in use for at least 30 years previously	
Viaduct in Okehampton area, Devon	1988–89	H. G. Hurrell
Britannia Bridge, Menai Straits, Gwynedd	'In recent years'	G. Kaczanow
Lunedale, Westmorland	1995–96	Lovegrove et al. (1994) C. Armitstead

PYLONS AND MASTS

Location	Date	Reference
Radio transmitter mast, Hensbarrow Beacon, Cornwall	1961	Penhallurick (1978)
Electricity pylons in two localities, Devon	Pre-1988	H. P. Sitters & P. J. Dare
Electricity pylon, North Cornwall	1978	Dingle (1980)
Electricity pylons, Uskmouth and north-west moorlands	1980s	Tyler et al. 1987
Roadside telegraph pole in Llyn, Gwynedd	1992	Lovegrove et al. (1994)
Electricity pylon, Glen Muck, Ayrshire	1969	J. Hutchinson, A. D. Watson, R. Roxburgh
Radio mast, Shetland	Early 1980s	Ewins et al. (1986)
TV transmitter, Mynydd Baeden, Mid-Glamorgan	1985	L. Thomas, P. Tagor, S. M. Squires
Electricity pylon, Dee estuary, Cheshire	1995	S. Marsden
Electricity pylons in 10 different territories, six regularly used, with nests only on pylons; Glamorgan–Breconshire	1993–95	A. Dixon

TABLE 10: *Vertical height of Raven nesting cliffs in several inland regions of Britain*

Region	Height of nesting cliff (m)				
	<10	11–20	21–50	51–100	>100
N Wales	1	21	51	16	6
Lake District	–	33	44	14	7
Pennines	1	14	2	–	–
Cheviots	5	16	2	–	–
Southern Uplands	4	35	18	3	–
C Highlands	–	15	41	11	4
NW Highlands	1	15	28	9	3

TABLE 11: *Pattern of use of alternative Raven rock-nest sites*

Territory number and location	Site	Dates used
1 LAKE DISTRICT		
Crag 1: sites spread over 280 m laterally, 60 m vertically	A	1945, 1946, 1962, 1967
	B	1946 (R), 1966, 1974
	C	1947, 1983
	D	1948, 1949, 1951 (R), 1952
	E	1951
	F	1953–56, 1959, 1960, 1964, 1969, 1970, 1973, 1976, 1977, 1979, 1984 (R), 1986 (R), 1987, 1987 (R), 1988, 1989, 1992 (R), 1993, 1993 (R), 1994 (R), 1995
	G	1957, 1958
	H	1963, 1968
	I	1961, 1968 (R), 1978, 1990 (R), 1991, 1996
	J	1972, 1975, 1992, 1994
	K	1965
	L	1984
Crag 2: 870 m from crag 1	M	1976 (R), 1990
Crag 3: 3 km from crag 1; 2.1 km from crag 2	N	1971, 1972 (R), 1981 (R). This locality has held a separate pair since 1983
2 LAKE DISTRICT		
Crag 1: sites spread over 400 m laterally and 50 m vertically	A	1946, 1952
	B	1948, 1954, 1958 (R), 1959–62, 1965 (R), 1967–72, 1974, 1975 (R), 1978 (R), 1989 (R), 1990, 1994
	C	1951, 1957, 1973
	D	1953, 1955 (R), 1956, 1958, 1963–65, 1975–85, 1987, 1991–93, 1995, 1996 (Photograph 19, p. 145)
	E	1986
	F	1987 (R), 1988, 1989
		Plus 2 other sites never known to contain eggs.
Crag 2: 650 m from crag 1	G	1955

continued overleaf

Table 11 (continued)

Territory number and location	Site	Dates used
3 LAKE DISTRICT		
Crag 1: sites spread over 300 m laterally and 120 m vertically	A	1926, 1946–49, 1959, 1960, 1963, 1964, 1969–74, 1988, 1989, 1992 (R), 1993
	B	1956
	C	1924, 1935, 1957, 1990, 1994–96
		Plus 4 other sites known to contain eggs before 1930
Crag 2: 700 m from crag 1	D	1954, 1958, 1967, 1968, 1975, 1976, 1977 (R), 1978–80, 1981 (R), 1983, 1986, 1987, 1991
4 LAKE DISTRICT		
Crag 1: sites spread over 150 m (Photograph 17, p. 141)	A	1945, 1947, 1948, 1949 (R), 1951, 1959, 1963, 1968, 1975, 1977, 1979, 1981, 1989, 1990 (R), 1991, 1992 (R) (Photograph 23, p. 156)
	B	1949, 1951 (R), 1965, 1966, 1970, 1975 (R), 1976, 1978, 1980, 1984, 1987, 1988, 1991, 1996 (Photograph 18, p. 143)
	C	1959 (R), 1961, 1971–74, 1986
Crag 2: 600 m from crag 1	D	1980 (R), 1987 (R)
5 SOUTHERN UPLANDS		
Crag 1: sites spread over 700 m (Photograph 11, p. 63)	A	1947–49
	B	1947 (R), 1949 (R), 1951, 1956 (R), 1965, 1986
	C	1952, 1954, 1959, 1960, 1967, 1987 (R)
	D	1953, 1990
	E	1954 (R), 1955
	F	1958 (R), 1968, 1969, 1974
	G	1961, 1988
	H	1984, 1985
	I	1975, 1977, 1978
	J	1982
Building 1: 1 km from crag 1		
Building 2: 4 km from crag 1		

6 SOUTHERN UPLANDS
  Crag 1: sites spread over 200 m

A	1924–26, 1928, 1931, 1932, 1947, 1960 (R), 1977–79
B	1936, 1937, 1949, 1956, 1962, 1968, 1970, 1973, 1976
C	1946, 1952, 1954
D	1948, 1951, 1959 (R), 1961, 1963, 1965
E	1957, 1958, 1971 (R), 1972, 1974, 1975
F	1959
G	1934, 1941, 1953, 1966, 1971, 1969
H	1955
I	1956 (R)
J	1982

  Crag 2: 350 m from crag 1
  Crag 3: 1.5 km from crag 1; 1.15 km from crag 2
Tree

NOTES:
R = repeat nest after failure of first laying.

TABLE 12: *Use of alternative Raven nest sites according to success or failure*

Region	Nesting sequence	No. of records	Moved to different site	Using same site	Source
N England, S Scotland & N Wales	Successful in previous year	80	35	45	D. A. Ratcliffe & G. Horne
	Failed in previous year	30	17	13	
	Repeat laying in same year	45	37	8	
C Wales	Repeat laying in same year	11	6	5	Davis & Davis (1986)
Shetland	Successful in previous year	18	5	13	Ewins, Dymond & Marquiss (1986)
	Failed in previous year	12	11	1	
	Repeat laying in same year	11	8	3	

TABLE 13: Altitude of Raven nesting cliffs in some higher mountain regions of Britain

Region	No. of cliffs	<100	101–200	201–300	301–400	401–500	501–600	601–700	701–800	>800	Mean
Snowdonia	94	1	20	21	22	11	11	6	2	–	347
Lakeland	106	1	7	12	31	30	20	3	2	–	412
Pennines	21	–	–	3	6	6	5	1	–	–	437
Southern Uplands	80	–	4	16	30	16	11	3	–	–	377
Perthshire	58	–	1	5	11	14	15	9	2	1	496
Ross-shire	28	1	6	14	4	2	1	–	–	–	269
Sutherland	39	5	21	9	3	1	–	–	–	–	188

TABLE 14: Size of Raven eggs for various regions of Britain and Ireland

Region	Years	No. of clutches	No. of eggs	Length of eggs (mm)	Standard error	Breadth of eggs (mm)	Standard error
SW England	1945–60	21	111	48.27	0.221	33.19	0.122
Wales	1899–1942	12	58	48.69	0.276	33.51	0.155
Mid-Wales*	1992	24	103	49.09	0.298	33.75	0.100
Mid-Wales*	1993	28	139	49.42	0.210	33.73	0.093
Mid-Wales*	1994	28	137	49.34	0.232	33.59	0.097
N England	1926–61	52	300	48.76	0.141	33.36	0.071
S Scotland	1911–60	6	24	48.86	0.522	33.63	0.187
Highlands	1898–1955	25	134	50.58	0.201	34.18	0.091
Ireland	Pre-1950	9	52	50.01	0.405	32.90	0.123

NOTES:
*Measured in the nest by A. V. Cross. All other measurements are of blown-eggs in collections.

Table 15: Raven clutch size for various regions of Britain and Ireland

Region	Period	No. of clutches	\multicolumn{7}{c}{Number of eggs in clutch}	Mean	Source
			1	2	3	4	5	6	7		
BRITAIN											
SW England	1983–96	100	–	–	9	33	37	21	–	4.70	G. Kaczanow
S Wales	1993–95	108	–	5	8	21	43	28	3	4.83	A. Dixon
C Wales	1975–86	143	2	6	18	37	45	32	3	4.57	Davis & Davis (1986), Cross & Davis (1986)
N Wales	1946–73	194	–	4	7	54	70	55	4	4.91	E. K. Allin and helpers
N England	1946–95	133	–	3	7	37	41	42	3	4.91	D. A. Ratcliffe, G. Horne
Southern Uplands	1946–94	204	1	4	16	47	86	48	2	4.79	D. A. Ratcliffe, D. Cross, J. Hutchinson
Southern Uplands	1974–94	51	–	1	4	14	17	13	2	4.84	M. Marquiss, R. Mearns, C. Rollie
Orkney	1972–77	51	–	–	4	9	17	19	2	5.11	C. J. Booth
Shetland	1984–93	43	1	2	6	8	13	12	1	4.63	Ellis et al. (1994)
Total		1027	4	25	79	260	367	270	20	4.81	
% of total			0.4	2.4	7.7	25.3	35.9	26.3	1.9		
BRITAIN & IRELAND											
England	1949–93	77	–	3	12	16	20	24	2	4.73	BTO
Wales	1947–93	214	–	7	22	51	67	51	16	4.85	BTO
Scotland	1959–93	46	–	3	9	5	16	11	2	4.63	BTO
Northern Ireland	1956–92	11	–	–	1	3	5	2	–	4.73	BTO
Republic of Ireland	1954–88	36	–	5	3	8	9	8	3	4.58	BTO
Total		384		18	47	83	117	96	23	4.77	BTO
% of total				4.7	12.2	21.6	30.5	25.0	6.0		

NOTES:
1. As some data from the first series (mainly from north Wales) were contributed to the BTO Nest Records Scheme, the two series of records are not independent.
2. The BTO records omit clutches of 1 egg and include only nests with eggs visited again to check that incubation had begun.

TABLE 16: *Raven clutch size for first and repeat layings in northern England and northern Scotland*

Region	Date	No. of clutches	Laying	No. of eggs in clutch							Mean	Source
				1	2	3	4	5	6	7		
N. England, S Scotland	1945–95	388	1st	1	8	27	98	144	103	7	4.84	See Table 17
		54	2nd	–	4	6	24	17	3	–	4.17	D. A. Ratcliffe, G. Horne
N England, S Scotland	1935–94	28	1st	–	–	–	5	10	13	–	5.29	D. Cross, J. Hutchinson
		28	2nd	–	–	1	9	14	4	–	4.75	D. Cockbain

NOTES:
1. In the first set of data, 1st and 2nd clutches are significantly different in size; Mann-Whitney test, $P = <0.001$ (A. Fielding).
2. The second set of data consists of instances where the sizes of the 1st and repeat clutches from the same birds in the same year were known. They are mainly egg-collectors' records and show an upward bias in clutch size because of selection for large 1st clutches. 1st and 2nd clutches are significantly different; Mann-Whitney test, $p = 0.02$ (A. Fielding).

TABLE 17: Date of Raven first egg for various regions of Britain and Ireland

Region	Period	No. of nests	Terrain	Mean date of 1st egg	Other observations	Source
England	1949–92	69	Coastal and inland	6 Mar		BTO
Wales	1946–93	220	Coastal and inland	5–6 Mar		BTO
Scotland	1961–93	28	Coastal and inland	20 Mar		BTO
Northern Ireland	1957–92	10	Coastal and inland	8 Mar		BTO
Republic of Ireland	1954–88	35	Coastal and inland	2 Mar		BTO
Total Britain and Ireland	1946–93	362	Coastal and inland	6–7 Mar	68% begin laying between 26 Feb and 14 Mar	BTO
SW England	1983–91	89	Inland	5 Mar	Laying spread 14 Feb to 28 Mar	G. Kaczanow
S Wales	1993–95	29	Inland	10 Mar	Laying spread 22 Feb to 24 Mar	A. Dixon
C Wales	1975–78	98	Inland	5 Mar		Davis & Davis (1986)
C Wales	1986	29	Inland	9 Mar	Very cold weather in Feb and Mar	Cross & Davis (1986)
C Wales	1992–93	50	Inland	4–5 Mar		A. V. Cross
N Wales	1951–79	42	Coastal and inland	5 Mar		D. A. Ratcliffe
N Wales	1946–67	167	Coastal and inland		76.5% of clutches completed by 10 Mar below 305 m. 37.5% of clutches completed by 10 Mar above 305 m.	Allin (1968)
N England	1933–88	68	Inland	8 Mar		D. A. Ratcliffe, G. Horne, R. J. Birkett, D. Cockbain
Southern Uplands	1948–85	54	Inland	9 Mar		D. A. Ratcliffe, D. Cross, J. Hutchinson
Highlands	1956–71	32	Coastal and inland	11 Mar		D. A. Ratcliffe
Orkney	1972–77	42	Coastal and inland	10 Mar	77% of clutches completed before 21 Mar	Booth (1979) and unpubl.
Fetlar, Shetland	1977–83	24	Coastal	29 Mar	Laying spread 24 Feb to 5 May	Ewins, et al. (1986)

304   Table 18

TABLE 18: *Raven brood size for various regions of Britain and Ireland*

Region	Period	No. of broods	No. of young in brood						Mean	Source
			1	2	3	4	5	6		
BRITAIN										
S Wales	1993–95	30	3	8	11	6	2	–	2.87	A. Dixon
C Wales	1975–86	202	27	52	60	48	15	–	2.86	Davis & Davis (1986); Cross & Davis (1986)
N Wales	1951–79	37	12	12	7	4	1	1	2.27	D. A. Ratcliffe
N Wales	1946–67	154	7	34	47	41	24	1	3.29	Allin (1968)
N Wales	1978–81	32	3	16	9	3	1	–	2.47	Dare (1986a)
N England	1935–75	91	11	25	27	27	1	–	2.80	D. A. Ratcliffe, E. Blezard, G. Horne
N England	1976–95	256	22	35	110	71	18	–	3.11	G. Horne, D. Hayward, P. Stott, S. King, G. Fryer
Southern Uplands	1946–75	118	11	34	32	33	7	1	2.95	D. A. Ratcliffe, M. Marquiss, R. Roxburgh
Southern Uplands	1976–95	74	5	14	21	23	9	2	3.31	R. Roxburgh, C. Rollie, M. Marquiss, R. Mearns
Perthshire	1982–95	101	11	27	37	23	3	–	2.80	P. Stirling-Aird, D. MacCaskill
Western Highlands	1956–71	52	6	17	20	9			2.62	D. A. Ratcliffe
Orkney	1983–95	205	28	51	60	46	17	3	2.91	C. J. Booth
Shetland	1984–92	176	23	33	44	48	22	6	3.18	P. M. Ellis
Total		1528	169	358	485	382	120	14	2.98	
% of total			11.1	23.4	31.7	25.0	7.9	0.9		

Table 18 (continued)

BRITAIN & IRELAND									
England	1943–93	294	44	56	89	68	34	3	3.00	BTO
Wales	1946–93	933	96	188	296	245	99	9	3.10	BTO
Scotland	1953–93	314	47	50	83	86	42	6	3.14	BTO
Northern Ireland	1947–62	53	3	7	13	16	11	3	3.64	BTO
Republic of Ireland	1954–88	164	25	30	43	44	22	–	3.05	BTO
Total		1758	215	331	524	459	208	21	3.10	BTO
% of total			12.2	18.8	29.8	26.1	11.8	1.2		

NOTES:
1. As some data from the first set were contributed to the BTO Nest Records Scheme, the two series of records are not independent. The first series is mainly records of older or fledged young, whereas the BTO series includes broods of all ages: the higher frequency of broods of 5 to 6 in the BTO series probably reflects this difference.
2. The Allin data for brood size in N Wales are significantly larger than those of Ratcliffe and Dare, but included many lowland and coastal nests whereas the other two sets were from uplands.
   N England broods were significantly larger in the later period: Jonckheere-Terpstra test, $p = 0.02$ (A. Fielding).
   Southern Upland broods were marginally significantly larger in the later period: JT test, $p = 0.04$ (A. Fielding).
3. J. H. Wells (1996, in press) found that 237 broods in Northern Ireland, 1984–96, averaged 3.54 young.
4. G. Kaczanow (unpublished) found that 33 broods in the Dartmoor area of Devon, 1983–96, averaged 2.10 young, coinciding with a high rate of egg infertility (p. 182).

TABLE 19: *Raven brood size for first and repeat layings in northern England and southern Scotland*

Region	Date	No. of broods	Laying	No. of young in brood						Mean	Source
				1	2	3	4	5	6		
N. England, S Scotland	1935–95	539	1st	49	108	190	154	35	3	3.05	See Table 18
N England, S Scotland	1926–95	37	2nd	8	9	14	6	–	–	2.49	G. Horne, E. Blezard, D. A. Ratcliffe, M. Marquiss, C. Rollie

NOTES:
Broods from repeat layings are significantly smaller than those from 1st layings: Mann-Whitney test, $p = 0.005$; χ^2 test, $p = 0.035$ (A. Fielding).

Table 20   307

TABLE 20: Raven breeding performance for various regions of Britain and Ireland

1	2	3	4	5	6	7	8	9	10	11
Region	Period	No. of territories visited	No. of territorial pairs	%	No. of pairs laying eggs	%	No. of pairs rearing young	%	Clutch size	Fledged brood size per successful pair	Productivity of breeders (young/pair)	Productivity of territorial pairs (young/pair)	Source

Region	Period	No. territories visited	No. territorial pairs	%	No. pairs laying eggs	%	No. pairs rearing young	%	Clutch size	Fledged brood/successful pair	Productivity breeders	Productivity territorial pairs	Source
S Wales	1993–95	109	100	92	83	83	~58	70	4.63	2.87	2.01	1.66	A. Dixon
C Wales	1975–80	323	304	94	269	88	181	67	4.50	2.83	1.90	1.68	Davis & Davis (1986)
C Wales	1986	72	61	85	52	85	34	65	4.79	3.03	1.97	1.69	Cross & Davis (1986)
C Wales	1992–94	345	329	95	*319	97	218	68	5.03	3.09	2.10	2.05	A. V. Cross
N Wales	1951–79	143	136	95	+53	–	#43	81	–	2.27	1.84	1.68	D. A. Ratcliffe
N Wales	1946–67	–	–	–	171	–	137	80	4.91	3.29	2.63	–	Allin (1968)
N Wales	1978–85	290	265	91	+108	–	#79	73	–	2.47	1.80	–	Dare (1986a)
N England	1974–95	–	–	–	567	–	413	73	4.91	3.10	2.26	–	G. Horne, P. Stott, D. Hayward, D. A. Ratcliffe
Southern Uplands	1989–94	156	108	69	79	73	61	77	4.84	3.31	2.54	1.87	C. Rollie, R. Roxburgh
Orkney	1983–95	–	–	–	368	–	205	56	5.11	2.91	1.63	–	C. J. Booth
Shetland	1982–84	–	–	–	133	–	69	52	4.70	3.19	1.66	–	Ewins et al. (1986)
Shetland	1987–93	301	240	80	221	92	128	58	^4.63	^3.18	1.84	1.70	Ellis et al. (1994)
Northern Ireland	1984–96	767	670	87	592	88	505	85	4.73	3.53	3.00	2.66	J. H. Wells

NOTES:
~ estimated from a smaller sample which gave percentage breeding success.
* nest-building pairs.
+ a sample of territorial pairs was monitored.
\# includes presumed successful large broods.
^ data for 1984–93.

1. Percentages are the figures in the column as a proportion of the figures in the previous column.
2. Figures in column 9 are obtained by multiplying those in column 8 by percentages in column 6.
3. Figures in column 10 are obtained by multiplying those in column 8 by the proportions of column 6 to column 4.

TABLE 21: Success rate of individual Raven territories since 1945

Territory	Brood reared		1st clutch robbed	Total failure	Repeat outcome unknown	Overall success	
	1st laying	Repeat laying				Total broods reared	% success
Lakeland 1	17	9	7(2)	11	44	26	59
Lakeland 2	24	4	5(2)	9	42	28	68
Lakeland 3	22	1	3(1)	5	31	23	74
Lakeland 4	14	4	8(6)	3	29	18	62
Lakeland	12	2	7(3)	2	23	14	61
Lakeland	10	4	1	1	16	14	88
Southern Uplands 5	7	–	5(4)	9	21	7	33
Southern Uplands 6	15	1	–	5	21	16	76

NOTES:
1. Figures are for number of occasions.
2. Total failure includes 1st and repeat clutches taken, or no repeat laid after 1st clutch taken, or failure for other reasons.
3. Bracketed figures after *1st clutch robbed* are for repeat clutches seen, but not followed up. Some of these could have been successful.
4. *% success* is therefore a minimum figure.
5. Territory numbers correspond to those in Table 11.

TABLE 22: Raven nest-spacing and breeding density for various regions of Britain and Ireland

Region	Period	Terrain	No. of pairs laying	Mean nearest neighbour distance (km)	Total area occupied (km²)	Mean density (km²/pair)	Mean density (pairs/100 km²)	Maximum density (pairs/100 km² grid square)	Source
Devon	1983–89	Stock-rearing and marginal farmland, upland sheepwalk	55	2.32	520	9.5	10.5	11	G. & J. Kaczanow
N Cornwall–Devon	1932	Coastal	15	1.81	–	–	–	–	D. Nethersole-Thompson
S Wales	1993–95	Marginal farmland and upland sheepwalk	40	2.00	399	10.0	10.0	10	A. Dixon
C Wales	1976	Marginal farmland	10	2.00	160	16.0	6.3	–	Newton, Davis & Davis (1982), Davis & Davis (1986)
C Wales	1976–7	Upland sheepwalk and forest	*50	1.70	315	6.3	15.9	–	Newton, Davis & Davis (1982), Davis & Davis (1986)
C Wales	1986	Previous two combined	52	1.74	475	9.1	11.0	17	Cross & Davis (1986)
NE Wales	1953	Upland sheepwalk and grouse-moor	11	3.60	288	26.2	3.8	–	Simson (1966)
Snowdonia	1951–61	Upland sheepwalk	38	3.15	690	18.2	5.5	7	D. A. Ratcliffe
Snowdonia	1978–85	Upland sheepwalk	#88	2.00	926	10.5	9.5	13	Dare (1986a)
Migneint-Hiraethog	1978–85	Moorland and marginal farmland	#18	3.55	477	26.5	3.8	6	Dare (1986a)
Pennines	1955–60	Upland sheepwalk and grouse-moor	10	10.40	3920	392.0	0.3	2	D. A. Ratcliffe
Lake District	1980–95	Upland sheepwalk	75	3.07	1235	16.5	6.1	9	G. Horne, P. Stott, D. A. Ratcliffe, D. Hayward
Cheviots	1967	Upland sheepwalk, grouse-moor and conifer forest	11	8.50	2040	185.0	0.5	2	Galloway & Meek (1983)

continued overleaf

TABLE 22 (continued)

Region	Period	Terrain	No. of pairs laying	Mean nearest neighbour distance (km)	Total area occupied (km²)	Mean density (km²/pair)	Mean density (pairs/ 100 km²)	Maximum density (pairs/ 100 km² grid square)	Source
Isle of Man	1987	Coastal and upland sheepwalk	40	–	565	14.1	7.1	–	Cullen & Jennings (1986)
Moffat Hills	1949–70	Upland sheepwalk	9	4.10	305	33.9	3.0	5–6	D. A. Ratcliffe
Galloway & Carrick	1946–70	Upland sheepwalk and conifer forest	29	3.40	627	21.6	4.6	6–7	D. A. Ratcliffe, D. Cross, J. Hutchinson
SW Scotland	1961–62	Coastal	21	4.45	–	–	–	–	R. Stokoe, Mearns (1983)
Perth-Stirling-Dunbarton	1992–95	Upland sheepwalk, grouse-moor and deer forest	35	5.23	1910	54.6	1.8	4	P. Stirling-Aird, D. MacCaskill, J. Mitchell
Mid-Argyll	1990	Sheepwalk, deer forest and conifer forest	62	3.60	1689	27.2	3.7	–	Thomas (1993)
N Mull	1993–95	Coast, upland sheepwalk and deer forest, conifer forest	20	3.20	359	18.0	5.6	–	P. Haworth
C/S Mull	1993–95	Coast, upland sheepwalk and deer forest	19	4.50	583	30.7	3.3	–	P. Haworth
Ross of Mull	1993–95	Coast, crofting farmland and upland sheepwalk	16	3.20	175	10.9	9.1	–	P. Haworth
Wester Ross	1956–70	Upland deer forest	18	6.10	1115	61.9	1.6	3	D. A. Ratcliffe, R. Balharry
Sutherland	1952–85	Upland deer forest	16	7.20	1396	87.3	1.2	3	D. A. Ratcliffe, S. Rae
Orkney	1972–77	Coastal and marginal farmland and moorland	90	3.50	974	10.8	9.2	–	Booth (1979)

Table 22 (continued) 311

Shetland	1984–93	Coastal and marginal farmland and sheepwalk	150	2.44	1468	9.8	10.2	–	Ellis et al. (1994) Ewins, Dymond & Marquiss (1986)
SE Ireland	1972–87	Coastal and upland sheepwalk	57	–	2025	35.7	2.8	–	G. C. Noonan in Hutchinson (1989)
Waterford	1976–96	Coastal	22	3.13	–	–	–	–	D. McGrath
Waterford	1976–96	Upland sheepwalk and conifer forest	10	2.20	95	9.5	10.5	–	D. McGrath
Connemara	1985–86	Upland sheepwalk	26	7.00	1690	65.0	1.5	3	P. Haworth

NOTES:

1. For measurements of breeding density to be comparable, they must be based on the number of pairs known to lay eggs in 1 year (or the average for a run of years).

   * therefore shows numbers of pairs which have been corrected downwards by using source figures for maximum number and % of pairs laying eggs.

   # indicates territory-holding pairs, as the sources do not give figures for % of pairs laying eggs.

   For some regions only segments of the population are included in this table.

2. Comparability also requires that measurements of breeding density are calculated by the same method and this is not the case for some of the above figures. I have followed the method (Ratcliffe, 1993) whereby the nearest-neighbour distance (NND) is used to delineate the outer boundary of study areas by drawing intersecting arcs with this radius around the outermost pairs. Simson (1966) used this method, but with a radius of half the NND, and I have recalculated his area; applying the full NND allows for the tendency for this parameter to underestimate the average spacing between pairs.

3. The figures for Orkney and Shetland are based on the total land areas of these island groups.

Overseas measurements of breeding density (pairs/100 km²)

(a)

Schleswig-Holstein, Germany	4.7	Simson (1966)
Mecklenburg, Germany	4.7	Prill (1982)
N Germany	0.7–2.1	Looft (1983)
Wolgast, Germany	18.7	Sellin (1987)
Switzerland (Wallis)	2.9–3.0	Oggier (1986)
Granada, Spain	5.8	Zuniga et al. (1982)
Botosani, Romania	9.6	Andriescu & Corduneau (1972)
Tula Forest, Russia	7.5	Likhachev (1951)
Tenerife, Canary Islands	3.4–3.9	Martin (1987)
El Hierro, Canary Islands	34.2–35.6	Nogales (1994)
Iceland	1.5–6.8	Skarphedinsson et al. (1990)
Utah (Eastern Great Basin), USA	1.9	Smith & Murphy (1973)
Malheur, Oregon, USA	4.0–4.6	Stiehl (1985)
North-West territories, central arctic Canada	0.35–0.6	Poole & Bromley (1988)
Snake River, Idaho, USA	72.6	Kochert, in Nogales (1994)
Los Esesmiles, El Salvador	6.0	Dickey & Van Rossem (1938)

TABLE 23: *Unusually close nesting and high breeding density in Ravens*

Region	Density	Date	Source
Cornwall, coast near Morwenstow	3 eyries in 1400-m cliff section, including 2 only 400 m apart	1932	D. Nethersole-Thompson
Dartmoor, Devon, inland	2 eyries 400 m apart	1994	G. & J. Kaczanow
	2 eyries 500 m apart	1994	
	both in different levels of two quarries		
Central Wales, inland	2 eyries 600 m apart	1975–79	Davis & Davis (1986)
	2 eyries 400 m apart	1993–95	A. V. Cross
Anglesey, coast at Carmel Head	2 eyries 650 m apart, 4 eyries in 3200 m	1951–52	D. A. Ratcliffe
Snowdonia, inland	2 eyries 350 m apart	1977–78	P. Dare
Lake District, inland	2 eyries 700 m apart	1987–92	G. Horne
Galloway, inland	3 eyries within circle of 1500-m diameter	1946–47	D. A. Ratcliffe
Orkney, coastal Mainland	3 eyries in 1300-m cliff section	1980	
	2 eyries (both successful) 400 m apart	1984	C. J. Booth
	5 eyries (3 successful) in 2.8-km cliff section, including NNDs of 500, 500 and 700 m	1985	
Orkney, coastal Rousay	2 eyries 800 m apart	1977	C. J. Booth
Lundy Island, Devon, 419 ha	9 pairs (0.47 km²/pair); more usually 2–4 pairs (1.04–2.09 km²/pair)	1966–67	Moore (1969)
Pembrokeshire Islands, *c.* 700 ha	12 pairs (0.58 km²/pair)	1970s, 1980s	Donovan & Rees (1994)
Colonsay, Inner Hebrides, 4550 ha	12 pairs (3.79 km²/pair)	1994	Argyll Bird Report (1994)
Coll, Inner Hebrides, 7326 ha	12 pairs (6.1 km²/pair)	1987	Stroud (1989)
Fetlar, Shetland, 3962 ha	9 pairs (4.4 km²/pair)	1984, 1987	P. M. Ellis
Fair Isle, 830 ha	6 pairs (1.38 km²/pair)	1983–84	Dymond (1991)
St Kilda (852 ha)	5–8 pairs (1.07–1.70 km²/pair)	1974–77	Harris & Murray (1978)
Mingulay & Berneray, Outer Hebrides, *c.* 750 ha	4 pairs (1.88 km²/pair)	1880s	Harvie-Brown & Buckley (1888)
Clear Island, Co. Cork, 602 ha	6 pairs (1.00 km²/pair)	1986	Cape Clear Bird Observatory Report
Rathlin Island, Co. Antrim, 1420 ha	6 pairs (2.37 km²/pair)	1996	J. H. Wells

NOTES:
NND Nearest-neighbour distance.

TABLE 24: *Life table for Ravens ringed in Britain and Ireland as nestlings*

Year of ringing	Total no. ringed	Age-class of recovered birds													Total deaths
		0	1	2	3	4	5	6	7	8	9	10	11	12	
1923–40	213	3	9	1	1										14
1941–50	182	13	5	1		1	2							1	23
1951–55	215	13	5	2	4		1								25
1956–60	242	8	6	5	7										29
1961–65	495	30	11	9	3	2			1						55
1966–70	496	15	12	6	3		2								37
1971	89	3	3									1			6
1972	120	4	4	1	1	2									12
1973	119	9	6	1	1	1	1	1		1			1		21
1974	118	12		2	1	2		1					1		18
1975	173	7			2		1								12
1976	206	12	2	2	3	1	1								20
1977	275	9	5	5	3	2	1	1		1				1	19
1978	178	2	1	2	1	1	1								8
1979	151	3	2	1		1	1								10
1980	189	9	2		1	1		1	1	1					18
1981	122	8	5	2	2	2			1				1		13
1982	313	3	5	2		1	1	2	1						14
1983	261	8	2	2	3				1						18
1984	284	9	2	5	1				1						18
1985	251	5	3	2		2	1								13
1986	417	11	4	2	2										19
1987	402	12	4	6	1										23
1988	319	6	6	3		2									17
1989	328	7	2	1	2	1									13
1990	331	5		2											7
1991	360	2	1												3
1992	362	2	2												4
1993	401	3													3
Total 1923–81	3583	160	78	32	33	15	9	3	2	2			2	4	340
Overall total	7612	233	109	57	42	21	12	5	5	2			2	4	492

NOTES:
1. Age 0 = up to 12 months; Age 1 = up to 24 months; and so on.
2. No records after 1993 have yet been processed by the BTO. Recoveries include live birds taken captive but later released, but exclude sight records of birds still at large in the wild (7) and records reported as long dead, ring only, or leg and ring only (13 up to 1981, 7 after 1981).

TABLE 25: *Increase in sheep numbers (a) on the Carneddau, north Wales; (b) in Northern Ireland*

(a)

Year	No. of breeding ewes	Total no. of sheep*	Area of rough grazing (ha)
1952	22,660	53,487	7373
1962	22,836	56,800	7239
1972	20,675	49,334	5671
1982	26,619	66,535	6919
1992	33,563	84,092	6642

NOTES:
* Includes breeding ewes, yearlings, rams, draft and cast ewes, wethers and lambs <1 year old.
Data are for Aber, Bethesda, Llan Llechid and Capel Curig communities and are from *A Habitat Management Survey of the unenclosed portion of the Carneddau SSSI* (1995): a report by A. J. Turner to the Countryside Council for Wales.

(b)

Year	Total no. of sheep
1980	1,061,000
1982	1,234,000
1984	1,450,000
1986	1,716,000
1988	2,093,000
1990	2,534,000
1992	2,656,000
1994	2,530,000
1996	2,475,000

NOTES:
Data from Department of Agriculture censuses provided by Wells (1996, in press).

TABLE 26: *Increase in sheep numbers in various regions of Britain: (a) England and Wales; (b) Scotland*

(a)

	Total number of sheep by regions (England and Wales)					
Year	SWENGLA	N'HUMBR	POWYS	SWALES	DYFED	NORTHW
1977	1,633,594	1,111,225	2,750,591	727,856	1,227,550	1,245,390
1978	1,777,610	1,171,819	2,897,708	771,861	1,357,425	1,267,836
1979	1,882,411	1,152,450	2,865,985	775,948	1,371,040	1,292,559
1980	1,969,433	1,223,453	3,019,463	825,675	1,445,363	1,343,062
1981	2,006,114	1,253,460	3,078,920	854,480	1,472,923	1,393,400
1982	2,034,732	1,301,317	3,126,386	853,042	1,516,116	1,433,959
1983	2,145,021	1,350,584	3,253,353	887,017	1,584,948	1,472,255
1984	2,105,055	1,369,555	3,332,635	920,365	1,678,550	1,496,440
1985	2,132,461	1,407,621	3,355,539	927,559	1,721,686	1,548,089
1986	2,250,776	1,452,855	3,451,847	952,488	1,788,670	1,588,994
1987	2,392,542	1,512,292	3,542,426	990,230	1,893,911	1,651,773
1988	2,253,192	1,575,960	3,694,779	1,033,669	2,044,747	1,710,442
1989	2,691,566	1,627,712	3,825,974	1,105,837	2,173,910	1,756,039
1990	2,648,016	1,673,989	3,857,029	1,124,993	2,235,680	1,752,816

(b)

	Total number of sheep by regions (Scotland)					
Year	HIGHLAN	D&GALLO	BORDERS	WISLS	SHETISL	ORKNEYS
1977	1,331,820	1,081,420	1,365,321	70,325	237,130	70,325
1978	1,327,489	1,091,605	1,412,668	74,381	233,589	74,381
1979	1,291,895	1,095,918	1,403,015	75,242	230,229	75,242
1980	1,375,808	1,144,691	1,487,008	76,913	242,668	76,913
1981	1,383,037	1,151,826	1,478,601	75,014	253,366	75,014
1982	1,384,306	1,194,027	1,514,117	84,724	254,397	84,724
1983	1,399,142	1,230,283	1,506,881	92,397	261,228	92,397
1984	1,444,684	1,232,292	1,519,841	103,110	267,094	103,110
1985	1,483,558	1,286,404	1,527,640	109,918	270,347	109,918
1986	1,487,558	1,353,700	1,539,521	115,627	285,726	115,627
1987	1,396,426	1,386,586	1,594,957	115,852	248,042	115,852
1988	1,427,376	1,458,454	1,663,474	126,507	258,287	126,507
1989	1,403,432	1,494,860	1,719,345	133,310	264,973	133,310
1990	1,380,649	1,547,591	1,767,567	222,262	343,001	135,447

NOTES:
(a) SWENGLA SW England – Devon & Cornwall
    N'HUMBR   Northumbria/Northumberland
    POWYS   Montgomery, Radnor & Brecknock/Powys
    SWALES   Monmouth, Glamorgan/Gwent & the Glamorgans
    DYFED   Cardigan, Carmarthen & Pembrokeshire/Dyfed
    NORTHW   Flint, Denbigh, Môn, Caernarvon, Merionedd/Clywd, Gwynedd

(b) HIGHLAN   Caithness, Sutherland, Ross & Cromarty, Inverness/Highland region.
    D&GALLO   Wigtown, Kirkcudbright & Dumfries/Dumfries & Galloway region
    BORDERS   Lothians, Berwick, Roxburgh, Selkirk & Peebles/Borders & Lothian regions
    WISLS   Outer Hebrides/Western Isles area
    SHETISL   Shetland Islands
    ORKNEYS   Orkney Islands

1. Sheep totals for Cumbria were 1,649,912 in 1977 and 2,621,028 in 1990.
2. Compiled by R. J. Fuller from data provided by the MAFF/DAFS June Agricultural Census Returns.

TABLE 27: Analyses of organo-chlorine (o/c) residues in Raven eggs (with Golden Eagle and Buzzard for comparison)

Species	Year	pp′DDE	pp′DDT	HEOD	Heptachlor epoxide	BHC	PCBs	Total o/c
Golden Eagle (10)	1963	1.0	0.1	1.3	trace	0.1	–	2.5
Buzzard (4)	1963	0.4	0.1	2.0	trace	trace	–	2.5
Raven (8)	1963	1.0	0.1	0.8	trace	0.3	–	2.1
Raven (10)	1970	0.23	–	–	–	–	–	–
Raven (6)	1975	0.19	–	0.04	–	–	1.0	1.23

NOTES:
1. pp′DDE is the principal metabolite of DDT in avian tissues.
2. DDT (dichlorodiphenyltrichloroethane) is an insecticide.
3. HEOD is the active ingredient in the insecticide dieldrin and a metabolite of the active ingredient in the insecticide aldrin in avian tissues.
4. Heptachlor epoxide is the metabolite of the insecticide heptachlor in avian tissues.
5. BHC is now known as HCH (hexachlorocyclohexane) and is an insecticide (also called lindane).
6. PCBs are the complex of polychlorinated biphenyls, organo-chlorine pollutants from industrial sources.
7. All measurements in parts per million (wet weight).

Index

Scientific names of plants and animals are given in Appendix 4.

Aberdeenshire 36, 65, 216, 262
Adder 89
Adolescence 102–4, 180–2, 199, 211–2, 215
Afforestation, effects of 35–6, 168, 186, 206, 225–32
Africa
  north 166, 241, 243, 246
  south 246–7
  tropical 245–7
Aggression
  interspecific 127–37
  intraspecific 97–102, 111–4, 209
Agriculture, influence of *see* Farming
Ailsa Craig 38, 88
Alaska 126
Albinism 269
Algal growths, below nests 157
Alpine zone 37, 158–9, 245
Altitude, effects of 158–9, 167, 169, 186, 206
America
  Central 241, 245
  North 214, 245
Amphibians, as food 89
Anglesey 50, 107, 186, 258
Angus 36, 65, 216, 262
Antrim, Co. 35, 74
Appearance, of Raven 9–10, 245, 268–70
Armagh, Co. 74
Arran 36, 64, 260
Arctic regions 7, 8, 29, 125, 134, 159, 166, 170, 193, 241
Argyll 25, 36, 43, 66, 68, 82, 84, 92–3, 101, 104, 128, 131, 136, 163, 166, 193–5, 199, 201, 211, 216, 229, 262
Asia 243
Aspect, effects of 158–9
Asynchronous hatching *see* Hatching
Atlases of birds in Britain and Ireland
  The Atlas of Breeding Birds 42, 72, 74
  The Atlas of Wintering Birds 118
  The New Atlas of Breeding Birds 42, 44–5, 53, 72, 74
Augury, Raven's supposed power of 9, 13, 265

Australia 152, 241, 247–8
Ayrshire 30, 36, 61, 63, 137, 223–4, 234, 260

Baits, use to attract Ravens 92, 104, 126, 238
Banffshire 36, 65, 262
Bare parts, of Raven 28, 174–5, 270
Barra 70
Bass Rock 60
Beachy Head 41
Bears 8, 138
Beetles, as food 79, 80, 89
Benbecula 70
Bergman's Rule 165, 245
Berkshire 41
Berneray 312
Berries, as food 90
Berwickshire 16, 36, 60–1
Berwyn Mountains 35
Biblical references to Ravens 12, 85
Bill, of Raven 82, 94, 175, 268
Birds, as food 81, 86
Blanket bog 34–5, 194
Bodmin Moor 46, 237, 256
Bones, remains of Raven 8, 13
Borders, Scottish 1, 2, 4, 14, 28, 150, 206
Borders, Welsh 14, 34, 224
Boreal Forest 8, 208, 225
Bounty schemes for Raven destruction 20–2, 60, 88
Bowland Fells 35, 56–7, 220
Breckland 47, 86, 201
Breconshire 49, 99, 159, 182, 232, 257
Breeding of Ravens
  age at first 124–5, 199, 214–5
  annual onset of 140, 168–70
  causes of failure 189–193
  cycle 194
  density 5, 62, 204–11, 229, 236, 309–12
  distribution 40–74, 207, 210–1
  performance 188–95, 228–9, 232, 239, 307–8
  see also Eggs of Raven, Clutch size, Young, Territory: breeding
British Trust for Ornithology (BTO) 119, 166–8, 186, 233, 240

317

318 *Index*

Brood
    reduction 184–5
    size 182–8, 228, 240, 304–6
Brooding behaviour 175–7
Brown trout, as food 89
Buildings, nest-sites on 46–8, 140, 151–2, 294–5
Buzzard 24, 26, 34–5, 76, 89, 92, 130–1, 133, 135, 157, 197, 205, 218, 223–5, 238–9
    Rough-legged 243

Caching of food 95, 175, 254
Caernarvonshire 35, 50–3, 163, 258
    see also Snowdonia
Cairngorms 65–6
Caithness 22, 36, 38, 68, 120, 232
Calls of Raven 161, 172, 178, 265–7
Cambridgeshire 46–7
Canada 7, 86, 88, 125, 241
Canary Isles 105, 208, 242–3
Cape Clear Island 72, 207, 264
Captive Ravens 10–3, 25, 109, 115–6, 142, 249–54, 265–7
Cardiganshire 49, 257
Caribou *see* Reindeer
Carlow, Co. 72
Carmarthenshire 48, 257
Carneddau 52–3, 233, 258
Carrion, importance as food 75–82, 92–3, 96, 168, 205, 226–34, 238–9, 246–7
Carrying capacity, of land 30–9, 204–9, 226–36
Castings (pellets) 78–9, 87, 90, 228, 231
Cattle 32–4, 75–6, 80, 205, 207
Caves
    paintings of Ravens 12
    Raven remains in 8
Cetaceans 77, 104, 242
Chalk downs 33
Chases between Ravens 98–9
Channel Islands 42
Cheshire 21, 55, 118
Cheviots 2, 29, 34–5, 57, 59, 61, 76, 87, 100, 107, 134, 186, 190, 197, 206, 210, 218, 226–31, 233
Chough 35, 72, 103, 136
Churchwardens' records 20–1, 43, 57
Clackmannanshire 285
Clare, Co. 72, 264
Cliff-nesting, of Ravens 2, 20, 23, 25, 30, 35–9, 130–2, 139–40, 143, 150–9, 210
Cliffs, used by Ravens
    alternative 155
    altitude 158–9, 300

aspect 159
availability 36, 152–4, 197, 210–1
height 152, 296
Climate, influence of 7, 8, 158–9, 205, 225, 241–8
Climbing disturbance 236–7
Clutch size 165–8, 194, 229, 301
    repeats 167, 302
Clyde Islands 64
Coastal haunts 29–32, 37–9, 150, 155, 158, 234
Cod, as food 89
Cold, effects of 29, 158–9, 187
Coll 68
Colonsay 68, 207
Colour-marking of Ravens 119, 123–5
Comeragh Mountains 72
Communal feeding 93, 96, 102–7
Communal roosts 96–7, 101–7, 125–6, 215, 256–64
Communities, of birds 32–9
Competition
    interspecific 127–37
    intraspecific 97–102, 202–9
Connemara 73
Conservation 25–6, 215–8, 225–6, 232, 234, 236–8, 240
Cooperative feeding 95
Copulation 115, 161, 172
Corby Messenger 12, 14
Cork, Co. 35, 37, 72, 81, 264
Cormorant 88
Cornwall 13, 25, 30, 43, 46, 83, 86–7, 93–4, 99, 100, 104, 107–8, 160, 171, 201, 207, 256
Courtship behaviour 108–15, 140, 160–1
Cover concept (Hickey) 152–3
Coyote 8, 94, 117, 138
Creation stories 11, 16
Crossfell range 220–1
Crow
    Black 163
    Carrion 14–5, 24, 26, 32, 34, 39, 76, 84–6, 103, 135–6, 165, 167, 211, 223–4, 238, 246–7, 250–2
    Hooded 16, 26, 39, 84–6, 93, 96, 103, 135, 167, 250–2
    House 248
    Little 248
    Pied 247
    Torresian 248
Crustaceans, as food 79, 81
Cumberland *see also* Lakeland 57, 75, 119, 133, 163, 259, 269
Cumbria 15, 55, 57
Curlew 36, 87–8, 138, 252
Czechoslovakia 242

Dartmoor 34, 42–3, 153, 207, 237, 256
DDT/DDE 165, 238–9
Deer, as food 75–6, 79–80, 104, 138, 228
Deer forests 36–7, 76, 205, 251
Deer, Red 36–7, 75–6, 80, 104, 142
Deer stalking 14, 36–7, 76
Deeside (Aberdeenshire) 21, 65, 238
Deluge stories 11, 16
Demonstrativeness, to humans 172, 178
Denbighshire 32, 35, 53, 153, 192, 210, 223
Denbigh Moors (Mynydd Hiraethog) 35, 258
Denmark 242
Density, breeding
 in Britain and Ireland 197, 204–8, 309–12
 in other countries 208, 311
 decrease in 206
 increase in 206–7
 measurement of 311
Derbyshire *see also* Peak District 54
Desert habitats 242–7
Desertion
 of nests *see under* Nest
 of territories *see* Territory
Devon 8, 30, 32, 42–3, 105, 107, 137, 146, 153, 207, 234, 256
Dickens, Charles 9, 254, 267
Dieldrin 238–9
Diet *see* Food
Dipper 4
Diptera, as food 90
Disease, unknown role of 189, 211
Dispersion, of nesting Ravens 197–8, 205–11
Dispersal, of young Ravens see Young
Display, ground 109–15
 advertising 111
 appeasement 114
 defensive threat 113
 ear-tuft intimidation 112–3
 frontal threat 113
 pre-copulatory 114
 self-assertive 111
 sexual significance 111–5
 soliciting 114–5
 thick-head intimidation 113
Display, flight 99, 107, 140
 alarm 265–6
 diving 108, 140
 rolling and tumbling 107–8, 116, 140
 unison 108, 140
Distribution
 England 40–7, 53–60, 284
 Europe 241–5

Ireland 71–4, 285
Scotland 60–71, 284–5
Wales 47–53, 284
World 7, 8, 241–8
Disturbance, effects of 172, 189, 192, 236–8
Diver
 Black-throated 36
 Great Northern 88
Dogs 80, 86, 116, 138, 205, 251, 253–4
Domestic fowl 86–7, 89, 253
Dominance relations 102–5, 109, 111–4, 204, 209, 266
Donegal, Co. 37, 73, 264
Dorset 25, 42, 100, 147
Dotterel 30, 88, 137
Down, Co. 35, 74
Dublin, Co. 72, 264
Dumfriesshire 61, 238
Dunbartonshire 64, 92, 224
Dung, as food 86, 90, 138
Dunlin 36, 208
Durham, Co. 59, 220, 222

Eagle
 Golden 26, 36–7, 76, 94–6, 127–30, 152, 159, 165, 180, 185, 189, 193, 205, 208, 225, 229, 238–9
 White-tailed 128
Earthworms 90, 205
Edinburgh 20, 60
Eggs of Raven 25, 161–73
 coloration 25, 161–3, 214
 erythristic 25, 163, 201, 214
 infertile 182
 shape 161
 size 163, 165, 300
 weight 163
Eggs, other birds', as food 80–1, 86–9, 95
Egg collecting and collectors 24–5, 168, 189–93
Eggshell
 thinning, lack of 165, 239
 disposal 174
Eider 88–9
Energy balance 95–6
England 74, 80, 121, 166, 224
 East Anglia 46–7
 Midlands 53–5, 258
 northern 165, 169, 217, 234, 259
 south-east 40–2
 south-west 42–3, 46, 120, 133, 151, 165, 190, 210, 218, 220, 240, 258
Essex 8, 47, 147
Exmoor 22, 34, 42–3, 46, 207, 256

Index

Eyes
  as expression of mood 115
  attacks on other animals' 85, 94, 114
  avoidance of damage to congeners' 114
  coloration 248, 270
Eye-sight (visual acuity) 91, 105
Eyrie *see* Nest

Faeroes 13, 205, 245, 269
Fair Isle 71, 120, 137, 207
Family behaviour 100–1, 180–2
Farming
  arable 31–3, 39, 77
  sheep 34–7, 75–6, 80, 82–5, 233–4, 236
  stock-rearing 32–4, 43, 75
Fear of carcasses 93
Feeding habits
  of female by male 172
  of young 175–6, 178, 180
  range 92–3, 125, 204, 226–32
  routine 96, 101
Fennoscandia 125, 208, 242
Fermanagh, Co. 74
Fetlar 71, 167
Fidelity
  to mate 201, 214
  to nest site 152, 214
Fighting, between Ravens *see* Aggression
Finland 144, 166
Fish, as food 76–7, 89
Fledging
  behaviour 178
  period 177
  timing of 180
Fledglings 177–8, 180–2
Flight characteristics 1, 98, 105–9, 116
  displays *see* Display: flight
Flintshire 53
Flocking of Ravens 101–7
Food
  intake 95
  range of diet 75–82, 286–7
  study of 78–82
  supply 92–3, 96, 208–9, 216
Foraging methods 91–6
Forestry Commission 225
Fostering behaviour 200–1
Fox, Red 76, 79, 94, 134, 138, 142, 153–5, 180, 189, 207, 211, 223
France 12, 242–3
Frogs, as food 89
Frost, effects of 29, 159
Fruits, as food 79, 81, 90, 246
Fulmar 137, 189

Galloway 2, 4, 29, 30, 35, 61–3, 107, 128–30, 142, 170, 206, 224, 232, 234, 238, 252, 260
Galway, Co. 73, 264
Gamekeepers 16, 21, 49, 103, 189, 198, 218–25, 238
Game-preserving 21–3, 218–25, 240
Gannet 88, 136, 208
Gastroliths 80, 91
Geographical races *see* Raven sub-species of *Corvus corax*
Geology, influence of 34–6, 159, 193
Germany 125, 144, 154, 204, 208, 242
Glamorgan 48, 171, 257
Gloucestershire 8, 46
Goat, feral 4, 80, 228
Goshawk 135
Gower 8, 48
Grain, as food 81, 90–1
Grassholm 48, 208
Grasshoppers, as food 90, 242
Great Glen 36, 66
Greeks, Ancient 13
Greenland 7, 94, 125, 241, 245
Greenshank 36, 138
Grey Wagtail 4
Ground-nesting, by Ravens 153–4
Grouse
  Black 35
  Red 35–6, 76, 80, 86–7, 208, 218, 251
Grouse-moors 23, 35–6, 76, 122–3, 131, 200, 218–24
Guillemot
  Brunnich's 86, 88, 95
  Common 88, 136
Gull
  Common 88
  Great Black-backed 88, 95, 136
  Herring 136, 211
  Lesser Black-backed 136
Gyr Falcon 243

Habitat needs, of Ravens 7–8, 27–39, 144–59, 210–1, 217–38
Hampshire 8, 41, 146
Handa 39
Hang-gliding 237
Hare
  Brown 80
  Mountain 80
Harris 70, 269
Hatching 171, 174–5
  asynchronous 184–5
Heathland 31, 34–6, 195, 234, 236
Heather 34–5, 194, 236

Hebrides 68–70, 104, 146, 155, 200, 234, 263, 269
Hen Harrier 16, 138
Herefordshire 32, 53, 258
Heron 88, 137
Hertfordshire 41
Highlands, Scottish 21–2, 30, 35, 92, 100, 120, 128, 158, 163, 189, 223, 234, 251, 260–3
  eastern and central 65–6, 88, 197, 210, 218, 223–4, 239
  northern 66–8
  southern 35, 64–5, 158–9, 205
  western 133, 154, 159, 186, 205
Hiking, effects of 237
Himalayas 125, 245
Hobby 134
Home range 195
Homing Pigeon 86, 224
Horse, as food 75–6
Howgill Fells 34, 55, 57, 170
Hoy (Orkney) 38, 151
Human associations
  Anglo-Saxon 13
  18–19th century 20–5
  Greco-Roman 13
  Medieval 14, 16, 20
  Norse 13
  prehistoric 8–9
  20th century 25–6, 217–40
Human corpses, as food 9, 14–5
Hungary 242

Ice Age 8
Iceland 13, 16, 77–8, 88–9, 151, 170, 193, 215, 245, 269
Idaho 90, 96, 126, 152, 166, 176, 182, 208, 210, 215–6, 243
Illegal killing, of Ravens 26, 211, 220–6
Incubation
  by female 171–3
  period 171, 185
India 170, 241, 246
Indians, North American 14
Individual variation 154, 172, 214
Information centre, theory of flocking 106
Instinct 250
Intelligence, of Ravens 249–55
Inuit 14–5
Inverness-shire 22, 66, 107, 262
Invertebrates, as food 77, 79–81, 89, 175, 205, 242, 246
Ireland 14, 20, 37, 71–4, 97, 120–1, 210–1, 240, 264
  northern 73–4, 151, 166, 217, 224, 232, 234
  southern 72–3, 151, 210, 217
  western 35, 39, 72, 92, 120
Islay 68, 107, 263
Isle of Man 30, 35, 38, 60, 88, 119–21, 135–7, 151, 260
Isle of Wight 30, 41
Israel 242
Italy 243

Jackdaw 86, 103, 136, 161, 252
Japan 245
Jura, Isle of 68
Juvenile
  appearance 180
  mortality 180
  movements 181–2, 120–5

Kent 41, 147, 216
Kerry, Co. 35, 37, 72, 264
Kestrel 35, 134, 253
Kildare, Co. 72
Kilkenny, Co. 72
Killing of Ravens, methods of 20–3, 221–5
Kincardineshire 36, 65, 262
Kirkcudbrightshire see also Galloway 61, 260
Kittiwake 86, 94

Lagomorph (rabbits and hares) 77, 80–1, 85, 90, 205, 228
Lakeland (Lake District) 2, 20–1, 23, 34–5, 57–9, 78, 100, 102, 107, 119, 123, 126, 130, 139, 142, 150, 153–4, 159, 163, 186–90, 196–7, 200–2, 206–7, 210, 214, 220, 229, 233, 236, 239
Lamb-killing by Ravens, allegations of 82–5, 247
Lammermuirs 60, 223
Lancashire 56–7, 118, 222
Land improvement 33, 205, 234
Land use change 226–36
Langholm Hills 16, 149, 162, 223, 227
Late nesting 170–1
Laying, interval 171
  date of 168–70, 303
  period 168–70
Learning ability, of Ravens 250–2
Legal protection, of Ravens see Protection
Leicestershire 54
Leitrim, Co. 73

322 *Index*

Lemming, as food 86
Lepidoptera, as food 90
Lewis, Isle of 70, 263
Life table, for Ravens 213, 313
Limerick, Co. 72
Lincolnshire 54
Lizards, as food 89, 246
Locusts, as food 90
London 16, 40
Londonderry, Co. 72, 74
Longevity, of Ravens 214
Lundy Island 43, 207, 256

Magpie 103, 135, 225
Maine 92, 106–7, 126
Mallard 138
Mate selection 117, 209
Maturity, breeding *see* Reproductive age
Mayo, Co. 73
Meadow Pipit 35, 86
Meath, Co. 8, 72
Mediterranean region 243, 246
Merioneth 21, 23, 35, 50, 79, 93, 101, 135, 223, 258
Merlin 4, 36, 134
Middle East 243, 246
Middlesex 41
Mimicry, of human voice 10, 249, 254, 266–7
Mingulay 70
Mink 154
Miraculous powers, attributed to Ravens 9–14, 16
Mischievous ways, of Ravens 116, 249
Moffat Hills 2, 35, 107, 238
Mole 32, 47, 79, 86, 205
Molluscs, as food 79, 81, 90
Monaghan, Co 72, 285
Monmouthshire 48
Montgomeryshire 35, 50, 223
Moorfoot Hills 60, 220, 223
Morayshire 36, 65
Mortality
  causes of 211, 224–5
  in adults 211–6
  in young 184–5, 213
  seasonal bias 211
Moult 269–70
Mouse, unspecified, as food 79, 85, 95
Movements 118–26
  age-related 121, 291
  daily 96, 125–6
  distance 120–1, 288–9
  long distance 120, 288
  migratory 120–5

orientation 121–3, 290
  relation to age 120–1
  seasonal 118–20, 291
Mull, Isle of 68, 125, 128, 131, 153, 216, 263
Mute Swan 89, 95
Mythology, of Ravens 9–16, 214

Nairn 36, 65
Names of Raven 274
Nature Conservancy 238
Nearest neighbour distance 195, 197, 204–10
Nest
  building 140–2, 157, 203
  defence 98–100, 178
  desertion 237, 254
  materials 140–2
  sanitation 154, 173, 175, 177
  structure 140–2, 171
  visiting, by other Ravens 100
Nest sites
  accessibility 144, 146–53, 195
  choice 144–54
  traditional nature 139, 146–7, 152, 157–8
  use of alternatives 154–7, 297–9
Nesting territory *see* Territory
Nestlings
  appearance 174–80
  behaviour 174–8
  food and feeding 175–8
  growth 175–80
  period in nest 177
Netherlands 242
Non-breeding
  flocks 144, 197, 200, 216, 245–8, 256–64
  pairs 99, 189, 197, 200, 202–4, 206, 228, 232
  population 103, 199–200, 216
Non-laying 202–4, 232
Norfolk 47
Northamptonshire 54, 147, 195
Northumberland *see also* Cheviots 59, 60, 133, 147, 153, 166, 168, 220, 236, 260
North Rona 39, 70
North York Moors 35, 55–6, 220
Norway 7, 16, 92, 170, 193, 252
Nottinghamshire 54
Numbers, of Ravens *see* Population: size

Omen, bird of 10–6
Oregon 88, 126, 152, 155, 166, 170–1, 178, 182, 201, 210

Organochlorine residues 165, 238–9, 316
Orkney Isles 70, 81, 85, 88, 120–1, 137, 142, 151, 153, 166, 168, 205, 215, 234, 263
Owl,
  Eagle 8
  Short-eared 36, 95, 197
  Snowy 134
Oxfordshire 42
Oystercatcher 138

Pair-bonding 109, 140
Pair formation 103, 105, 109, 111–5, 117, 140, 202
Parkland, nesting habitat 33, 40–2, 46–7, 53–5, 59
Partridges 218
PCBs 239
Peak District 8, 54–6, 139, 220, 236
Peeblesshire 61
Pembrokeshire 48, 83, 93, 99, 105, 107, 135, 207, 257
Pennines 34, 55–7, 101, 106–7, 119, 123, 133, 135, 142, 155, 190, 197, 210, 218–22, 224, 229, 236, 259–60
Pentland Hills 60, 223
Peregrine 5, 16, 23, 46, 72, 95, 98, 131–4, 137, 151–2, 154, 180, 189, 193, 196, 200, 211, 223–4, 237–8, 243, 266
Persecution, effects of 20–6, 178, 189, 193, 195, 212, 215–6, 218–25, 236, 238–40, 247, 255
Perthshire 22, 64–6, 80–1, 85, 101, 105, 107, 133, 159, 261
Pet, Raven as 2, 13, 25, 267
Pheasant 86, 218, 224
Pine Marten 138, 154
Piracy, by Ravens 95
Place-names, containing Raven 40
Play 108, 115–7
Plover
  Golden 4, 36, 88
  Killdeer 138
Plumage 111, 269–70
Poe, Edgar Allen 9, 14
Poison, use against Ravens 23, 26, 189, 195, 201, 211, 216, 223–4, 234, 238–9, 242
Poland 144
Polecat 138, 154
Population
  age-structure 213–4, 313
  breeding 40–74, 284–5
  decline 20–6, 40–3, 45–7, 53–7, 59–61, 63–6, 71–2, 197, 217, 226–9, 234
  increase 42–3, 46–55, 58, 60–1, 63–4, 72, 74, 206, 217, 233–5
  limitation 196–216
  regulation 196–216
  size 74, 216, 284–5
  stability 58, 196–8, 202, 206, 215
  status 23–6, 44–5, 74, 240
  turnover 213
Portugal 170
Post-glacial period 8
Pouch, food 95, 140, 171, 175
Power-line towers, use by Ravens 130, 140, 151, 210–1, 216, 243
Predator, Raven as 20, 82–9, 136–8
Preening, mutual 115
Prehistory 8–9
Prey, Raven as 128, 131, 133
Productivity 189–95, 215
Protection laws 25–6, 82, 224–5
Ptarmigan 8, 37, 88, 94, 208, 252
Puffin 82, 86, 94, 136

Quarry nesting haunts 42–3, 46, 53, 57, 74, 151, 153, 210, 234, 237

Rabbit, as food 32, 47, 79–81, 85, 95, 142, 234
Radio-tracking 96, 126
Radnorshire 49, 53, 137, 223, 257
Rain, effects of 159, 187, 242, 247
Ramsey Island 48
Rathlin Island 312
Rats, as food 79, 254
Raven, species of 241–8
  American White-necked 247
  Australian 247
  Brown-necked 245–6
  Common or Northern 241–5
  Dwarf 245
  Fan-tailed 246
  Forest 248
  Little 247
  Thick-billed 246–7
  White-necked 246
Raven, sub-species of *Corvus corax* 243–5
  C. c. canariensis 243
  C. c. corax 243
  C. c. hispanus 243
  C. c. kamtschaticus 245
  C. c. laurencei 243
  C. c. principalis 245
  C. c. sinuatus 245
  C. c. tibetanus 245
  C. c. tingitanus 243
  C. c. varius 245

Raven Crags 139, 141, 143, 156
Razorbill 136
Reactions, of other birds to Ravens 130–8
Reasoning ability 250–3
Recreation, effects of 236–7
Recruitment, to breeding population 125, 191, 212–6
Re-directed aggression 171, 178
Red Kite 16, 20, 24, 26, 34–5, 76, 130–1, 172, 205
Reed Bunting 86
Refuse, as food 91
Reindeer 8, 86, 242
Re-mating, rapid 197, 200–2
Renfrewshire 61
Repeat nestings 155, 167, 171, 188, 190
Replacement output, of young 213, 215
Reproductive age 103, 124–5, 214–5
Reptiles, as food 89
Ringing (banding), BTO scheme 119–25
  recoveries 119–25, 133, 180, 211–3, 220, 224, 288–91
Ring Ouzel 36
Risk-taking, by Ravens 95, 116–7
Road victims, as food 77, 85
Rock Dove 86, 93–4
Rodents, as food 78–81, 85
Romans 13, 97, 195
Rook 103, 135, 165, 171
Roosting 173
  communal 101–7
Ross-shire 22, 66–8, 128, 133, 136, 159, 205
Roxburghshire 61
Royal Society for the Protection of Birds 221, 224
Rubbish tips, as Raven feeding places 78, 80, 89, 91–2, 104, 205, 215, 236–8, 257
Russia 144, 242–5

Sandhill Crane 88
Scavenger, Raven as 8–9, 16, 20, 134, 138, 236, 242, 245
Scilly Isles 46
Scotland 21–3, 34, 74, 165–7, 188, 199, 218, 239
Scrub 31
Seabirds, as food 38, 77, 136–7
Sea-cliffs *see* Coastal haunts
Seals, as food 77
Seaweed, as food 79, 81, 90
Sedentary habits 118–21, 245
Seeds, as food 79, 81, 90
Selkirkshire 61

Sentinel, role of male 172
Separation, of young from parents *see* Family behaviour
Sexual dimorphism 96, 268
Shag 87–8, 136–7
Shakespeare, William 9, 14
Sheep, importance to Ravens as food 23, 34–5, 75–85, 93, 205–7, 226, 230–6
  increase in numbers 206, 233–4, 314–5
  killing, allegations of 82–5
Sheep-dips 238–9
Sheepwalks 34–7, 80, 82–5, 96, 220–6, 228–36
Shepherds 23, 82–5, 189, 192, 198, 234
Shetland Isles 8, 23, 38, 71, 76–7, 81, 85, 101, 104, 107, 119–21, 137, 151, 153–4, 163, 167–8, 187, 193, 195, 199, 202, 205, 264
Shiant Islands 39, 70
Shooting of Ravens 22, 84, 180, 201, 211, 223
Shropshire 32–3, 53–4, 107, 124, 215
Siberia 241, 243–5
Size 268
Skokholm 48
Skomer 48, 257
Skye, Isle of 22, 25, 68–9, 82, 107, 130, 211, 263
Slaughterhouses, as feeding places 78, 104, 205, 234, 236, 256–7
Sligo 73
Small Isles 69
Smell, sense of 92, 246–7
Snakes, as food 89, 246
Snow, effects of 28–9, 159
Snowdonia 22, 35, 50–3, 78, 93, 98, 100, 107, 119, 134, 136, 153, 155, 158–9, 186, 191–2, 196–7, 199, 206–7, 232–3, 236
Soaring 92, 109, 116, 246
Social behaviour *see also* Displays 105
Soil, in castings 80
Somerset 30, 32, 46, 256
Southern Uplands 2, 34, 60–4, 80, 87, 89, 107, 135, 141, 150, 153–6, 163, 166–9, 186, 190, 196–7, 202, 205–6, 214, 218, 223, 226–9, 233–5, 239
Spacing behaviour *see also* Territory 197, 204, 208–9
Spain 12, 170, 242–3, 253
Sparrowhawk 208–9, 238
Spey Valley 66, 223
Spiders, as food 90
Spread of Raven *see* Population: increase
St. Kilda 39, 82, 88, 151, 207, 263

Index 325

Stability of numbers *see* Population: stability
Staffordshire 8, 54
Starling 86
Starvation, possible 96
Stirlingshire 64, 224
Stoat 86
Stonechat 36
Suffolk 47, 134
Surrey 41
Sussex 41, 216
Sutherland 22–3, 36, 67–8, 120, 128, 155, 159, 205, 208, 232, 262
Sweden 144, 214
Symbolism of Raven 9

Taiga *see* Boreal Forest
Tameness, of Ravens 10, 13, 25, 254–5
Territory, breeding 93, 97–102, 107, 197
  constancy
  defence 97–101, 204, 209
  desertion 23, 40–8, 53–7, 59–61, 64–7, 72, 202–4, 206, 228–33, 236–7
  doubling 155, 204, 207, 209
  expansion/merging 206, 209, 232
  purpose of 204–5
  quality 186, 193–5
  size 204–8, 309–12
Territory, non-breeding 202–4
Thermals, soaring in 109
Tibet 242, 245
Tipperary 264
Tiree 68, 153, 263
Toad, as food 89
Tors, as nesting places 43, 46, 237
Tower of London 10, 25, 214
Traditional nesting haunts 40, 139, 164–7, 156–8, 297–9
Trapping, of Ravens 22, 211, 224, 254
Tree-line, climatic 35, 37, 158
Tree-nesting habit 23, 35, 42–3, 46–7, 140, 142, 144–50, 153, 157–8, 210, 220, 223, 225, 229, 232, 237
Tree species, used by nesting Ravens 144–50, 292–3
Trossachs 35
Trout, Brown, as food 89
Truce with neighbours 93
Tundra 8, 134
Twa Corbies, ballad of The 15, 17–9
Tyrone, Co. 74

Uist, North 70, 263
Uist, South 70, 263

United States 89, 92–4, 100–1, 104–7, 116, 125–6, 130, 135, 138, 152, 166, 170–1, 178, 182, 201, 208, 210, 242–5, 247
Upside-down hanging 116
Urban habitats 75, 236, 238, 247

Vegetable matter, as food 79–82, 90–1, 246–7
Vegetation, of Raven habitats 30–9
Vermin lists 21–2
Vikings (Norsemen) 13
Vision 91
Voice, vocalisations *see* Calls of Raven
Vole, Short-tailed Field 80, 86, 95, 228, 230–1
Vulture
  Turkey 95
  unspecified 105, 242

Wales 15–6, 21–4, 28–30, 32–5, 43, 47–53, 74, 92, 98, 121, 153, 166, 217–8, 220, 224–5, 232, 234, 240, 256–8
  central 48–9, 80, 84, 92, 96, 119, 123–5, 130, 145, 150, 153, 158, 166–8, 171–2, 185–7, 192–3, 195, 202–3, 206, 208, 229–32
  north 22, 49–53, 94, 100, 134–6, 157, 166–9, 184, 186
  see also Snowdonia
  south 47–8, 92, 119–20, 135, 140, 151, 159, 234
Warwickshire 54
Water, Raven's need for 171, 175–6
Waterford, Co. 35, 72
Weasel 86
Weather, effects of 4, 28–9, 76, 119, 143, 150, 168, 187
Weight 268
West Country 13, 22, 30, 32, 34, 145, 153, 207, 217–8, 225
  see also England, south-west
Western Isles *see* Hebrides
Westmorland 20–1, 57, 75, 106, 142, 163, 221, 259
  see also Lakeland
Wexford, Co. 72
Whales
  Killer 104
  unspecified 75, 77
Wheatear 35
Whimbrel 138
White, Gilbert 146
Wicklow, Co. 35–6, 72, 264

Wigtownshire 61, 63, 223, 260
Wildcat 154
Wiltshire 41, 46
Wind, effects of 29–30, 116, 144
Wing-loading 269
Wing-tagging, of Ravens *see* Colour-marking of Ravens
Wolves 1, 8, 117, 138, 242, 251
Woodland habitat 32, 34–5, 43, 106, 131, 144–50, 157, 225–32, 237, 247–8
Worcestershire 54
Wordsworth, William 20
World distribution 241–8

Yorkshire 15, 21, 55–6, 190, 216, 220, 222, 260
Young
  appearance 174–80
  dispersal of 100–1, 121, 180–2, 200
  feeding of 175
  growth of 175–80
  relationships to adults 100–1